Fabric testing

The Textile Institute and Woodhead Publishing

The Textile Institute is a unique organisation in textiles, clothing and footwear. Incorporated in England by a Royal Charter granted in 1925, the Institute has individual and corporate members in over 90 countries. The aim of the Institute is to facilitate learning, recognise achievement, reward excellence and disseminate information within the global textiles, clothing and footwear industries.

Historically, The Textile Institute has published books of interest to its members and the textile industry. To maintain this policy, the Institute has entered into partnership with Woodhead Publishing Limited to ensure that Institute members and the textile industry continue to have access to high calibre titles on textile science and technology.

Most Woodhead titles on textiles are now published in collaboration with The Textile Institute. Through this arrangement, the Institute provides an Editorial Board which advises Woodhead on appropriate titles for future publication and suggests possible editors and authors for these books. Each book published under this arrangement carries the Institute's logo.

Woodhead books published in collaboration with The Textile Institute are offered to Textile Institute members at a substantial discount. These books, together with those published by The Textile Institute that are still in print, are offered on the Woodhead web site at: www.woodheadpublishing.com. Textile Institute books still in print are also available directly from the Institute's web site at: www.textileinstitutebooks.com.

A list of Woodhead books on textile science and technology, most of which have been published in collaboration with The Textile Institute, can be found on pages xv-xix.

Woodhead Publishing in Textiles: Number 76

Fabric testing

Edited by
Jinlian HU

The Textile Institute

CRC Press
Boca Raton Boston New York Washington, DC

WOODHEAD PUBLISHING LIMITED
Cambridge England

Published by Woodhead Publishing Limited in association with The Textile Institute
Woodhead Publishing Limited, Abington Hall, Granta Park,
Great Abington, Cambridge CB21 6AH, England
www.woodheadpublishing.com

Published in North America by CRC Press LLC, 6000 Broken Sound Parkway, NW,
Suite 300, Boca Raton, FL 33487, USA

First published 2008, Woodhead Publishing Limited and CRC Press LLC
© Woodhead Publishing Limited, 2008
The authors have asserted their moral rights.

British Library Cataloguing in Publication Data
A catalogue record for this book is available from the British Library.

Library of Congress Cataloging in Publication Data
A catalog record for this book is available from the Library of Congress.

Woodhead Publishing ISBN 978-1-84569-297-1 (book)
Woodhead Publishing ISBN 978-1-84569-506-4 (e-book)
CRC Press ISBN 978-1-4200-7988-3
CRC Press order number: WP7988

The publishers' policy is to use permanent paper from mills that operate a
sustainable forestry policy, and which has been manufactured from pulp which is
processed using acid-free and elementary chlorine-free practices. Furthermore,
the publishers ensure that the text paper and cover board used have met
acceptable environmental accreditation standards.

Typeset by SNP Best-set Typesetter Ltd., Hong Kong
Printed by TJ International Limited, Padstow, Cornwall, England

Contents

6   Fabric appearance testing   148

X Binjie and J Hu, The Hong Kong Polytechnic
University, China

7   Fabric permeability testing   189

X Ding, Donghua University, China

8   Testing for fabric comfort   228

P Bishop, University of Alabama, USA

S Nazaré and AR Horrocks, University of Bolton, UK

Contributor contact details

(* = main contact)

Chapter 1

Professor Jinlian Hu
Institute of Textiles and Clothing
The Hong Kong Polytechnic
   University
Hung Hom
Kowloon
Hong Kong
China
E-mail: tchujl@inet.polyu.edu.hk

Chapter 2

Dr Ka Fai Choi
Institute of Textiles and Clothing
The Hong Kong Polytechnic
   University
Hung Hom
Kowloon
Hong Kong
China
E-mail: tcchoikf@inet.polyu.edu.hk

Chapter 3

Professor Z. Zhong* and Professor
   C. Xiao
School of Textiles
Tianjin Polytechnic University
63, Cheng Lin Road
Tianjin 300160
China
E-mail: zhong_zhili@yahoo.com.cn
zhongzhili@tjpu.edu.cn

Chapter 4

Professor X. Wang*, Dr X. Liu and
   C. Hurren
Centre of Material and Fibre
   Innovation
Deakin University
Geelong
Victoria
Australia 3217
E-mail: xwang@deakin.edu.au

Chapter 5

Dr Qinguo Fan
Department of Materials and
   Textiles
University of Massachusetts
   Dartmouth
285 Old Westport Road
North Dartmouth, MA 02747-2300
USA
E-mail: qinguo.fan@umassd.edu

Chapter 6

X. Binjie*
Institute of Textiles and Clothing
The Hong Kong Polytechnic
   University
Hung Hom
Kowloon
Hong Kong
China
E-mail: Binjie.XIN@inet.polyu.
   edu.hk

Professor Jinlian Hu
Institute of Textiles and Clothing,
The Hong Kong Polytechnic
   University
Hung Hom
Kowloon
Hong Kong
China
E-mail: tchujl@inet.polyu.edu.hk

Chapter 7

Dr Xuemei Ding
No. 1882, Yan'an West Road
(Near Yan'an Road Gate)
Donghua University
Shanghai 200051
China
E-mail: fddingxm@dhu.edu.cn

Chapter 8

Dr P. Bishop
The University of Alabama
PO Box 870312
Tuscaloosa
Alabama 35487-0312
USA
E-mail: pbishop@bama.ua.edu

Chapter 9

C. Hurren
Centre for Material and Fibre
   Innovation
Deakin University
Geelong
Victoria
Australia 3217
E-mail: cjhurren@deakin.edu.au

Chapter 10

Professor Jinlian Hu*
Institute of Textiles and Clothing,
The Hong Kong Polytechnic
   University
Hung Hom
Kowloon
Hong Kong
China
E-mail: tchujl@inet.polyu.edu.hk

Dr K. Murugesh Babu
Institute of Textiles and Clothing,
The Hong Kong Polytechnic
   University
Hung Hom
Kowloon
Hong Kong
China
E-mail: kmb6@rediffmail.com

Chapter 11

Dr R.V. Mahendra Gowda*
Department of Textile-Fashion
    Technology
Bannari Amman Institute of
    Technology
Sathyamangalam
Tamil Nadu 638 401
India
E-mail: rvm_gowda@rediffmail.
    com

Dr K. Murugesh Babu
Institute of Textiles and Clothing
The Hong Kong Polytechnic
    University
Hung Hom
Kowloon
Hong Kong
China
E-mail: kmb6@rediffmail.com

Chapter 12

Shonali Nazaré* and Professor
    A.R. Horrocks
University of Bolton
Centre for Materials Research and
    Innovation
Deane Campus
Bolton BL3 5AB
UK
E-mail: S.Nazare@bolton.ac.uk

Woodhead Publishing in Textiles

Preface

Textile fabrics are manufactured for many different end uses, each of which has different performance requirements. The chemical and physical structures of textile fabric determine how it will perform, and ultimately whether it is acceptable for a particular use. Fabric testing plays a crucial role in gauging product quality, ensuring regulatory compliance and assessing the performance of textile materials. It provides information about the physical or structural, chemical and performance properties of the fabrics.

As consumers become more aware and more demanding of products, the number of tests required for textile materials has grown. As a result the testing of fabrics is increasingly varied, in constant flux and full of the unprecedented challenges of globalization. With the onset of new types of fabrics for the apparel industry and of technical textiles for functional applications, and with the increasing number of innovations taking place in the garment sector, fabric testing procedures have undergone tremendous changes and there is a need to understand all the procedures before a testing system is adapted to investigate the performance of fabrics.

It is very important to predict the textile fabric's performance by testing. Fashion merchandisers, apparel designers, interior designers and textile scientists who have an understanding of textile properties and testing are equipped to make decisions that will benefit their clients and enhance profits for their businesses. Knowledge of fabric testing and its performance analysis can contribute to efficiency in solving consumer problems with textile products, and to the development of products that perform acceptably for consumers. As indicated above, retail buyers and producers of apparel and textiles are among those who use the fabric testing data and results in making decisions about their products. Most textile or apparel manufacturers will use either test methods or performance specifications that are published by testing organizations.

A number of textile research and testing organizations have published data on fabric testing and their procedures. There exists a great variety of textile testing procedures for different fabrics for different end uses. Researchers all over the world have been constantly involved in developing newer methods of fabric testing so as to meet the ever-growing globalization and quality requirements. Their researches have resulted in an enormous quantity of data and testing procedures for fabrics. These results should be providing the industry, fabric suppliers, apparel manufacturers, exporters, fashion designers and retailers with an enormous amount of information about the testing aspects of fabrics and apparel to meet the international standards. It appears that coverage of the existing literature in textbooks on fabric testing procedures and results is insufficient, although there have been a great number of research achievements by scientists, researchers and industry experts in the areas of apparel, industrial fabrics such as technical fabrics, intelligent fabrics for special applications, nanotechnology applications, medical textiles, etc. Hence, a systematic approach towards integrating the knowledge available in the literature on fabric testing and developments in different aspects of fabric testing and the achievements of researchers and industry experts would help all those who are involved in quality assessment and evaluation of textile products to a great extent.

Based on the above considerations, it was thought desirable to compile a book on testing principles and procedures of various aspects of fabrics. Hence an effort has been made in this book to include the latest procedures of testing of fabrics for their comfort, appearance, intelligence, damage analysis, etc. Wide coverage of advanced topics on composition testing, chemical testing, physical and mechanical testing, statistical testing, flammability analysis, testing for colour and dye analysis, and permeability will help readers to understand these tests in detail.

Finally, this book is a compilation of research works on fabric testing by experienced researchers worldwide. I sincerely feel that a complete book on fabric testing of this scope will help all those involved in the fabric, garment and fashion industries and the import and export businesses to adopt new testing procedures to meet the international standards and to maximize their profits. In addition, research and academic organizations can benefit from this book in exploring the possibilities of new test methods and testing procedures for new types of fabrics, including smart and intelligent fabrics.

Professor Jinlian Hu
Hong Kong Polytechnic University

Acknowledgements

This book is a result of research contributions in the area of fabric testing by experienced researchers all over the world. I would like to take this opportunity to acknowledge my sincere thanks to various people who contributed in successfully making this book a reality.

I am extremely grateful to Dr K. Murugesh Babu for his consistent help and hard work during the editing and preparation of this book. His outstanding reviewing and editing skills combined with sincere efforts have made this book a meaningful piece of work.

I wish to thank all the authors of this book for their valuable contributions in presenting their chapters in a befitting manner. Their efforts in making this book a reality are greatly appreciated.

1

Introduction to fabric testing

J HU, The Hong Kong Polytechnic University, China

Abstract: Fabric testing plays a crucial role in gauging product quality, assuring regulatory compliance and assessing the performance of textile materials. It provides information about the physical or structural properties and the performance properties of the fabrics. Today more and more countries and markets have a stake in the treatment and testing of fabric. As consumers become more aware and more demanding of products, the number of tests required for textile fabrics has grown. As a result the testing of fabrics is increasingly varied, in constant flux and full of the unprecedented challenges of globalization. This introductory chapter describes the importance, scope, current status and future trends in fabric testing.

Key words: scope of fabric testing, importance of fabric testing, future of fabric testing, standards for tests.

1.1   Introduction

Testing of textiles refers to numerous procedures for assessing myriad fibre, yarn and fabric characteristics such as fibre strength and fineness, yarn linear density and twist and fabric weight, thickness, strength, abrasion resistance, colour fastness, wrinkle resistance and stiffness. It is the application of engineering knowledge and science to the measurement of the properties and characteristics of, and the conditions affecting, textile materials. It involves the use of techniques, tools, instruments and machines in the laboratory for the evaluation of the properties of the textiles (Grover and Hamby, 1960). Textile testing has become more important in recent years as a result of the new demands placed upon the products of textile manufacturers. Advances in textile technology, combined with the rise in the number of knowledgeable consumers with firm demands for specific performance behaviour, have made it essential that the properties of a material must be well understood and must be maintained over a long period of time (Slater, 1993). An understanding of the principles of these procedures, a certain degree of skill in carrying them out and the expertise to interpret reported results are important steps in developing the ability to correlate structure with performance.

The main reasons for testing of textiles are control of product, control of raw materials, process control and analytical information. Testing is actually a two-way process, in which the incoming raw materials that will be needed to manufacture the company's products will be scrutinized to ensure that they meet the specifications. That is, any manufacturing problems will be minimized while also ensuring that the textile item thus made will not result in problems for the customers; namely, that the item being manufactured is a quality product (Adanur *et al.*, 1995).

Testing is important, mainly for customer satisfaction of the textile product as well as to ensure product quality for the market in which the textile manufacturer competes. Testing is also important in order to control the manufacturing process and cost. In the textile industry, it is very important to use testing to control the manufacturing process for cost and other reasons. The importance of testing cannot be disregarded for product satisfaction and control of manufacturing cost. There are additional reasons such as customer relations, reputation, employee satisfaction and sales. Proper testing programmes are a very important ingredient of the efficient manufacturing business. Testing informs us whether the product will be saleable or not (McCullough, 1978).

Quick response and just-in-time delivery have become increasingly important as textile suppliers and purchasers like to shorten the supply-side pipeline. Quality considerations, mandated by the International Organization for Standardization (ISO), have forced suppliers to update testing methods, explore opportunities for more rapid testing and develop entirely new test methods. One of the most compelling reasons for the rise of rapid testing of textile products is the increasing globalization of the textile industry (Mock, 2000). Materials for an individual garment or fabric are often sourced today from a variety of suppliers, literally from around the globe. This necessitates the testing procedures to be highly competitive and accurate to analyse the textile product's characteristics to meet a particular end use. The test procedures today need to be more objective than subjective. Instrumentation may definitely help in this regard. A key issue in modern testing is to understand the complexity of the instruments and their working principles and finally to interpret the results in a systematic and scientific way.

1.2   Fabric testing for innovation and commercial needs

Textile fabrics are manufactured for many different end uses, each of which has different performance requirements. The chemical and physical structures of textile fabric determine how it will perform, and ultimately whether it is acceptable for a particular use. Fabric testing plays a crucial role in gauging product quality, assuring regulatory compliance and assessing the performance of textile materials. It provides information about the physical or structural properties and the performance properties of the fabrics.

Physical properties include those that characterize the physical structure of the fabric and tests that measure these properties are sometimes called characterization tests. Physical properties include fabric thickness, width, weight and the number of yarns per unit fabric area (i.e. fabric count). Performance properties are those properties that typically represent the fabric's response to some type of force, exposure or treatment. These include properties such as strength, abrasion resistance, pilling and colour fastness. Performance properties are mostly influenced by their physical properties. Although performance properties are often the primary factors in product development, aesthetic properties are equally important such as the way a fabric feels or drapes in design and development decisions. In some cases, trade-offs occur between performance characteristics and aesthetics, while in others, decisions based on aesthetic factors can also enhance product performance (Collier and Epps, 1999).

Throughout the textile supply chain, from distributors and textile mills to dyers and finishers, speciality textiles continue to grow in complexity. Today, more and more countries and markets have a stake in the treatment and testing of fabric (Hildebrandt, 2006). As consumers become more aware and more demanding of products, the number of tests required for textile materials has grown. As a result the testing of fabrics is increasingly varied, in constant flux and full of the unprecedented challenges of globalization. With the onset of new types of fabrics for the apparel industry, the development of technical textiles for functional applications, and the increasing number of innovations taking place in the garment sector, fabric testing procedures have undergone tremendous changes and there is a need to understand all the procedures before a testing system is adopted to investigate the performance of fabrics. For example, photochromic textiles change colour when exposed to UV light and revert to their original colour in the absence of UV. Ultraviolet radiation, with wavelengths ranging from 280 to 400 nm, has significant detrimental effects on both synthetic and natural fibre fabrics. Also, various stab and ballistic resistant garments are worn by police, soldiers, prison correction officers and other types of security, military and law enforcement personnel. The new fabric testing programmes must include techniques that determine the fabric's quick photochromic response, colour fastness and impact resistance properties. It is required to develop improved ways for UV protection of a range of fabric substrates and their evaluation, through systematic approaches to analyse the efficacy of such a finish.

It is difficult generally to describe what is meant by the term 'performance'. One may say that it has to do with how well the fabric 'holds up' in its intended end use, or we often use another equally ambiguous term, 'durability'. Although 'performance' is not easily defined directly, there is seldom any doubt in describing poor performance. A fabric may be deemed unacceptable because it fades, wrinkles, tears or shrinks or because it is too stretchy, or for numerous other reasons that are obviously important factors in the fabric's performance. The desirable level of fabric performance is defined in terms of the intended end use, ultimately by the user.

It is very important to predict a textile fabric's performance by testing. Fashion merchandisers, apparel designers, interior designers and textile scientists who have an understanding of textile properties and testing are equipped to make decisions that will benefit their clients and enhance profits for their businesses. Knowledge of fabric testing and its performance analysis can contribute to efficiency in solving consumer problems with textile products, and to the development of products that perform acceptably for consumers. As indicated above, retail buyers and producers of apparel and textiles are among those who use the fabric testing data and results in making decisions about their products. Most textile or apparel manufacturers will use either test methods or performance specifications that are published by testing organizations.

Innovations in fabric development have taken a new path. The applications of high performance and functional fabrics have been expanding rapidly. These fabrics have enhanced performance attributes and functionalities over commodity fabrics, and are used in areas such as:

- Protective garments (e.g. ballistic and stab resistant fabrics, UV protective wear)
- Functional fabrics (e.g. photochromic textiles)
- Textiles for acoustic applications (e.g. in automotives)
- Smart and electronic textiles (e.g. fabric sensors and actuators).

According to textile intelligence, 'performance textiles' represent one of the fastest growing sectors of the international textile and clothing industry. It has been estimated that in the European Union, the sports market alone is worth over 37 billion euros. In the USA it is worth around US$46 billion.

1.3   Need for integration of fabric testing literature

A number of textile research and testing organizations have published data on fabric testing and their procedures. There is a wide range of textile testing procedures for different fabrics for different end uses. The current literature on fabric testing is available in the form of textbooks and published articles. Researchers all over the world have been constantly involved in developing newer methods of fabric testing so as to meet the ever growing globalization and quality requirements. Their researches have resulted in extensive quality data and testing procedures for fabrics. These results should provide industries, fabric suppliers, apparel manufacturers, exporters, fashion designers and retailers with an enormous amount of information about the testing aspects of fabrics and apparel to meet the international standards. It appears that coverage of the existing literature in textbooks on fabric testing procedures and results is insufficient, although there have been many research achievements by scientists, researchers and industry experts in the areas of apparel, industrial fabrics such as technical fabrics, intelligent fabrics for special applications, nanotechnology applications, medical textiles, etc. Hence, a systematic approach towards integrating the knowledge available in the literature on fabric testing and developments in different aspects of fabric testing and the achievements of researchers and industry veterans would greatly benefit all those involved in quality assessment and evaluation of textile products.

1.4   Scope of fabric testing

The performance of a fabric is ultimately related to the end-use conditions of a material. The physical, chemical, physiological and biological influences on fabrics affect their end-use performance (Saville, 1999). Although all agents affect textile performance at the fibre, yarn and fabric levels, emphasis is generally given to fabrics since they represent the largest class of textile structures in a variety of applications.

Thus, a fabric is usually the most complex and representative form of a textile structure that is subjected to these agents and influences in most end uses. Testing of fabrics and quality control is broad in its scope (Fig. 1.1). It

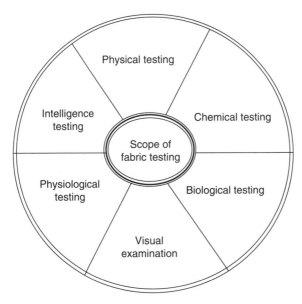

1.1 Scope of fabric testing.

can include, for instance, the means for determining and controlling the quality of a manufactured product. It can be used to measure the outside factors that influence the test results. Testing of fabrics to the above influences of a physical, chemical and biological nature would be of great help to manufacturers in adjusting their process control parameters to produce the right material. An understanding of the visual examination of fabrics for their surface characteristics, shape, texture, etc., would be useful to designers in making proper selection of fabrics for a particular end use. These days, a number of fabrics are being modified to act as 'smart fabrics' in various industrial applications. These fabrics incorporate wearable electronic devices and gadgets to serve a specific function such as heat resistance, breathability, protection against external conditions, biomimicry, etc. The testing of the intelligent properties of these fabrics is increasingly important in the modern textile world.

1.4.1   Physical testing

The first broad class of factors that affect the performance of fabrics are physical agents and influences. These may be further subdivided into mechanical deformation and degradation, tactile and associated visual properties of fabrics (such as wrinkling, buckling, drape and hand) after their use and manufacture, and their response to heat, liquids and static charge. The testing of fabrics to mechanical deformation is very important and refers to fabrics that are subjected to variable and complex modes of deformation. They include tensile behaviour, compression, bending or flexing, shrinkage, abrasion resistance, frictional rubbing, torsion or twisting, and shear. Fabrics with special features or constructions require either additional tests or modification of existing tests for conventional fabrics to characterize adequately their mechanical and related properties. Coated fabrics must be evaluated not only for their mechanical integrity and behaviour but also for their bonding integrity of the coating to the fabric (Vigo, 1994).

The development of new fabrics for industrial and functional purposes has introduced a new set of physical testing procedures for these fabrics. Conductive fabrics used in conjunction with metals for detecting electromagnetic signals need to be tested for their electromagnetic behaviour. High performance clothing demands the testing of parameters such as impact resistance, thermal resistance, moisture vapour transmission, etc. Comfort and aesthetic properties of speciality fabrics is another area that needs to be addressed. Wear resistance of protective fabrics has to be considered in physical testing.

1.4.2 Chemical testing

Chemical and photochemical exposure of textiles may lead to yellowing or discolouration of undyed fabrics, to fading of dyed fabrics, and/or to degradation of dyed and undyed fabrics. These adverse results are due to depolymerization of the polymer chain in the fibre that may occur by hydrolysis, oxidative processes and/or crosslinking. Textile fabrics have varying degrees of resistance to chemical agents such as water and other solvents, to acids, bases and bleaches, to air pollutants and to the photochemical action of ultraviolet light. Resistance to chemical agents is dependent on fibre type, chemical nature of the dyes, additives, impurities, finishes present in the fibre, and to a lesser extent on the construction and geometry of the fabric. The testing of fabrics towards the above influences is very important in assessing the performance of the fabrics for various end uses (Jones, 1981). The development of chemically resistant protective clothing and textile filtration media has led to test methods relevant to these end uses.

1.4.3 Biological testing

Textile fabrics may be adversely affected by various microorganisms and insects. The effect of biological agents on textile fabrics is important for enhancing their end-use performance in many areas. Fabrics will have desirable aesthetic qualities if they can suppress odour-causing bacteria and other types of odour-causing microorganisms. The hygienic and medical effectiveness of fabrics is required to prevent the growth of dermatophytic fungi (those that cause skin disease), pathogenic and potentially lethal microorganisms on fabrics and to prevent their infestation by insects. Finally, prevention of fibre discolouration and degradation, usually by fungi and insects, prolongs the useful life of the material. Testing of fabrics and evolution of specific test methods for the above biological influences would help manufacturers, retailers and users of fabrics to develop strategic ways to maintain and protect their fabrics in storage and transportation. These tests would be useful for rapid screening of various modified and unmodified fabrics for their ability to withstand biological attack.

1.4.4 Visual examination

Fabrics can be evaluated for a variety of attributes to assess their performance by visual assessment either manually (subjective assessment) or by objective evaluation techniques. Visual examination of fabrics includes evaluating the texture, surface characteristics, dye shade variations, design details, weave patterns, construction particulars, pilling assessment, etc.

Subjective or objective measurement techniques may be employed to determine the above properties of fabrics. New types of fabrics and garments require careful examination of their surfaces for change in functional groups due to the application of finishes such as shape memory polymers or plasma treatments. Detailed microscopic examination can reveal distinctive microscopic and macroscopic features of fabrics. SEM, TEM and other microscopic tests may be used to identify the details. Objective measurement using image analysis is a new area in which a fabric's surface is analysed for texture, change in colour and surface modification due to a special finish application. High performance scanning can be used to visualize the colour aspects of a dyed and printed fabric, and colour coordinates can be measured using advanced colour physics principles.

Defect analysis is another major area today and is widely discussed in the textile industry. Defects are bound to occur in fabric during manufacture for a number of reasons. The quality of the final fabric is assessed based on the minimum number of defects present in it. Visual examination is a preliminary tool to detect faults in the fabric before they are being processed further. A good visual examination technique would help identify the faults and ensure that necessary corrective measures are undertaken by the respective departments to reduce the incidence of faults. Computerized image processing techniques are now widely used in the textile industry. The complex problem of fabric quality control through defect analysis may be solved by means of computer vision using advanced digital signal and image processing tools. Many of these image processing applications aim at detecting textural characteristics and textural defects of fabrics, including colour detection and dye shade variations.

1.4.5 Intelligence testing

In the last decade, research and development in smart/intelligent materials and structures have led to the birth of a wide range of novel smart products in aerospace, transportation, telecommunications, homes, buildings and infrastructures. Although the technology as a whole is relatively new, some areas have reached the stage where industrial application is both feasible and viable for textiles and clothing.

Intelligent textiles are fibres and fabrics with a significant and reproducible automatic change of properties due to defined environmental influences. They represent the next generation of fibres, fabrics and articles produced from them. They can be described as textile materials that think for themselves. This means that they may keep us warm in cold environments or cool in hot environments or provide us with considerable convenience and even fun in our normal day-to-day lives, for example through the incorporation of electronic devices or special colour effects. The most important intelligent materials at present are phase change materials, shape memory materials, chromic materials and conductive materials. Many intelligent textiles already feature in advanced types of clothing, principally for protection and safety and for added fashion or convenience. The testing programmes must include the testing of these fabrics to meet the ever growing demand for hi-tech fabrics and garments.

A shift in consumer values has occurred; instead of wanting the finest natural materials, people look at the engineered beauty, innovative design and intelligent aspects of products. Working closely with the clothing industry will develop the base that is needed to offer developments in intelligent clothing with huge commercial potential at minimum risk. At a later stage of development, such cooperation is likely to create more solid product assortments. Veterans expect that smart clothing technologies will be launched in the market within the next five to ten years. Hence there exists an absolute necessity to understand these new fabrics and their technologies. Testing of fabrics hitherto limited to traditional fabrics such as apparel, home furnishings and some varieties of technical textiles may not help in the long run to understand the properties of these new fabrics. New methods of testing and evaluation for intelligent and smart fabrics will become extremely important in the industry as the future relies more and more on these textiles.

We are still far from taking full advantage of the potential of information technology services, but the future for fully soft electronic products is very attractive and requires a different, but interesting, design approach. The geometric and mechanical properties of textiles (large flexible area) differ strongly from those of conventional electronics and can create new computer designs and architectures.

1.4.6 Physiological testing

Fabric physiology deals with the physiological characteristics of fabrics that are expressed in the well-being, performance and health of the wearer. It covers the areas of physics, chemistry, medicine, physiology, psychology and textile technology. Three important physical parameters that are instrumental in the physiological processes of fabrics are heat transmission, moisture transport and air permeability (Welfers, 1978). The physiological properties of fabrics relate to what the fabric or garment feels like when it is worn next to the skin, such as too warm, too cold, sweaty, allergic, prickly, etc. The psychological properties include mainly the aesthetics of the fabric such as colour, fashion, prejudice, suitability for an occasion, garment style, fabric finish, etc. (Smith, 1986).

Clothing is designed to maintain a hygienic and comfortable zone about the human body in which one feels well, even if inner or outer influences change rapidly. The zone in which the temperature, moisture and air circulation are properly matched is called the 'comfort zone'. The so-called microclimate that prevails there is defined by definite physical and physiological conditions. There are physiological and psychological positive comfort sensations but these tend to be more individualistic and less frequently noticed in the wearer of the garment. Therefore, in the assessment of a fabric or garment for a particular end use, the comfort of that product is considered to be very important. Fabric testing therefore needs to address the comfort properties of fabrics.

1.5   Importance of fabric testing

Globalization of the clothing industry and increased competition in the world market have encouraged consumers to expect high-quality garments at affordable prices. The quality of a garment, as normally perceived by a customer, depends on its aesthetic appeal, its ability to drape gracefully, its 'handle' and durability. These depend largely on the quality of the fabrics used and the making-up process (Potluri et al., 1995). However, a more expensive fabric does not necessarily result in a better-quality garment. The colour and design of a fabric, along with drape, contribute to the aesthetic appeal of a garment.

During the past few years, the demand for quality textiles has increased globally, with the steady growth in population and income resulting in a rise in production and usage of different types of fibres, yarns and fabrics. To compete, especially in the world market now, there is greater demand for consistency in quality rather than for quantity of products. This implies production processes that guarantee overall quality – 'built-in' instead of merely 'inspected-on'. Today, quality assurance programmes have become necessary for survival not only in textiles but in every branch of industry.

Testing of fabrics has attained an important position in the textile industry due to the development of new types of fabrics for various apparel, furnishing and industrial applications. It has become almost mandatory for any textile manufacturing activity to carry out testing and evaluation of their textile products such as yarns and fabrics to meet international standards and the customer's satisfaction. Fabrics undergo a number of deformations during their use and a systematic analysis of the defects occurring in the fabrics is a key role played by the testing department.

The importance of fabric testing lies in the fact that in order to control the product and its cost, testing the performance of the goods becomes absolutely necessary. Fabric testing would benefit many in the industry and those involved in the export business. Professionals developing new fabrics use results from testing in selecting the right raw materials (Shaw, 1985). Decisions based on the accurate results of testing result in fewer rejections and customer complaints. Designers who create new fashions for the high-fashion industry are sometimes confronted with the problem of selection of the right quality of fabric for their garments. Fabric testing results would help them to understand the construction, properties and behaviour of fabrics for a particular end use. The textile scientist also stands to gain from a thorough understanding of fabric testing and analysis. Although one may be a specialist in textile chemistry or textile engineering, an understanding of how physical tests relate to fabric performance and consumer expectations is a necessary prerequisite to successful development of new textile fabrics. There are various stages in which fabric has to undergo quality inspection and this would reduce the burden on management to supply a fabric consignment with minimum defects. Proper testing on a regular basis can make the difference in the success or failure of a product and indeed the whole business.

There are a number of points in the production cycle at which testing may be carried out to improve the product or to prevent sub-standard fabric progressing further in the cycle. Fabric testing becomes important from the point of view of the following considerations.

1.5.1 Quality control for manufacturing

The meaning of the term quality is elusive: everybody has their own idea of what is meant by it, but it is difficult to express the idea in a concrete form. However, in order to produce a quality product, manufacturers need to have a definition of quality which will allow them to measure how far their products meet the requirements. Quality can be defined in two broad dimensions: perceived quality and functional quality. The perceived quality supports the corporate image, creates interest and generates an initial purchase. The buying is done on the basis of three basic parameters – the item looks good, feels good and offers good value for money – whereas the functional quality is related to the supplier and involves make, size and performance.

The buyer is responsible for selecting the best product on the basis of perceived quality (Garner, 1977). The supplier must produce this product and also achieve the functional aspects of quality. The key to all quality is the need for a 'standard'. A customer (buyer) in total expectation of quality demands a 'standard' quality and therefore this must be defined as 'an agreed' standard between the customer and the supplier.

Quality control refers to the performance on a periodic basis of certain tests designed to measure the characteristics of the raw or processed material. Regular controls on the quality of fabric produced at every stage of manufacture become most important for the fabric to live up to international standards. Such controls must conform with the following:

- The testing standards established by an individual organization, for example most of the big importers and retail store-chains have their own standard specifications and test methods for various clothing items.
- Established scientific specifications, such as those laid down by ASTM, AATCC, ISO, etc., and by other authorities of various countries.
- Market requirements or standards, such as meeting the requirements for width, ends/picks and weight for certain staple fabrics listed periodically in trade journals.
- Consumer needs or demands, for example to forecast the effectiveness of a material to meet consumer needs for wear or dye-fastness or to assess its satisfactory performance for the end use of the fabric.

1.5.2 Selection of fabrics

The production cycle as far as testing is concerned starts with the delivery of raw material. If the material is incorrect or sub-standard then it is impossible to produce the required quality of final fabric. Proper selection of fabrics becomes highly important for fashion designers, retailers, exporters and scientists and a thorough checking of fabrics for defects and construction particulars is a key step in preventing rejection of final consignments. Fabric manufacture consists of a number of separate processes such as winding, warping, sizing and weaving before the final fabric is produced. Knitted and non-woven fabrics form a separate section and all these processes demand the checking of quality at every stage of their production. The final product has to be checked for the required properties so that unsuitable material can be rejected or appropriate adjustments made to the production conditions.

1.5.3 Production monitoring

Production monitoring, which involves testing fabric samples taken from the production line, is known as quality control. Its aim is to maintain, within known tolerances, certain specified properties of the product at the level at which they have been set. A quality fabric for these purposes is defined as one whose properties meet or exceed the set specifications. Besides the need to carry out the tests correctly, successful monitoring of production also requires the careful design of appropriate sampling procedures and the use of statistical analysis to make sense of the results.

Online production monitoring systems monitor fabric quality and raise product packing efficiency. Using these systems it is possible to inspect the final product and collect all fabric defect data. This data provides precise information on the fabric rolls to optimize packing, to give access to all fabric defect data which gives the true picture of fabric quality, and to make snap cost-saving decisions in the manufacturing. In addition, online production monitoring tools provide an organization with a complete inventory of defects, while the fabric is being inspected on the folding table. They record all the defects, their location on the roll and their severity. Based on the final product specifications, it is possible to present an optimal cutting–mending master plan.

1.5.4 Assessing the quality of the final fabric

In this process the bulk production is examined before delivery to the customer to see if it meets the specifications. By its nature this takes place after the fabric has been produced. It is therefore too late to alter the production conditions. In some cases selected samples are tested and in other cases all the material is checked and steps are taken to rectify faults. For instance, some qualities of fabric are inspected for faulty places which are then mended by skilled operatives; this is a normal part of the process and the material would be dispatched as first quality.

1.5.5 Investigation of faulty material

If a faulty material is discovered either at final inspection or through a customer complaint, it is important that the cause is isolated. This enables steps to be taken to eliminate faulty production in future and so provide a better quality product. Investigations of faults can also involve the determination of which party is responsible for faulty material in the case of a dispute between a supplier and a user, especially where processes such as finishing have been undertaken by outside companies. Work of this nature is often contracted out to independent laboratories that are then able to give an unbiased opinion.

1.5.6 Product development and research

Product development is an important aspect in any manufacturing activity. The need for new product development arises from aggressive global competition, rapidly changing technologies, increasing complexity of markets and diversifying consumer trends. Today's consumer in the textile industry wants more variety in fabrics and garments, and better quality products at a lower price. Before the 1980s, competition in the business was in manufacturing capacity, whereas during the 1980s competition arose in product development and since the 1990s product development has been the battleground in any textile and garment business. The product development process converts ideas (inputs) into products (outputs) using the company's

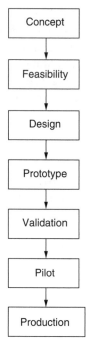

1.2 Product development cycle.

workforce of designers, marketers, production personnel and accountants. A product development cycle can be represented as shown in Fig. 1.2.

Textile fabrics are evaluated during the development process. This helps textile scientists determine how to proceed at each stage of development. In the textile industry technology is changing all the time, bringing modified materials or different methods of production. Before any modified product reaches the marketplace it is necessary to test the material to check that the properties have been improved or have not been degraded by faster production methods. In this way an improved product or a lower-cost product with the same properties can be provided for the customer.

A large organization will often have a separate department to carry out research and development; otherwise it is part of the normal duties of the testing department. This section also includes testing in order to study theories of fabric or fibre behaviour. With advances in research and development, new products and processes may require testing procedures that are not provided through standard test methods. Test methodologies developed for a specific research application within one laboratory often gain wider acceptance and eventually are developed into industry-wide standard test methods.

1.5.7 Ecological considerations

In the past, the quality requirement of consumers for textile and garment products has concentrated on the comfort and performance of fabrics. Accordingly, regular testing usually focuses on the dimensional stability test, colour fastness, fabric construction and composition analysis, fabric performance test, fibre and yarn tests, etc. In developed Western countries, this perception is changing and a new element, ecological concern, is also becoming important. Especially in northern Europe, there is an increased awareness of environmental protection and the potentially harmful effects to human health from chemical processes and ingredients in food and clothing (Chan, 2007). This has resulted in an extra demand on testing and product certification of textile products. New laws and directions require manufacturers to make their products safer by minimizing the use of chemicals with hazardous substances and heavy metals such as cadmium, azo dye, chromium, lead and arsenic. While improving the quality of textile fabrics, ecological factors cannot be overlooked. The textile industry uses many chemical pollutants, allergens and carcinogens. These have to be severely restricted by laying down ecological requirements (Bhattacharya and Varadarajan, 1993). Only limited use of various chemicals such as azo dyes, heavy metals, harmful odours, etc., should be permitted. The textile industry also needs to address the problem of indiscriminate disposal of waste water loaded with toxic chemicals. Added quality control not only covers the materials purchased, but also includes all stages of production, quality control, final processing, garment wash, even the packaging and stock management. It thus provides much greater assurance and peace of mind. From the point of view of ecological requirements, testing of parameters such as prohibited azoic dyestuffs, carcinogenic and allergy-inducing dyestuffs, formaldehyde, pesticides, chlorinated phenols, chloro-organic carriers, extractable heavy metals, nickel, phthalates in baby articles, butyltin compounds (TBT and DBT), emission of volatile components, odours, etc., becomes extremely important.

1.5.8 Teaching and scientific institutions

A large number of technical and fashion institutes and universities are today offering quality education in textile engineering, technology and fashion-related subjects. Textile testing is a major subject of study in their courses. The testing of fabrics needs to become a regular subject in the curriculum, as fabrics play the major role in the textile and garment industry. Textile fabric testing should include laboratory experience for students, although there is hardly time in one semester of laboratory work for a student to perform all or even most of the tests that are included in fabric performance specifications. Students need to become comfortable reading actual standard specifications and test methods as they carry out their laboratory work. A systematic approach towards teaching of testing of fabrics using subjective and objective techniques will help the students and the faculty to understand the newer methods of testing.

Today's textile industry is in the transition zone between traditional textile production and the realization of highly focused design and production of added-value textiles. The innovative field of smart and intelligent fabrics is becoming increasingly popular and commercially successful because it combines product use with new material properties. More and more instrumental analysis techniques and the evolution of new test methods for testing intelligent textiles, for example testing the thermoregulatory response of a phase change material, or the shape memory effect of a shape memory garment, photo-responsive testing of fabrics, self-cleaning tests, testing of medical textiles and implants, and testing of technical textiles, would really help both students and the faculty to enrich their knowledge of testing procedures and standards. It is necessary to introduce future-oriented design concepts and textiles with high quality and added value that are focused on people's needs and that integrate high technology and design in a sensible manner. It is important to educate fashion and textile designers within the area of advanced fabrics and state-of-the-art technologies to prepare students for situations they will face once they have entered the job market. In addition, interaction between academic institutions and the industry needs to be improved by coordinating the testing activities, and they need to help each other in developing new testing procedures and standards for establishing constant quality standards for the industry.

1.6   Current status of fabric testing

Concern about product quality is universal in business and industry. Everyone in the industry needs his product to be tested according to the international standards so as to ensure quality in all respects. This will largely depend on the testing programme adopted by the testing house to evaluate the quality of the fabrics. Currently, fabrics are tested at different testing houses for their quality particulars before being exported or sold in the market. There are large commercial organizations that have set up their own laboratories and standards to assess the quality of their products to satisfy their customers. On the other hand, there exist a vast majority of private testing organizations that take up testing of fabrics on a commercial basis for the industries, testing their products as per the standards. In addition, there are a number of governmental and approved research organizations throughout the world that test and certify the fabric testing results to meet the standards. Hence, the whole scenario of fabric testing is not integrated to a particular area and the testing is not completely standardized and integrated.

1.6.1 Commercial testing

Globalization has clearly led more and more companies to produce newer products to meet the ever-growing demands of the textile industry. A company in the USA buying products from Asia and Europe may use a number of external testing laboratories and these companies are looking for more testing methods that will satisfy everyone. Today there are many laboratories for testing fabrics. They include in-house company or manufacturers' laboratories for testing. Thorough testing of fabrics is carried out in these in-house laboratories before being released to the market. On the other side, the buyers who buy the products may have their own testing laboratories for testing the incoming goods for their quality and standards.

Commercial testing of fabrics by private and governmental nodal agencies has made tremendous progress as these agencies have set up their laboratories throughout the world to undertake testing of fabrics according to the international standards. Laboratories such as AATCC, SGS, TRI, the Atlas Material Testing Laboratory in the USA, the Hohenstein Institutes in Germany, the Centro Controllo Tessile in Italy, the Korea Textile Inspection and Testing Institute in Korea, Covitex in Europe, Contexbel in Belgium, Inotex in the Czech Republic, BTTG in the UK, the Textiles Committee and SGS in India, etc., have been continuously catering to the needs of the industry and the exporters. The laboratories set up by the research organizations also cannot be ignored. The contribution of these non-profit testing organizations is tremendous in developing new sets of testing standards and procedures for various types of fabrics. These laboratories incorporate advanced testing instruments and objective techniques to carry out testing of fabrics to meet customer requirements.

Most competent fabric manufacturers have their own test procedures in addition to the recommendations of the above professional organizations. New and innovative fabrics that are coming on to the market, such as intelligent fabrics, nano-textiles, medical implants and various technical fabrics, must be watched carefully and new testing schemes must be introduced in order to ensure performance of these fabrics and their acceptance by customers. It is most important to test the fabrics in order to control their cost.

1.6.2 Research and development

The role of the research organizations involved in testing of fabrics and other products is to collect relevant data or facts and invent falsifiable hypotheses about relationships among these facts. Experiments or tests are then conducted to verify the predictions of the hypotheses. In addition, the organizations set up standards for the testing of fabrics and garments to be eventually followed by the industries. A great amount of time and effort has been spent by researchers and scientists to develop new fabrics to cater to the needs of the industry. The direction of research suggests that these new fabrics will make headway in the coming years and a revolutionary change might be expected in areas such as industrial textiles, intelligent fabrics and medical textiles. The credit goes to the scientists and researchers of various textile institutes of great repute and the research community in the industry and several other nodal organizations. A wealth of information is available on their research and it is time for industries, exporters and buying houses to borrow their scientific findings and research results and to incorporate them into their manufacturing and testing programmes. Such an effort to integrate the results of researchers and scientists with the test procedures followed by the industry would go a long way towards understanding the procedures for testing new types of fabrics and their characteristics in order to alleviate the problems of misunderstanding among manufacturers, retailers and exporters. The integration of industry practices and the scientific results would help standardize the testing practice for existing as well as newer types of fabrics in the long run.

1.7   Standards for tests

The term 'standard' is used often in regard to testing of products. It may be ambiguous at times as it can have several different meanings. It can refer to the actual test methods or to the minimum acceptable level of performance on a particular test. Any testing which is to be done on a product will need to be done by employing standard test methods. In this way, every possible variable within the test method will be precisely controlled. The reason for this is that reproducibility must be absolutely assured. That is, the test results in a plant or laboratory will need to be the same as those obtained within the customer's laboratory; otherwise, lawsuits would abound.

Test methods are developed for textiles and textile products by several different organizations. They are typically developed in response to a need expressed by an individual manufacturer, a product user, or occasionally by a consumer group. In most organizations that develop standard test methods, once the test procedure is clearly defined, the proposed method then undergoes interlaboratory trials. Interlaboratory testing can reveal problems with procedures that must be corrected and they can also be used to determine whether the test method is applicable to a particular type of product; for example, does the method work only on woven fabrics, or can it also be used to determine the precision of the test? Precision indicates whether the tests will repeatedly produce the same results on the same fabric specimen. Interlaboratory tests determine the reproducibility of the test from one lab to another and from one operator to another. A test which has a high level of precision has good interlaboratory reproducibility and good between-operator reproducibility.

Throughout the world there are numerous organizations that develop standard test methods and performance standards for textiles. Because of the increasing global market in textiles and apparel, a growing need exists for uniformity of standards on an international basis. This would alleviate some of the problems faced by manufacturers that export, and countries that import from foreign manufacturers. For example, products that are imported to the United States must meet standards set by the United States, regardless of the standards or methods that exist in the country where the products are manufactured.

1.7.1 International organization for standardization

The International Organization for Standardization (ISO), based in Geneva, Switzerland is an organization that serves member organizations throughout the world. There are three categories of membership in ISO. These are member body, correspondent member and subscriber member. A member body is the national organization that is most representative of standardization in its country. The member body which represents the United States is the American National Standards Institute (ANSI). Other countries have comparable organizations that are member bodies, such as the Standards Council of Canada (SCC), the British Standards Institution (BSI), Standards Australia (SAA), the Bureau of Indian Standards (BIS), the China State Bureau of Technical Supervision (CSBTS) and Ente Nazionale Italiano di Unificazione (UNI), which represents Italy. Member bodies are responsible for informing potentially interested parties in their respective countries of relevant international standardization initiatives and assuring that a concerted view of each country's interest is represented during international negotiations leading to standards agreements.

1.7.2 Major American organizations

American National Standards Institute

ANSI represents the United States member body in the ISO. The purpose of ANSI is to coordinate voluntary standards development and use in the United States and to serve as liaison between standards organizations in this and other countries, through the ISO.

American Society for Testing and Materials

The purpose of this organization is to develop standards on characteristics and performance of materials, products, systems and services. The standards developed by ASTM include test methods, specifications and definitions and usually deal with physical properties of materials. ASTM writes standard tests not only for textiles but also for virtually every other product such as steel, plastics, lumber, etc. For textiles, ASTM writes primarily physical-type tests such as methods for testing the tensile strength, abrasion resistance, twist determination, fibre maturity, denier and yarn count, among many others.

American Association of Textile Chemists and Colourists

AATCC was founded to promote greater knowledge of textile dyes and chemicals and therefore is concerned specifically with textile products. This organization works very closely with ASTM but writes chemical-type tests. In addition to the development of test methods, AATCC sponsors scientific meetings and promotes textile education. The activities are concerned primarily with the chemical properties of textiles in contrast to ASTM's emphasis on physical properties.

1.7.3 Other national organizations

Apart from international organizations and major American standards organizations, many countries have their own national standards organizations, for example BSI (Britain), BIS (India), JIS (Japan) and DIN (Germany) standards. The same arguments that are used to justify national standards can also be applied to the need for international standards to assist worldwide trade, hence the existence of International Organization for Standardization (ISO) test methods and, within the European Union, the drive to European standards.

Japanese Industrial Standards

Japanese Industrial Standards (JIS) specifies the standards used for industrial activities in Japan. The standardization process is coordinated by the Japanese Industrial Standards Committee and published through the Japanese Standards Association. The JIS in many ways has been Japan's answer to ISO. The JIS is extremely sophisticated and complex and goes beyond the requirements of the ISO 9000 series but essentially performs the same quality management function. The JIS is more rigorous and comprehensive in standards, making it extremely challenging for an organization to successfully implement. This fact has led to its adoption almost exclusively in Japan and makes its requirement outside Japan very rare. Organizations that have a JIS certification can be considered to be at least as good as if not better than an organization that has an ISO 9000 certification, with most JIS systems being closer in scope to ISO 9001.

Bureau of Indian Standards

The Bureau of Indian Standards (BIS), the national standards body of India, is involved in the development of technical standards (popularly known as Indian Standards), product quality and management system certifications and consumer affairs. Apart from setting standards for textiles, the organization sets standards for other major industries in the country. It resolves to be the leader in all matters concerning standardization, certification and quality.

British Standards Institution

The British Standards Institution (BSI) is the world's leading standards organization, and facilitates the setting of standards, inspections and quality management. It is a non-profit distributing organization, independent of government, industry or trade associations, whose operating divisions (product certification, quality assurance, standards, testing and training services) are designed to further the use of standards. Manufacturers, importers and retailers rely upon BSI testing to independently assess the performance of their products for safety, reliability and quality. Its certification trade mark, the Kitemark, has been established for more than 90 years.

German Institute for Standardization

With its 81 standards committees the German Institute for Standardization (DIN) is the responsible standardization body of the Federal Republic of Germany. Within the framework of its terms of reference, DIN is the German member in the European and international standardization organizations. DIN represents the interests of German standardization at both the national and international levels. Maintaining a collection of more than 38000 documents, DIN offers standards covering everything from textiles to standard methods for analysis of contaminants in water, to screws and bolts. Standards offered by DIN provide companies and individuals doing business in Germany or with German businesses a solid basis for quality, safety, and minimum functionality expectations.

1.8   Future trends

The textile industry is poised for a phenomenal growth in the coming years. With the introduction of new types of fibres, yarns and fabrics for high-end applications, the testing for the performance of these new materials has to undergo rapid changes and take new dimensions. Fabrics are no longer meant only for traditional apparel and the demand for new types of fabrics such as technical fabrics and medical textiles for applications in industrial segments is increasing. The traditional way of testing fabrics for their performance properties may not be critical for the sustainable growth and requirement of standards for fabrics to meet the international challenges. New methods of testing and procedures are the need of the hour because fabric testing plays a crucial role in gauging product quality, assuring regulatory compliance and assessing the performance of textile materials for future requirements. In the wake of globalization and faster assessment of fabric quality parameters and to meet the international challenges, the future of fabric testing relies on the following considerations.

1.8.1 Manual to automatic testing

Fabric testing relies more on the manual methods of testing. There is always an element of human error in the testing process, which leads to spurious results and affects the final product evaluation. In addition, the reproducibility of the results may not be possible with manual methods to reconfirm the results of testing. There is a limitation on the testing speeds that one can achieve during manual testing. The increased stress factor on the human mind also cannot be ignored. All these factors have resulted in the introduction of automatic testing procedures using high-speed instruments to evaluate the performance of the fabrics.

The trend is definitely changing. Many fabric and garment tests that once required manual preparation and expert judgement have adapted to accommodate devices that take over part or all of the testing processes (Thiry, 2002). Instrumental testing is fulfilling the needs in quality testing labs, in the production environment and in research and product development. One of the most compelling reasons for the rise of instrumentation in fabrics and garment testing is the increasing globalization of the textile industry. Materials for an individual garment or a line of garments are often sourced today from a variety of suppliers, literally from around the globe. It is important to retailers to increase the amount of testing offshore before the garments arrive at their destination. However, they find that it is impossible to standardize testing conditions around the world. Appropriate instruments help disparate labs to produce reliable and repeatable results. In addition to the several reasons mentioned above, the move towards instrumentation in the testing of fabrics has come as a result of the need for repeatable and reliable results, the lack of educated manpower to perform manual tests, the savings in time and money and the globalization of the textile industry. In some areas, instrumentation is well established. In others, researchers are developing new technologies. Despite compelling needs, the movement towards instrumentation is progressing in fits and starts, and as instrumentation moves more into fabric and garment testing, test methods will have to be adapted to include instrumental measurements for accurate and reliable results.

With the development of new sensors and faster computers at cheaper prices, online measurement and quality control is becoming a reality in the textile industry. Online measurement of length, speed and shrinkage of a fabric, using the latest online equipment, is widely used in automotive testing. Online fabric fault detection using digital image processing, opto-electronic processing and online monitoring for uniformity of non-woven fabrics has to be considered in future. Fabric mechanical properties vary along the length of a fabric roll. The properties measured on a few fabric samples do not adequately represent the entire length of the fabric roll. Therefore, it is desirable to measure properties continuously along the length of a fabric (Potluri *et al.*, 1995). Very little progress has been reported in this area. The ultimate objective of automation should be to achieve online continuous measurement of all the fabric properties, during finishing or the final inspection process. The fabric property data can be supplied with each roll of a fabric, which may be utilized during the making-up process.

Based on extensive research, it has been well established that garment quality and its making-up process can be controlled based on fabric mechanical and surface properties. Widespread use of this technology in industry would depend on the availability of totally automated test equipment, to eliminate human errors and improve productivity, in terms of the number of samples tested in a given duration. The development of robotic systems, capable of conducting all the fabric tests on a single sample without operator intervention, would completely eliminate the human handling and operative errors to achieve the best results of fabric testing.

1.8.2 Artificial intelligence

Artificial intelligence is a branch of computer science dedicated to the study of computational activities that require intelligence when carried out by human beings. Artificial neural networks and expert systems technologies have been particularly helpful in solving an array of problems in the textile industry and are one of the promises for the future in computing. Offering an ability to perform tasks outside the scope of traditional processors, they can recognize patterns within vast data sets and then generalize those patterns into recommended courses of action (Liu and Mandal, 2006a). Artificial neural networks offer the textile and apparel industry numerous opportunities, with potential applications in fabric inspection, fabric colour fastness grading, fabric comfort control and layout planning in apparel manufacture. The application of artificial intelligence through neural networks will become crucial in future testing of fabrics and garments. Such techniques can prove to be very useful in correlating the fabric structure with its properties and hence predicting the performance of the fabric. Simulation of fabric shapes and patterns for a particular body configuration can be studied through artificial intelligence techniques and a variety of fabric data can be handled at a great pace.

1.8.3 Subjective to objective measurement

In the clothing industry traditionally, the experts relied on their subjective estimations of the fabric properties without using any testing equipment. They examined the fabric by performing certain physical movements, such as stretching, bending, shearing and rubbing, and expressed their feelings in terms of subjective sensations, such as stiffness, limpness, hardness, softness, fullness, smoothness and roughness. These expressions formed the basis for fabric selection, though such subjective evaluation is not very accurate. This manually oriented estimation is being followed even today in the majority of clothing industry production lines and particularly in fabric handling for sewing.

These days, fabric testing is changing from subjective to objective measurement techniques. The objective measurement of fabric properties has attracted the interest of researchers from the beginning of the previous century. Fabric objective measurement of mechanical, geometrical, surface and large deformation properties represents a very powerful tool for quality control of fabric manufacturing, finishing and refining operations. It enables the setting up of an integrated computerized scientific database incorporating in objective terms the enormous wealth of experience of numerous experts who have worked in the textile and clothing industries over many years in different countries throughout the world. Test methods that require an advanced level of expertise and skilled judgement are today being performed by a workforce with diminishing abilities and experience. In the research community as well as in the clothing industry, there is wide interest in the objective measurement of fabric properties.

1.9   Conclusions

Textile testing has attained an important position in the textile industry. It has become absolutely necessary for any textile manufacturing activity to subject its products for scientific testing before being released to the market.

Fabric testing, a part of textile testing, is gaining importance as a variety of fabrics are being introduced in the market for specific end uses. The main objective of any textile operation is to make a profit. The past two decades have seen the marketplace change dramatically, making this objective harder to reach. Competition around the world in every area of textiles has become more difficult. In order for a company to make a profit, it has to change its strategy.

Fabrics are used in a wide range of applications in apparel, domestic and industrial areas. Applications include not only clothing and accessories, bedding and interior decoration, but also textile structures that are used to make cables, cords, parachutes, hot-air balloons, tents, etc. Further, new fabric composites and other textile fabrics can be traced in space structures, aircraft, machine parts, civil engineering, marine engineering, artificial limbs, sports goods, musical instruments and so on. The ever increasing application of textile fabrics in various fields is making textile companies emphasize value. In order to thrive in the marketplace they have to gain a sustainable competitive advantage. In order for many companies to achieve targeted competitive advantage, manufacturing facilities have to undergo a major transition in their testing programmes. This transition requires new tools and technologies for managerial decisions. More new and advanced fabric testing techniques such as automatic testing using high-tech instruments, and objective measurements of fabric sensory and mechanical properties must be introduced. Artificial intelligence techniques for quality and process control must become the routine test procedures for the testing houses and research organizations. The future certainly lies with the ability to test fabrics reliably and cost-effectively. This is only possible with the availability of totally automated test systems, to enable their use in an industrial environment.

The tedious process of sample preparation should be eliminated in future developments. A test system capable of measuring fabric properties on a large fabric sample should be developed, to eliminate the need for sample preparation and the difficulties associated with free curling edges. The programme of quality control in any organization must include instrumental analysis of data, neural networks for prediction of fabric properties, and assessment of colour and defect analysis through image processing techniques to streamline and simplify test methods. Finally, it is envisaged that online measurement of fabric mechanical, surface and dimensional properties would become a reality in the near future.

1.10   References

Adanur S, Slaten L B and Hall D M (1995), Textile testing, in *Wellington Sears Handbook of Industrial Textiles*, ed. Sabit Adanur, Technomic Publishing, Lancaster, PA, 231–271.

Bhattacharya N and Varadarajan S (1993), Importance of testing of textiles to eco-standard specifications, *Textile Dyer and Printer*, **18**, September, 1, 25–29.

Chan S (2007), Ecological textiles lead to change – Dynamic change in textile testing and requirements, CMA Testing and Certification Laboratories, *BFC Online*, China, 24 March.

Collier B J and Epps H H (1999), *Textile Testing and Analysis*, Prentice-Hall, Upper Saddle River, NJ.

Garner W (1977), Cloth quality–cost considerations of faulty materials, *British Clothing Manufacturer*, **13**, December, 25–26.

Grover E B and Hamby D S (1960), *Handbook of Textile Testing and Quality Control*, Interscience, New York.

Hildebrandt P (2006), Fabric testing: What's your score?, *Industrial Fabric Products Review*, March, **91**, 3, 44–47.

Jones E B (1981), Chemical testing and analysis, *Textile Progress*, **10**, 4, The Textile Institute.

Liu J and Mandal S (2006a), Artificial intelligence in apparel technology and management, *Textile Asia*, June, 30–35.

Liu J and Mandal S (2006b), Human v. artificial, *Textiles*, **33**, 3, 16–19.

McCullough T (1978), Importance of fabric testing, *Bobbin*, February, 158.

Mock G N (2000), Textile testing for quick response, *America's Textiles International*, **29**, 11, 48–51.

Potluri P, Porat I and Atkinson J (1995), Towards automated testing of fabrics, *International Journal of Clothing Science and Technology*, **7**, 2/3, 11–23.

Saville B P (1999), *Physical Testing of Textiles*, Woodhead, Cambridge, UK.

Shaw I P (1985), Fabric quality to meet modern garment making techniques, *Apparel International*, October, 15–17.

Slater K (1993), Physical testing and quality control, *Textile Progress*, 23, 1/2/3.

Smith J E (1986), The comfort of clothing, *Textiles*, **15**, 1, 23–27.

Thiry M C (2002), Objective and subjective measurement opinions, *AATCC Review*, December, 12, 10–14.

Vigo T L (1994), *Textile Processing and Properties, Textile Science and Technology Series – 11*, Elsevier, Amsterdam.

Welfers E (1978), Insights of garment physiology, *Melliand Textilberichte* (English Edition), June, 504–506.

2
Sampling and statistical analysis in textile testing

K F CHOI, The Hong Kong Polytechnic University, China

Abstract: The nature and the major reasons of sampling are described at the beginning of this chapter. In the study of sources of error, particular attention is put on the discussion of measurement precision of test methods in relation to repeatability and reproducibility. The essence of statistical analysis is on the treatment of variability. The reduction of variability is briefly discussed. In the application side, both point estimation and interval estimation are mentioned with special emphasis on the determination of sample size, i.e. number of tests. The methodology of comparing means using ANalysis Of VAriance (ANOVA) approach is illustrated with solid examples in the textile field.

Key words: statistical analysis, source of error, precision, accuracy, repeatability, reproducibility, point estimation, interval estimation, sample size, comparing means, hypothesis testing, analysis of variance, interaction effect, fabric sampling

2.1 Introduction: the requirement for sampling and statistics in textile testing

A sample is a subset of a population. The researchers have to decide on how to obtain the sample and, once the sample data are in hand, how to use them to estimate the characteristic of the whole population. Sampling involves questions of sample size, how to select the sample, and what measurements to record.

Sampling is usually distinguished from the closely related field of experimental design, in that in experiments one deliberately adjusts the parameters of a population in order to see the effect of that action. In sampling, one wants to find out what the population is like without disturbing it.

Sampling is also distinguished from observational studies, in which one has little or no control over how the observations on the population were obtained. In sampling one has the freedom to deliberately select the sample, thus avoiding many of the factors that make data observed by happenstance or convenience.

In many daily cases sampling is the only practical way to determine something about the population. Some of the major reasons for sampling include:[14]

- The destructive nature of certain tests. In the area of textile product development, yarns, fabrics and similar products must have a certain minimum tensile strength. To ensure that the product meets the minimum standard, the quality control department selects a sample from the current production. Each piece is stretched until it breaks, and the breaking point is recorded. Obviously, if all the fabrics or all the yarns were tested for tensile strength, none would be available for sale or use.
- The physical impossibility of checking all items in the population. The populations of fabrics are continuous pieces instead of isolated items. For each test specimen, the portions of the fabric not fixed by the jaws in the test machine are not involved in the tensile test, and in this sense not the whole population is evaluated even though all the fabric rolls in the population are cut into test specimens for testing.
- The cost of studying all the items in a population is often prohibitive.
- To include the whole population would often be too time consuming. In the commercial world, buyers will not wait for a year for test reports on their newly purchased fabric lots.
- The adequacy of sample results. Even if funds were available, it is doubtful whether the additional accuracy of 100% sampling (that is, studying the entire population) is essential in most problems.

2.2   Sampling and statistical techniques used

Sampling design[20] is the procedure by which the sampling units are selected from a population. Most inference problems in sampling are to estimate some summary characteristics of the population, such as the mean or the variance, from the sample data. Additionally, in most sampling and estimation situations, one would like to assess the confidence associated with the estimates; this assessment is usually expressed using confidence intervals.

Theoretically, if the sample size were expanded until all N units of the population were included in the sample, the population characteristic of interest would be known exactly. The uncertainty in estimates obtained by sampling is due to the fact that only part of the population is observed.

With careful attention to the sampling design and using a suitable estimation method, one can obtain estimates that are unbiased for population quantities, such as the population mean and variance, without relying on any assumptions about the population itself. The estimate is unbiased in that its expected value over all possible samples that might be selected with the design equals the actual population value. Additionally, the random or probability selection of samples removes recognized and unrecognized human sources of bias, such as conscious or unconscious tendencies to select units with larger (or smaller) than average values of the variable of interest. Such a procedure is especially desirable when survey results are relied on by persons with conflicting sets of interests. In such cases, it is unlikely that all parties concerned could agree on the purposively selected sample.

A probability design such as simple random sampling thus can provide unbiased estimates of the population mean and also an unbiased estimate of variability, which is used to assess the reliability of the sampling result. In addition to the unbiased or nearly unbiased estimates from the sample, researchers are looking for precise or low-variance estimates and procedures that are convenient or cost-effective to carry out.

In sampling theory it is assumed that the variable of interest is measured on every unit in the sample without error, so that errors in the estimates occur only because just part of the population is included in the sample. Such errors are referred to as sampling errors. But in real situations, non-sampling errors may arise as well. Errors in measuring or recording the variable of interest may also occur. Careful monitoring throughout every stage of a sampling process is needed to keep errors to a minimum.

2.3   Sources of error

The *Encyclopaedia Britannica* has the following description on measurement error. Measurement begins with a definition of the measurand, the quantity that is to be measured, and it always involves a comparison of the measurand with some known quantity of the same kind. If the measurand is not accessible for direct comparison, it is converted or 'transduced' into an analogous measurement signal. Since measurement always involves some interaction between the measurand and the observer or observing instrument, the effects of such interaction can become considerable in some types of measurement and thereby limit accuracy.

The problem of error is one of the major concerns of measurement theory.[11,16] A measurement is useless without a quantitative indication of the magnitude of that error. Such an indication is called the uncertainty. Without knowing the uncertainty,[9,10,13] the comparison of a measurement result with a reference value or with results of other measurements cannot be made. Among the various types of error that must be taken into account are errors of observation (which include instrumental errors, personal errors, systematic errors and random errors), errors of sampling, and direct and indirect errors (in which one erroneous measurement is used in computing other measurements). The above description of measurement error can be summarized through the following expression: measurement error = methodology error + instrumental error + personal error.

2.3.1 Precision and accuracy

A great deal of attention has been given to the measurement of the precision of test methods under ideal conditions.[7] Ideal conditions within a laboratory might be considered to involve one operator, using the same instrument, testing a sample of standard material, and duplicating the test within a short period of time (repeatability[5,15]). Performance under ideal conditions may have little relevance to the precision of the test method under routine conditions of use. In commercial tests – between buyer and seller, upon shipment and receipt, in national and international trade – it is important to know the routine but not the ideal precision so that decisions to accept or reject are not made on falsely optimistic premises. It is desirable to adopt a concept of the precision of a test as the inherent variation in results when a test method is used to compare materials in commerce. The inherent variation must be measured under conditions of actual use. Differences larger than the inherent variation, then, are real differences in material that have economic consequences in buying and selling.

Precision is a measure of the scatter of results when a test method is repeated. The standard deviation is a basic measure of scatter when the variable measurements follow a normal distribution. Within-laboratory precision limits would include random variation in material, different operators, different instruments and different days. Between-laboratory precision limits should be based on using samples of the same homogeneous material and a completely different set of operators and instruments and no restriction on time (reproducibility[5,15]).

Precision implies a state of control. Within a laboratory a state of control means that instruments are properly calibrated, operators are well trained and follow standard procedures, chemicals and reagents are fresh and environmental conditions are stable. A state of control indicates that variations in test results are random; departure from control means that some assignable, non-random cause has affected the result. Thus, precision involves random scatter only. The more precise a method, the smaller the random scatter.

Between laboratories the scatter is usually greater than within a laboratory because of differences in instruments, in calibration or operating procedures, in operator training, in reagents, or in environmental control. Differences between laboratories should be random differences rather than fixed biases. Inter-laboratory control[6] can be maintained by participation in check sample programmes (inter-laboratory calibration, or proficiency test).

Accuracy is the degree of agreement between the experimental result and the true value. Precision is the degree of agreement among a series of measurements of the same quantity; it is a measure of the reproducibility of results rather than their correctness. Errors may be either systematic (deterministic) or random (stochastic). Systematic errors cause the results to vary from the correct value in a predictable manner and can sometimes be identified and corrected. A popular cause of systematic error is improper calibration of an instrument. Random errors are the small fluctuations introduced in nearly all analyses. These errors can be minimized but not eliminated. Statistics is used to estimate the random error that occurs during each step of an analysis, and, upon completion of the analysis, the estimates for the individual steps can be combined to obtain an estimate of the total experimental error.

2.3.2 Reduction of test variability

It is stated in the ASTM Standard D4853[3] that there are three circumstances a test method would require to reduce test variability:

1. During the development of a new test method, ruggedness testing might reveal factors which produce an unacceptable level of variability, but which can be satisfactorily controlled once the factors are identified.
2. Another circumstance is when analysis of data from an inter-laboratory test of a test method shows significant differences between levels of factors or significant interactions which were not desired or expected. Such an occurrence is an indicator of lack of control, which means that the precision of the test method is not predictable.
3. The third situation is when the method is in statistical control, but it is desired to improve its precision, perhaps because the precision is not good enough to detect practical differences with a reasonable number of specimens.

Identifying probable causes of test variability

Sometimes the causes of test variability will appear to be obvious. These should be investigated as probable causes, but the temptation should be avoided to ignore other possible causes. To aid in selecting the items to investigate in depth, plot frequency distributions and statistical quality control charts. Make these plots for all the data and then for each level of the factors which may be causes of, or associated with, test variability. In examining the patterns of the plots, there may be some hints about which factors are not consistent among their levels in their effect on test variability. These are the factors to pursue.

2.4 Applications

The statistical theory has extensive applications in textile testing. Based on the theory, R&D personnel could determine the quantity of sampling units and estimate the range of their product's quality level under prescribed tolerance and chance factors. The details of the applications of the statistical theory are described in the following sections.

2.4.1 Point estimations of mean and variance

The estimations of population mean and variance are the major tasks in the general textile evaluation. The average quality is mostly recognized as the representation of the product quality. Sometimes uniformity of quality may be even more important than its average performance, e.g. for yarn strength, weavers would prefer more uniform yarn with slightly lower average strength instead of a high-strength yarn with occasional ultra-weak points.

In random sampling, the sample mean $\bar{x}$ and variance s^2 are unbiased estimators of the population mean μ and variance σ^2, i.e. $\hat{\mu} = \bar{x}$, $\hat{\sigma}^2 = s^2$.

$$\mu = \frac{\sum_{i=1}^{N} x_i}{N}$$

$$\sigma^2 = V(X) = \frac{\sum_{i=1}^{N}(x_i - \mu)^2}{N}$$

$$\bar{x} = \frac{\sum_{i=1}^{n} x_i}{n}$$

$$s^2 = \hat{V}(X) = \frac{\sum_{i=1}^{n}(x_i - \bar{x})^2}{n-1}$$

$$V(\bar{X}) = \begin{cases} \dfrac{\sigma^2}{n} & \text{for infinite population} \\[2ex] \dfrac{N-n}{N-1} \times \dfrac{\sigma^2}{n} & \text{for finite population with size } N \end{cases}$$

$$\hat{V}(\bar{X}) = \begin{cases} \dfrac{s^2}{n} & \text{for infinite population} \\[2ex] \dfrac{N-n}{N-1} \times \dfrac{s^2}{n} & \text{for finite population with size } N \end{cases}$$

The quantity $(N - n) / (N - 1)$ is the finite population correction factor, when N is much larger than n, i.e.

$$\lim_{N \to \infty} \frac{N-n}{N-1} = 1$$

On the other hand, when the sample size n is close to the population size N, the variance will become smaller until when $n = N$ the variance is zero, since there is only one possible way to take the sample and $\bar{x} = \mu$.

2.4.2 Interval estimation of mean and variance

With the point estimates of the population mean, it is desirable in addition to make an assessment regarding the precision of the estimate. This can be done by constructing a confidence interval within which it is sufficiently sure that the true value lies under the assumption that bias does not exist. The $100(1 - \alpha)\%$ confidence interval (I) of a population parameter θ is derived from the probability equation $\Pr(\theta \in I) = 1 - \alpha$. For the population mean μ, its $100(1 - \alpha)\%$ confidence interval I is given as

$$\left(\bar{x} - t_{\alpha/2, n-1} \times \frac{s}{\sqrt{n}} \sqrt{\frac{N-n}{N-1}}, \bar{x} + t_{\alpha/2, n-1} \times \frac{s}{\sqrt{n}} \sqrt{\frac{N-n}{N-1}} \right)$$

where n is the sample size, N is the population size, $\bar{x}$ is the sample mean, and $t_{\alpha/2, n-1}$ is the value of the Student's t-distribution with $n - 1$ degrees of freedom and right tail area $\alpha/2$ (see Fig. 2.1).

For the population variance σ^2, its $100(1 - \alpha)\%$ confidence interval I is given as:[12]

$$\left(\frac{(n-1) \times s^2}{\chi^2_{\alpha/2, n-1}}, \frac{(n-1) \times s^2}{\chi^2_{1-\alpha/2, n-1}} \right)$$

where s^2 is the sample variance as defined in Section 2.4.1, and $\chi^2_{n-1, \lambda}$ is the value of the χ^2 distribution with $n - 1$ degrees of freedom (see Fig. 2.2) and right tail area λ ($\lambda = \alpha/2$ or $\lambda = 1 - \alpha/2$).

2.4.3 Number of tests

For the purpose of calculating the number of specimens per laboratory sampling unit for testing, information is needed on the variability of individual observations made as directed in the method to be employed. The variability of individual observations depends upon the test method itself,

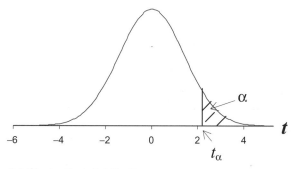

2.1 Student's t-distribution.

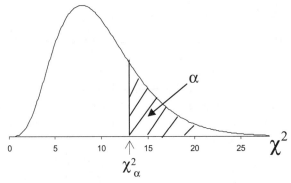

2.2 Chi-square distribution.

upon the experience and training of the operator, upon the calibration and maintenance of the apparatus used, and especially upon the variability of the property in the material being tested. For this reason, it is preferable for the user of the test to determine the variability of individual observations experimentally under the actual test conditions of interest and upon specimens from the types of material to be tested by use of the method.

In addition to the variability of individual observations, the required number of specimens depends upon the values chosen for the allowable variation and probability level. Both of these factors are based on engineering judgement.

For a normal population with mean μ and variance σ^2, when the population mean is to be estimated, the number of tests n is determined as follows:[20]

$$\Pr\left(|\hat{\mu} - \mu| > d\right) < \alpha$$

$$\Rightarrow z \times \sqrt{\frac{N-n}{N-1} \times \frac{\sigma^2}{n}} = d$$

$$\Rightarrow n = \frac{1}{\frac{1}{n_o} + \frac{1}{N}} \quad \text{where } n_o = \left(\frac{z_{\alpha/2} \times \sigma}{d}\right)^2$$

where $z_{\alpha/2}$ is the value of the standard normal distribution with right tail area $\alpha/2$, d is the acceptable error or the tolerance, and α is the probability that the estimated mean is within tolerance.

When σ is directly proportional to μ, the relative error d/μ is acceptable instead of the absolute error d. The number of tests n is also determined as above with n_o defined as:

$$n_o = \left(\frac{z_{\alpha/2} \times V}{P}\right)^2 \quad \text{where } V = \frac{\sigma}{\mu} \text{ and } P = \frac{d}{\mu}$$

Normally the variability of individual observations made under conditions of single-operator precision is considered. Other sources of variation, such as differences between instruments, operators or laboratories, or differences with time, may also have to be taken into account when certain test methods are being written or revised.

According to the standard ASTM D2906,[2] the components of variance are combined to obtain the total variance S_T^2:

$$S_T^2 = S_B^2 + S_W^2 + \frac{S_S^2}{n}$$

where S_S^2 = single-operator component of variance

S_W^2 = within-laboratory component of variance (all laboratory variation except single-operator variation)

S_B^2 = between-laboratory component of variance (calibration practice, etc.)

n = number of observations by a single operator in each average.

The contributions of the variability of lot sampling units, laboratory sampling units and test specimens to the variation of the test result of a sampling is fully explained in the ASTM Standard D4854.[4]

In the absence of necessary information, the standard advises a somewhat larger number of specimens than would ordinarily be obtained from an estimate of the variability of determinations in a user's laboratory, with 1.414 times the best estimate for a typical value of the variability,[1] i.e. twice the number of specimens.

Remarks

- When variability is directly proportional to the mean, use CV% to replace standard deviation, and % error to replace absolute error.
- When a property needs to be controlled in only one direction instead of both directions, use one-sided limits. (Yarn count is an example of a property that ought to be controlled in both directions. Fabric strength is an example of a property which needs to be controlled in only one direction, since fabric strength has no maximum limit.)
- When the frequency distribution of the data is markedly skewed or the standard deviation is correlated with the mean but not proportionally, consider making a transformation of the original data which will result in a normally distributed variate with standard deviation being independent of the mean. Please refer to the standard statistical texts for the choice of suitable transformations, e.g. *Statistical Theory with Engineering Application*, by A. Hald, John Wiley & Sons, New York, 1952.

- Selection of the probability level is largely a matter of engineering judgement. Experience has shown that for both one-sided and two-sided limits, a 95% probability level is often a reasonable one.
- The allowable difference d is the smallest difference of practical importance,[19] small enough to ensure that the variability of the test results will not exceed the normal needs of the trade but not so small that an unrealistically large number of observations are required.

2.4.4 Significance test

Hypothesis testing by performing a significance test is the inference of unknown population parameters, e.g. the average tensile strength of a fabric lot, or the variance of water-resistance capability of fabrics coming out from several coating machines. Random samples are taken from the populations to shed light on the values of the parameters. Based on the probability theory, whether the hypothetical statement (null hypothesis) is rejected or not will depend on whether evidence is sufficient or insufficient. When the null hypothesis is rejected, the alternative hypothesis will be accepted. The sufficiency of evidence is judged with reference to a fixed level of significance.

Testing for differences among many population means

One-way analysis of variance (ANOVA) is the extension of the pooled t-test to several samples. The pooled t-test compares two means, $\bar{x}_1$ and $\bar{x}_2$, and tests the null hypothesis H_0: $\mu_1 = \mu_2$, which can be written as H_0: $\mu_1 - \mu_2 = 0$. We could use $\bar{x}_1 - \bar{x}_2$ to infer $\mu_1 - \mu_2$.

When we are going to test the equality of the population means of more than two populations, we have to use another approach – the ANOVA approach using the F-test. The F-test for the one-way analysis of variance will tell you whether the averages of several independent samples are significantly different from one another. This replaces the pooled t-test when you have more than two samples and gives the identical result when you have exactly two samples.

The null hypothesis for the F-test in the one-way analysis of variance claims that the k populations (represented by the k samples) all have the same mean value. The alternative hypothesis claims that they are *not* all the same, that is, at least two population means are different. For the one-way study, the null hypothesis is

$$H_0: \mu_1 = \mu_2 = \ldots = \mu_k \quad \text{where } k \text{ is the number of populations} \quad 2.1$$

Let

$$\mu_1 = \mu + \tau_1, \ \mu_2 = \mu + \tau_2, \ \ldots, \ \mu_k = \mu + \tau_k \quad\quad\quad 2.2$$

Table 2.1 Data from *k* samples

	Sample				
	1	2	3	...	k
	x_{11}	x_{21}	x_{31}	...	x_{k1}
	x_{12}	x_{22}	x_{32}	...	x_{k2}
	x_{13}	x_{23}	x_{33}	...	x_{k3}
	...	...	...	...	...
	...	...	...	...	...
	...	...	...	...	...
	x_{1n}	x_{2n}	x_{3n}	...	x_{kn}
Mean	$\bar{x}_1$	$\bar{x}_2$	$\bar{x}_3$	...	$\bar{x}_k$

We have *k* samples from *k* populations which are independent and normally distributed with common variance σ^2.

$$x_{ij} = \mu_i + e_{ij} = \mu + \tau_i + e_{ij}$$

where e_{ij} is the random error of the *j*th observation in the *i*th sample.

The data set for the one-way analysis of variance consists of *k* independent univariate samples, as shown in Table 2.1. The sample sizes may be different from one sample to another. For simplicity, the sample sizes of all the *k* samples are set to *n*. These data values are different from one another due to the following two sources of variation here:

1. One source of variation is due to the fact that the populations are different from one another. This source is called the *between-sample variability* (S_A^2). The larger the between-sample variability, the more evidence you have on the differences among the populations.
2. The other source of variation is due to the diversity within each sample. This source is called the *within-sample variability* (S^2). The larger the within-sample variability, the more scattered are the data within each of the populations and the harder it is to tell whether the populations are actually different or not.

We define S_A^2 and S^2 such that they represent the between-group variation (the effect) and within-group variation (the error) respectively.

$$S_A^2 = \frac{n \times \sum_{i=1}^{k} (\bar{x}_i - \bar{\bar{x}})^2}{k-1} \qquad S^2 = \frac{\sum_{i=1}^{k} \sum_{j=1}^{n} (x_{ij} - \bar{x}_i)^2}{nk - k}$$

$$E(S_A^2) = \sigma^2 + \frac{n \times \sum_{i=1}^{k} \tau_i^2}{k-1}$$

2.3

$$E(S^2) = \sigma^2 \hspace{4cm} 2.4$$

where n is the size of each group (or sample), k is the number of groups, $\bar{\bar{x}}$ is the grand average (see below) and $E(X)$ stands for the expected value (mean value) of the random variable X.

Because the null hypothesis claims that all population means are equal, we will need an estimate of this mean value that combines all of the information from the samples. The *grand average* is the average of all of the data values from all of the samples combined. It may also be viewed as a *weighted average* of the sample averages, where the larger samples have more weight.

$$\text{Grand average: } \bar{\bar{x}} = \frac{\sum_{i=1}^{k} \bar{x}_i}{k}$$

To test	$H_0: \mu_1 = \mu_2 = \ldots = \mu_k,$
	$H_1:$ At least two μ_i's are unequal
is the same as to test	$H_0: \tau_1 = 0, \tau_2 = 0, \ldots, \tau_k = 0,$
	$H_1:$ At least two τ_i's are non-zero
or to test	$H_0: \sigma_A^2 = \sigma^2$ from eqns 2.1, 2.2
against	$H_1: \sigma_A^2 > \sigma^2$ from eqns 2.3, 2.4

Since under H_0, $\tau_i = 0$ for $i = 1, 2, \ldots, k$, $E(S_A^2) = \sigma_A^2 = \sigma^2$, the test statistic is $F = S_A^2/S^2$. We reject H_0 when $S_A^2/S^2 > F_{k-1,\, nk-k;\, \alpha}$ (critical F value – see below).

The F statistic for a one-way analysis of variance is the ratio of variability measures for the two sources of variation: the between-sample variability divided by the within-sample variability.

The *between-sample variability* is an estimator which measures the differences among the sample averages. This would be zero if the sample averages were all identical, and it would be large if they were very different. The number of degrees of freedom of this estimator is equal to $k - 1$. One degree of freedom is lost (as for an ordinary standard deviation) because the grand average was determined from the k averages.

The *within-sample variability* is an estimator which measures how variable each sample is. Because the samples are assumed to have equal variability, there is only one measure of within-sample variability. This would be zero if each sample consisted of its sample average repeated many times, and it would be large if each sample contained a wide diversity of numbers. The square root of the within-sample variability is in fact an estimator of the population standard deviation. The number of degrees of freedom of this estimator is equal to $nk - k$, since all nk data values are involved in the calculation of the within-sample variability but have lost k degrees of freedom because k different sample averages were estimated from the same set of data.

The F-table is a list of critical values, $F_{k-1, nk-k; \alpha}$ for the distribution of the F statistic when the null hypothesis is true. If the F-statistic as calculated from the sample data exceeds the critical F-value, the null hypothesis will be rejected with a level of significance of α. This indicates greater differences among the sample averages. To find the critical F-value, use your numbers of degrees of freedom to find the row and column in the F-table corresponding to the level of significance you are testing at. It is usually the case with hypothesis testing that when you can't reject the null hypothesis you have a weak conclusion in the sense that you should *not* believe that the null hypothesis has been shown to be true.

Nowadays we perform the ANOVA test by computer using statistical software, e.g. SPSS, SAS, Minitab, etc. The advantages of using computer processing are obvious: tedious calculation work can be eliminated, careless mistakes can be largely avoided, and more importantly the level of significance of a calculated F-value can be obtained with as many decimal places as required (this is called the significance of F; normally three decimal places are enough). We don't need to compare the calculated F-value with the critical F-value as obtained from an F-distribution table. Instead we could refer the significance of F to a level of significance of 0.1 (slightly significant), 0.05 (significant), 0.01 (highly significant) and 0.001 (very highly significant). In practice, the 0.05 level is the generally accepted significance level.

Example 1: One-way ANOVA

Table 2.2 shows the results of an experiment designed to compare the effectiveness of four types of shower-proof fabric. Five pieces of each type were chosen at random and subjected to a test in which the time for water to penetrate the fabric was measured.

Table 2.2 Sample data of four types of fabrics

	Type of fabric			
	1	2	3	4
Penetration time	103	127	149	123
(seconds)	110	141	113	118
	107	136	146	106
	100	127	134	137
	119	119	153	106
Mean	108	130	139	118

Preliminary examination of the data using a box plot

The analysis of variance might tell you that there are significant differences, but you would also have to examine ordinary statistical summaries (means and standard deviations) and a box plot to actually see those estimated differences. Box plots are particularly well suited to the task of comparing several distributions because unnecessary details are omitted, allowing you to concentrate on the essentials. Here is a checklist of things to look for when using box plots or histograms to compare similar measurements across a variety of situations:

- Do the medians (dark lines in Fig. 2.3) appear different from one box plot to another? This provides an initial, informal assessment for which the analysis of variance will provide an exact, formal answer. From Fig. 2.3, it can be observed that the specimens of fabric type 3 stand out to have the best water-resistance property, while specimens of fabric type 1 are consistently the worst.
- Is the variability reasonably constant from one box plot to another? This is important because the analysis of variance will assume that these variabilities are equal in the population. The variability is represented by the interquartile range (height of the boxes in Fig. 2.3). If, for example, the higher boxes (with larger medians) are systematically wider (indicat-

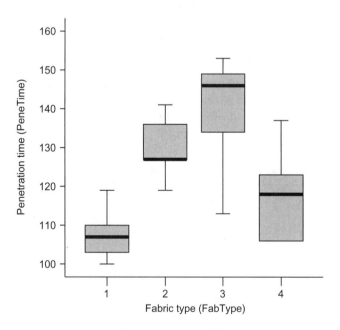

2.3 Box plot of water penetration data.

ing more variability), then the analysis of variance may give incorrect answers. This systematic change in variability cannot be observed in Fig. 2.3.

The one-way ANOVA table

The computer output in Table 2.3 shows an ANOVA table for this example, using a standard format for reporting ANOVA results. The sources include the factor (this is the fabric type, indicating the extent to which the four types of fabrics vary systematically from one another), the error (the random variation within a type of fabric) and the total variation. The sums of squares (SS) are in the next column, followed by the degrees of freedom (df). Dividing SS by df, we find the mean squares (MS), which are the between-sample and the within-sample variabilities. Dividing the factor MS by the error MS produces the F statistic in the next column, followed by the significance of F (also called the p-value) in the last column, indicating that the differences in water-resistance property among the four types of fabrics are highly significant. From Table 2.3, the significance F is equal to 0.004 which is less than 0.01, so we could conclude that the differences in water-resistance performance of the four types of fabrics are highly significant.

Example 2: Two-way ANOVA with replication

The tensile strengths (in N/cm) of three types of fabrics were measured in two independent testing laboratories. The results are shown in Table 2.4. Based on the results of the F-tests on main effects, we need to determine whether there is a difference in fabric strength due to fabric type, and to investigate the consistency of the fabric strength test between the two laboratories.

Table 2.5 shows the significance test results of the two main factors, fabric type (Fabric) and testing laboratory (Lab). The significance of the F-test on Fabric is equal to 0.000, which is less than 0.001. We can conclude that the difference in fabric strength among the three types of fabrics is very highly significant. For the other main factor Lab, the significance of the F-test is equal to 0.323, which is even larger than 0.1. We can conclude that the

Table 2.3 One-way ANOVA

Dependent variable: PeneTime

Source	Sum of squares	df	Mean square	F	Sig. F
FabType	2795.400	3	931.800	6.683	0.004
Error	2230.800	16	139.425		
Total	311060.000	20			

Table 2.4 Strengths (in N/cm) of three types of fabrics tested in two laboratories

	Fabric 1	Fabric 2	Fabric 3
Lab. A	27.6	64.0	107.0
	57.4	66.9	83.9
	47.8	66.5	110.4
	11.1	66.8	93.4
	53.8	53.8	83.1
Lab. B	49.8	48.3	88.0
	31.0	62.2	95.2
	11.8	54.6	108.2
	35.1	43.6	86.7
	16.1	81.8	105.2

Table 2.5 Analysis of variance

Dependent variable: Strength

Source	Type III sum of squares	df	Mean square	F	Sig. F
Fabric	19317.331	2	9658.665	51.172	0.000
Lab	192.027	1	192.027	1.017	0.323
Fabric * Lab	177.144	2	88.572	0.469	0.631
Error	4529.948	24	188.748		
Total	145959.890	30			

strength test results are not significantly different between the two testing laboratories.

Example 3: Two-way ANOVA with interaction effect

The tensile strengths (in N/cm) of three types of fabrics were measured under two atmospheric conditions: dry and wet. The results are shown in Table 2.6 with averages shown at the lower right-hand sides.

The data were processed using the SPSS statistical software and the results are shown in Table 2.7. The dependent variable is fabric strength (STRENGTH) and the two factors include fabric type (FABRIC) and testing condition (COND). The purpose of this study is to determine the effect of the two factors on the dependent variable. Table 2.7 shows the results of the significance tests of the two main effects – fabric type (FABRIC) and testing conditions (COND) – and also the interaction effect between FABRIC and COND. The significance *F* of the main effect FABRIC

Table 2.6 Strength of three types of fabrics in dry and wet

	Woollen fabric		Nylon fabric		Cotton fabric		Row average
		Mean		Mean		Mean	
Dry	77.6		64.0		64.0		
	72.0		66.9		72.0		
	65.0		62.5		78.0		
	62.0		58.8		75.4		
	59.0	*67.12*	53.8	*61.20*	68.0	*71.48*	**66.60**
Wet	60.3		58.3		88.0		
	51.0		62.2		80.5		
	41.8		44.6		89.3		
	35.1		65.0		86.7		
	66.1	*50.86*	51.8	*56.38*	79.4	*84.78*	**64.01**

Table 2.7 Tests of between-subjects effects

Dependent variable: STRENGTH

Source	Type III sum of squares	df	Mean square	F	Sig. F
FABRIC	2 468.051	2	1234.025	20.227	0.000
COND	50.440	1	50.440	0.827	0.372
FABRIC * COND	1 110.835	2	555.417	9.104	0.001
Error	1 464.244	24	61.010		
Total	133 029.3	30			
Corrected total	5 093.570	29			

is equal to 0.000, which is less than 0.001, so fabric type has a very highly significant effect on fabric strength. The significance F of the other main effect COND is equal to 0.372, which is larger than 0.1, so the testing condition does not have a significant effect on fabric strength. When looking further at the interaction effect, the significance F of the interaction effect FABRIC*COND is equal to 0.001, which is less than 0.01. Now we encounter a significant interaction effect.

The interpretation of the significant interaction effect can be visualized in the box plot as shown in Fig. 2.4. For the woollen fabrics, the dry strength is higher than the wet strength, while for cotton fabrics it is just the opposite. Nylon fabrics are similar in fabric strength in dry and wet. A significant interaction effect means that different levels of a main effect lead to a different behaviour of the other main effect, i.e. dry condition tends to increase

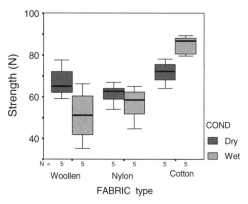

2.4 Box plot of two-way ANOVA.

Table 2.8 Independent samples test (fabric type 3)

	t-Test for equality of means			
	t	df	Sig. (two-tailed)	Mean difference
STRENGTH (equal variances assumed)	−4.125	8	0.003	−13.3000

Table 2.9 Independent samples test (fabric type 1)

	t-Test for equality of means			
	t	df	Sig. (two-tailed)	Mean difference
STRENGTH (equal variances assumed)	2.448	8	0.040	16.2600

the fabric strength only for a particular type of fabric (woollen fabric), not for the other types of fabrics. The positive effect of the dry condition on woollen fabrics is counterbalanced by the negative effect on cotton fabrics. That explains why the main effect COND does not have a significant effect on fabric strength. This can be proved by performing a pooled *t*-test with results as shown in Table 2.8.

Table 2.8 shows the significance test results for the effect of testing condition COND on fabric strength only for fabric type 3 (cotton fabrics). The significant *t* is equal to 0.003, which is less than 0.01. This demonstrates the fact that the testing condition has a significant effect on fabric strength, and similarly for fabric type 1 (woollen fabrics) as shown in Table 2.9.

This example illustrates the importance of studying the interaction effect and the use of a box plot on exploring the data set. Further in-depth analysis is necessary when there exist significant interaction effects in multi-factor analysis of variance.

2.5   Sampling of fabrics

According to British Standard BS EN12571,[8] theoretically correct methods of sampling from lengths or rolls of fabric require that laboratory samples be taken at intervals along the length of the fabric. In practice, rolls of fabric can be very long, and there are economic and practical handling difficulties in applying such methods. In most cases fabrics cannot be cut, except at the ends of the fabric piece, and therefore sampling cannot be carried out across the whole population.

If a particular fabric property is known to vary along the length of the roll, then tests carried out only at the ends of the roll will not be representative of the whole length of the fabric. In this case a statement to this effect is included in the test report, and if the results are to be used as a basis for quality standards, then this is agreed between the interested parties.

In special cases, for example where faults or changes in a property along the length of the fabric are to be examined, samples may need to be taken at intervals along the whole length of the roll. In this case, the number and arrangement of samples depend on the specific requirement and are agreed between the parties.

The number of cases (containers, etc.) and the number of rolls to be inspected are specified in the British Standard.[8] In practice, samples from the selected rolls are taken by cutting, not tearing, at a distance of 1 m from the end of the roll. The size of the sample taken is sufficient to carry out the required tests. Any visible irregularities, damage or colour differences are avoided. In addition, no two specimens should contain the same set of warp or weft threads.[18]

2.6   Future trends

The experimental measurement of sampling results can be carried out more satisfactorily through the improvement of the performance of measurement equipment according to particular requirements.[17] There is a continuous demand for measurement instruments that perform better with respect to speed (faster), dimensions (smaller), applicability (wider range, higher environmental demands), automation (to be embedded in automated systems), quality (higher accuracy, higher sensitivity, lower noise levels), energy (minimizing power consumption), reliability (more robust, long lifetime) and cost (cheaper).

Rapid growth in microprocessors and computer science has contributed to an expansion of software tools for the processing of measurement signals. The computer is gradually taking over more and more instrumental functions. The development of 'virtual instrumentation' is still expanding. The advancement of virtual instruments simplifies the set-up of a measurement; however, it also tends to mask the essences of a measurement: the transformation from quantifiable properties to an electrical signal and the associated measurement errors. A measurement can never be done by software alone, since measurements must be made by hardware.

2.7   References

1. ASTM Standard, D2905:2002, *Standard Practice for Statements on Number of Specimens for Textiles.*
2. ASTM Standard, D2906:2002, *Standard Practice for Statements on Precision and Bias for Textiles.*
3. ASTM Standard, D4853:2002, *Standard Guide for Reducing Test Variability.*
4. ASTM Standard, D4854-95 (reapproved 2001), *Standard Guide for Estimating the Magnitude of Variability from Expected Sources in Sampling Plans.*
5. ASTM Standard, E456:2006, *Standard Terminology Relating to Quality and Statistics.*
6. ASTM Standard, E691:2005, *Standard Practice for Conducting an Inter-laboratory Study to Determine the Precision of a Test Method.*
7. Bicking CA, 'Precision in the routine performance standard tests', *ASTM Standardization News*, 1979, 7(1), 12–14.
8. BSI Standards, BS EN12751:1999, *Textiles – Sampling of Fibres, Yarns and Fabrics for Testing.*
9. BSI Standards, PD6461-3:2004, *General Metrology, part 3: Guide to Expression of Uncertainty in Measurement* (GUM).
10. BSI Standards, PD6461-4:2004, *General Metrology, part 4: Practical Guide to Measurement Uncertainty.*
11. Bucher JL, *The Metrology Handbook*, ASQ Quality Press, Milwaukee, WI, 2004.
12. Freund JE, *Mathematical Statistics*, 5th edition, Prentice-Hall, Englewood Chiffs, NJ, 1992.
13. ISO/IEC 17025:1999, *General Requirements for the Competence of Testing and Calibration Laboratories.*
14. Lind DA, Marchal WG, Mason RD, *Statistical Techniques in Business and Economics*, 11th edition, McGraw-Hill, New York, 2002.
15. Mandel J, Lashof TW, 'The nature of repeatability and reproducibility', *Journal of Quality Technology*, 1987, 19(1), 29–36.
16. Rabinovich SG, *Measurement Errors and Uncertainties: Theory and Practice*, Springer, New York, 2005.
17. Regtien PPL, van der Heijden F, Korsten MJ, Olthuis W, *Measurement Science for Engineers*, Kogan Page Science, London, 2004.
18. Saville BP, *Physical Testing of Textiles*, Woodhead Publishing, Cambridge, UK, 1999.

19. Shombert G, 'The practical significance of the precision statements in ASTM test methods and a proposal to expedite their application to quality control in the laboratory', *ASTM Standardization News*, 1976, 4(4), 24–28.
20. Thompson SK, *Sampling*, 2nd edition, Wiley Series in Probability and Statistics, John Wiley & Sons, New York, 2002.

3

Fabric composition testing

Z ZHONG and C XIAO, Tianjin Polytechnic
University, China

Abstract: This chapter discusses fabric composition testing which is
necessary for organizations such as textile manufacturers, import and
export houses and R&D institutes, for carrying out applied research
and investigation. Much of the information presented in this chapter
focuses on some new testing methods such as quantitative fibre
mixture analysis for fabrics by scanning electron microscopy (SEM),
using the environmental scanning electron microscope (ESEM)
method to characterize the surface, interface and dynamic properties
of fabrics, computer image processing technology and so on. The
traditional tests such as burning, chemical, optical, staining and
density tests are systematically considered. The exposition in the
chapter indicates general procedures and only suggests interpretation.

Key words: fabric composition testing, environmental scanning electron
microscope (ESEM) technology, near infrared spectral image
measurement system, thermogravimetry (TG) analysis, computer image
processing technology.

3.1 Introduction: the importance of testing fabric composition

The testing of composition of fabric is necessary for various organizations
such as textile manufacturers, import and export houses, government agencies, R&D institutes and academic institutions, for carrying out applied
research, investigation such as archaeological studies for identification of
ancient textile fibres from various archaeological tracts, case analyses by
legal medical experts, and so on. In response to ever-changing governmental
regulations and the ever-increasing consumer demand for high quality,
regulatory testing of fabric composition is essential to minimize risk and
protect the interests of both manufacturers and consumers.

Most countries importing apparel and soft home furnishing products
require testing the fabric composition mainly for fibre identification labels
that indicate the fibre type and percentage of fibre components. Some
countries even use fibre composition to classify quota categories. All fabrics
contain textile fibres. The exact fibre composition of a specific fabric is
important because different fibre types have different characteristics, which
can significantly affect the properties and value of the item in question. This

is a particularly important consideration for clothing. A consumer paying a high price for a silk garment would be most unhappy to discover it was made of polyester, for example. There are also implications for duty on imported goods in some countries, such as the United States, while in others regulations may exist to protect consumers. These regulations may require fibre content to be stated, and impose legal penalties for infringement. For fabric manufacturers testing the fabric composition will play an important role in developing and/or revising quality assurance programmes and product specifications. In addition, it also helps to develop, monitor and/or improve manufacturing processes and assists fault-finding and problem-solving processes.

The trend of green consumerism has been extended to textile and apparel products. Major textile product buyers throughout the world, especially in Europe and the USA, have responded to this public awareness by viewing their textile products from an ecological viewpoint and are establishing relevant requirements. All finished fabrics contain different chemicals which may be hazardous or poisonous to both environment and consumer health. So current testing methods on fabric composition also include eco-testing such as chemical analysis for banned azo colourants, formaldehyde content, heavy metal residues, ozone-depleting chemicals, pesticide residues and so on.

The very survival of early humans was greatly aided by fabrics. In modern times, fabrics still perform many essential protective, decorative and social functions, as well as being the basis of a range of advanced structural and functional products. An important factor which is driving the changes in the fabric industry at the present time is the expanding markets created through increase in population and increase in disposable income. Competition between a wide range of similar fabrics, especially in the mass markets, demands a constant pressure for cost reduction in processing methods. On the other hand, new technologies create opportunities for new fabrics, particularly in niche markets. Demand for environmental friendliness will also be important in the future. Environmental concerns have already led to changes in processing methods to conserve energy, water and chemicals. Biodegradable products and recycling technologies are being demanded by consumers.

The fabric industry is becoming an increasingly competitive environment and new types of fabrics have been developed with the help of new technologies. Differentiating fabrics is therefore important and this can be facilitated through optimizing fibre composition, rationalizing fabric structure for more suitable end use, and improving the quality. Testing fabric composition can be done to legal standards, industry standards or custom standards to test whether a product complies with requirements. Testing fabric composition can be used to improve fabric quality and achieve compliance to international, regional or retailer-specific standards, especially to provide good support in product safety management.

The traditional way of testing composition for single-layered fabrics was to identify the fibres in the fabrics. But currently there are number of different new types of fabrics in the market, such as industrial fabrics and technical fabrics of high-performance fibres, multilayer fabrics with different combinations of materials such as non-wovens, wovens, films and paper, phase change fabrics, electrically conductive fabrics and so on. For these fabrics, testing fabric composition needs new methods besides traditional ones based on traditional textile fibre identification. Of course, when testing composition for any fabric, fibre identification is necessary first. Much of the information presented in this chapter focuses on some new testing methods for fabric composition such as 'quantitative fibre mixture analysis' for fabrics by scanning electron microscopy (SEM), using the environmental scanning electron microscope (ESEM) method to characterize the surface, interface and dynamic properties of fabrics. Other methods covered include micro IR spectroscopy, the near infrared spectral image measurement system for water absorbency of woven fabrics, the technique using capillary electrophoresis/mass spectrometry (CE/MS) for identification of dyed textile fibres, thermogravimetry (TG) analysis and computer image processing technology on fabric testing, as well as the traditional methods such as burning, chemical, optical, staining and density tests. The discussion included in this chapter indicates general procedures and only suggests interpretation.

3.2   Methods of testing fabric content and composition

Testing composition of fabrics is a skilled activity, especially for blended fabrics and multilayer fabrics, where the fibre mixture must first be visually identified by a microscope and then by chemical tests. The amount of each fibre is then determined either by physical separation – where the mixture is created by twisting yarns together or by sewing panels of fabric together – or by chemically extracting one fibre at a time using gravimetric analysis to give a composition result.

Qualitative identification of textile fibres of a specific fabric can be difficult and may require several tests. Simple tests that can be used in identification are described here along with brief comments about their use and importance. In addition to their use in fibre identification, some tests yield insight into problems of processing and care of textile products.

Testing of composition of fabric needs modern laboratories with sophisticated testing and analytical instruments. The methods of testing can be grouped under the following headings:

- Traditional methods
  - Optical test
  - Density test
  - Chemical test
  - Staining test
  - Burning test
- New methods
  - Environmental scanning electron microscope (ESEM) technology
  - Near infrared spectral image measurement system
  - Capillary electrophoresis/mass spectrometry (CE/MS) technique
  - Thermogravimetry (TG) analysis
  - Computer image processing technology

Microscopic evaluation by optical test is a little more specific and in some cases may be accurate enough to identify individual fibres in the fabrics; and the density test or physical separation or chemical extraction, especially chemical solubility, may be accurate enough to categorize fibres in the fabrics into generic groups. Instruments used in more precise and accurate identification are also mentioned above. Investigators who wish to be able not only to state what fibre type in the fabrics is involved but also to identify a specific fibre would usually need instrumental evaluations as well as those described herein.

3.3   Traditional testing methods

For single-layer fabrics having traditional textile fibres, testing fabric composition is almost testing for fibre identification. In other words, certainly, when testing composition for any fabric, fibre identification is necessary first.

The traditional ways of fibre identification can be divided into five smaller groups or tests as above. They are optical, density (Table 3.1), chemical, staining, and burning tests (Table 3.1). Each test has its own advantages and disadvantages. Most of them are cheap and simple identification techniques and are easy to use.

3.3.1   Optical tests: visual identification – microscopy

Optical tests are the simplest tests available, and the use of a microscope allows the observer to see the fabrics up close. This is valuable because certain fibres have particular shapes which can be identified when viewed under a microscope. However, not much information can be obtained from the longitudinal sections alone and viewing the cross-section helps greatly, but preparing cross-sectional samples takes great skill and time. Although

Table 3.1 Physical properties of fibres

	Density (g/cm³)	Melting point (°C (°F))
Natural fibres		
Cellulose	1.51	None
Silk	1.32–1.34	None
Wool and other hair	1.15–1.30	None
Artificial fibres		
Acetate (secondary)	1.32	260 (500)
Acetate (tri)	1.30	288 (550.4)
Acrylic	1.12–1.19	None
Modacrylic	1.30 or 1.36	188 (370.4) (not sharp) or 120 (248)
Nylon 6	1.12–1.15	213–225 (415.4–437)
Nylon 66	1.12–1.15	256–265 (492.8–509)
Polyester	1.38 or 1.23	250–260 (482–500) or 282 (539.6)
Polypropylene	0.90–0.92	170 (338)
Rayon	1.51	None

definite information about the fibres can be obtained from their shapes, for some fibres, especially artificial fibres with similar shapes, additional analysis is required for specific identification. Therefore, photomicrographs of selected fibres may be used for comparison. Microscopic examination is indispensable for positive identification of the several types of cellulosic and animal fibres, because the infrared spectra and solubility techniques will not distinguish between different species.

Analyses by infrared spectroscopy and solubility relationships are the preferred methods for identifying artificial fibres. The analysis scheme based on solubility is very reliable. The infrared technique is a useful adjunct to the solubility test method. The microscopic examinations are generally not suitable for positive identification of most artificial fibres and are useful primarily to support solubility and infrared spectra identifications.

The American Association of Textile Chemists and Colorists (AATCC) publishes a *Technical Manual* annually. It has test methods for the identification of textile fibres qualitatively and quantitatively along with a number of other test methods and evaluations. The photomicrographs are very helpful as are the various schemes for identification, including solubility. The ASTM *Identification of Fibres*, D276, is also very helpful but has no photographs.

3.3.2 Density tests: physical separation

The density test offers a simple test of preparing a liquid in which the fibres will either sink or float, but porous fibres and fibre blends will skew the density results. The density gradient column is now widely used and necessitates observing the level to which a test specimen will sink in a column of liquid, the density of which increases uniformly from top to bottom. The column should be about 40–50 mm high and normally graduated from 0 at the top to 100 cm at the bottom. Up to three such columns can be positioned in a water bath controlled at 23 ± 1°C. A valuable attribute of this method is that the columns, when prepared, can be accurately calibrated with glass beads of known density.

Various liquids are used to cover a spread of working densities (Table 3.1) depending on the density of the test material.

3.3.3 Chemical tests: chemical extraction

Chemical tests are cheap and simple methods, but the tests are not quantitative. Also, the number of elements that can be detected is limited. So for more accurate analysis, better and more expensive equipment is needed.

The solubility of a fibre extracted from a fabric in specific chemical reagents is frequently a definitive means of specific fabric identification. Frequently, however, this process identifies only generic groups or categories of fibres in the fabrics. If one combines the results of the microscopic evaluation, the burning test, which will be introduced in later sections, and the chemical solubility test, it is possible, in many cases, to positively identify specific fibres in the fabrics. The solubility behaviour of fibres in various solvents is given in Table 3.2.

3.3.4 Staining tests

Staining tests can help to show changes in fibre structure from one process to another. If the fibre structure is changed, the dye shade will be different from batch to batch. A characteristic that can be obtained from dyeing is whether the fibre is hydrophobic or hydrophilic, because hydrophilic fibres are easier to dye. Staining can be used to group fibres into three groups: cellulosic, protein-based, or artificial fibres, but the process is not good on deep-dyed samples, and chemical finishes can interfere with the process.

When fibres are white or off-white or when colour can be stripped from fibres, staining techniques may be used as a part of identification. Specially prepared mixtures of dyes are used to stain the fibres for a specified time and at a specified temperature. After staining, the fibre, yarn or fabric pieces are dried and then compared with a known sample.

3.3.5 Burning tests

The burning test is a good preliminary test for fabric identification. It provides valuable data regarding appropriate care and will help place a fibre

Table 3.2 Solubility behaviour of fibres in the fabrics

	Acetone 100%	Hydrochloric acid 20%	Sulfuric acid 60%	Sulfuric acid 70%	Chlorine bleach 5%	Formic acid 90%
Acetate	Soluble	Insoluble	Soluble	Soluble	Insoluble	Soluble
Acrylic	Insoluble	Insoluble	Insoluble	Insoluble depending on type	Insoluble	Insoluble
Cotton	Insoluble	Insoluble	Slightly soluble	Soluble	Insoluble	Insoluble
Hair	Insoluble	Insoluble	Insoluble	Insoluble	Soluble	Insoluble
Hemp	Insoluble	Insoluble	Slightly soluble	Soluble	Insoluble	Insoluble
Linen	Insoluble	Insoluble	Slightly soluble	Soluble	Insoluble	Insoluble
Modacrylic	Soluble or insoluble depending on type	Insoluble	Insoluble	Insoluble	Insoluble	Insoluble
Nylon	Insoluble	Soluble	Soluble	Soluble	Insoluble	Soluble
Olefin	Insoluble	Insoluble	Insoluble	Insoluble	Insoluble	Insoluble
Polyester	Insoluble	Insoluble	Insoluble	Insoluble	Insoluble	Insoluble
Ramie	Insoluble	Insoluble	Slightly soluble	Soluble	Insoluble	Insoluble
Rayon	Insoluble	Insoluble	Soluble	Soluble	Insoluble	Insoluble
Silk	Insoluble	Partially soluble	Soluble	Soluble	Soluble	Partially soluble
Wool	Insoluble	Insoluble	Insoluble	Insoluble	Soluble	Insoluble

in the fabrics into a specific category. A burning test can help determine the class to which a fibre belongs by observing its burning behaviour. Observing how things smell and char when they are burnt are qualities that can help. To identify a fabric that is unknown, a simple burn test can be done to determine whether the fabric is made of a natural fibre, an artificial fibre, or a blend of natural and artificial fibres. The burn test is used by many fabric stores and designers and generally requires practice to determine the exact fibre content. However, an inexperienced person can still determine the difference between many fibres to narrow the choices down to natural or artificial fibres. This elimination process will give information necessary to decide the care of the fabric.

The following procedure may be used to carry out the burning test for fabrics:

1. Select one or two yarns from the warp of a woven fabric or unravel a length of yarn from a knitted fabric.
2. Untwist the yarn so that the fibres are in a loose mass.
3. Hold the loosely twisted yarn in forceps; move them towards the flame from the side (i.e., approach the flame from its own level, not by bringing the sample down into the flame).
4. Observe the reaction as the yarn approaches the flame.
5. Move the yarn into the flame, and then pull it out of the flame and observe the reaction. Does the yarn start to burn as it nears the flame? Does it start to melt? Does it shrink away from the flame? Does it burn quickly or slowly? Does it have a sputtering flame, a steady flame, no flame at all? When removed, does it continue to burn? Is it bright red or coloured to indicate that it has reached a high temperature? Does the flame go out when removed from the source? What type of ash or residue, if any, is formed?
6. Notice any odour given off by the fibre both while it is in the flame and after it is removed.
7. Observe the ash or residue formed and what characteristics it has. Is it brittle? Is it bead-shaped? Is it fluffy? Is it the shape of the yarn? Or is there nearly no residue?
8. Repeat for the filling yarn of woven fabrics.
9. If the fabric does not have yarn structure, or if it is impossible to 'deknit' a length of yarn from complex knitted structures, a small sliver of fabric can be cut and used in place of the yarn.

3.4   Burning behaviour of fibres

All fibres will burn but asbestos-treated fibres are, for the most part, fire-proof. The burning test should be done with caution. A small piece of fabric is preferred. The fabric should be held with tweezers but not with fingers.

Some fabrics will ignite and melt. The result is burning drips which can adhere to fabric or skin and cause a serious burn.

3.4.1 Natural fibres

Cotton is a plant fibre. When ignited it burns with a steady flame and smells like burning leaves. The ash left is easily crumbled. Small samples of burning cotton can be blown out as you would with a candle.

Linen is also a plant fibre but different from cotton in that the individual plant fibres which make up the yarn are long where cotton fibres are short. Linen takes longer to ignite. The fabric closest to the ash is very brittle. Linen is easily extinguished by blowing on it as you would with a candle.

Silk is a protein fibre and usually burns readily, not necessarily with a steady flame, and smells like burning hair. The ash is easily crumbled. Silk samples are not as easily extinguished as cotton or linen.

Wool is also a protein fibre but is more difficult to ignite than silk as the individual 'hair' fibres are shorter than silk and the weave of the fabrics is generally looser than with silk. The flame is steady but more difficult to keep burning. The smell of burning wool is like that of burning hair.

3.4.2 Artificial fibres

Acetate is made from cellulose (wood fibres), technically cellulose acetate. Acetate burns readily with a flickering flame that cannot be easily extinguished. The burning cellulose drips and leaves a hard ash. The smell is similar to that of burning wood chips.

Acrylic, technically acrylonitrile, is made from natural gas and petroleum. Acrylics burn readily due to the fibre content and the lofty, air-filled pockets. A match or cigarette dropped on an acrylic blanket can ignite the fabric which will burn rapidly unless extinguished. The ash is hard. The smell is acrid or harsh.

Nylon is a polyamide made from petroleum. Nylon melts and then burns rapidly if the flame remains on the melted fibre. If you can keep the flame on the melting nylon, it smells like burning plastic.

Polyester is a polymer produced from coal, air, water and petroleum products. Polyester melts and burns at the same time; the melting, burning ash can bond quickly to any surface it drips on, including skin. The smoke from polyester is black with a sweetish smell. The extinguished ash is hard.

Rayon is a regenerated cellulose fibre which is almost pure cellulose. Rayon burns rapidly and leaves only a slight ash. The burning smell is close to that of burning leaves.

Blends consist of two or more fibres and, ideally, are supposed to take on the characteristics of each fibre in the blend. The burning test can be used but the fabric content will be an assumption.

In order to determine the fibre content in a fabric, a piece of fabric of area approximately 1 square inch (6.5 cm^2) is ignited using a butane lighter, holding it with a pair of tweezers over a non-flammable surface in a well-ventilated area. The quality and colour of the flame, the odour produced and the quality of the resulting ash or cinder are observed carefully. Table 3.3 may be used to determine the fabric's content.

The burning test is a simple test to start with when testing the fibre content of fabrics. However, there are many ways to interpret the results. Incorrect interpretation, of course, will lead to false information. Whether a fibre burns or self-extinguishes is not dependent upon whether it is a 'natural' fibre or a 'synthetic' fibre. For example, because of their chemical make-up, both silk and wool will self-extinguish when the flame is removed. Rayon is regenerated cellulose, so even though it is an artificial fibre, it burns extremely well and will continue to do so after the flame is removed. It will burn until the sample is exhausted if unhindered. The smell given off when rayon burns will be similar to that of paper (also cellulose based), but silk will smell like burning hair.

Silk fabrics will shrink away due to the heat given off even before they burn from contact with the flame. The best way to view this 'shrinking' of silk is to slowly bring it close to the flame but not into the flame. While a synthetic like polyester will also self-extinguish, it does not behave in the same manner as silk when exposed to the flame. Silk will 'shrink' away from the flame and form small beads on the ends of the fibres after the flame is removed. Polyester will also form beads but it does not shrink away from the flame (it does melt, but it doesn't 'jump back' like silk). The colour, shape and size of the beads at the end of comparable-sized threads of polyester and silk will also differ.

Certain artificial fibres are not easily identified by any of the testing procedures cited. Positive verification of some fibres depends on the use of one or more sophisticated instrumental techniques. These include testing for melting point, refractive index, index of birefringence, the use of X-ray diffraction machines, infrared spectrophotometers, chromatographs of various types, electron scanning microscopes and polarizing microscopes. These are standard equipment in many university laboratories, testing laboratories and research laboratories. Although they may not be available in departments where textile science is taught, they may be available in chemistry departments.

3.5   New testing methods

In this section a detailed discussion is provided on the principles and procedures of new testing methods such as environmental scanning electron microscope (ESEM) technology, near infrared spectral image measurement system, the capillary electrophoresis/mass spectrometry (CE/MS)

Table 3.3 Flammability behaviour of fabrics

Fabric	Flame quality	Odour	Ash quality	Comments
Wool	Orange colour, sputtery	Burning hair or feathers	Blackish, turns to powder when crushed	Flame will self-extinguish if flame source removed, no smoke
Silk	Burns slowly	Burning hair or feathers	Greyish, turns to powder when crushed	Burns more easily than wool but will self-extinguish if flame source removed
Cotton	Yellow to orange colour, steady flame	Burning paper or leaves	Greyish, fluffy	Slow-burning ember
Linen	Yellow to orange colour, steady flame	Burning paper or leaves	Similar to cotton	Takes longer to ignite than cotton but otherwise very similar
Rayon	Fast orange flame	Burning paper or leaves	Almost no ash	Ember will continue to glow after flame source removed
Polyester	Orange flame, sputtery	Sweet or fruity smell	Hard shiny black bead	Black smoke
Acetate	Burns and melts, sizzly	Acidic or vinegary	Hard black bead	Will continue to burn after flame source removed
Nylon	Burns slowly and melts, blue base and orange tip, no smoke	Burning celery	Hard greyish or brown-ish bead	Will self-extinguish if flame source removed
Acrylic	Burns and melts, white-orange tip, no smoke	Acrid	Black hard crust	Will continue to burn after flame source removed
Polypropylene (olefin)	Burns and melts	Not defined	Hard, round bead, maybe light brown	Shrinks quickly

technique, thermogravimetry (TG) analysis and computer image processing technology.

3.5.1 Environmental scanning electron microscope technology

Environmental scanning electron microscope (ESEM) technology was introduced in the mid-eighties. This technology has improved over time. The ESEM now can offer full functionality in three modes of operation: High Vacuum, Low Vacuum and ESEM mode. Conventional high vacuum SEM is also available on the ESEM. This mode can be used for the examination of vacuum-compatible or gold/carbon-coated non-conductive samples. A low vacuum mode is suitable for the examination of uncoated non-conductive samples. The ESEM mode allows very high chamber pressures of up to 50 torr. This is achieved by the differential pumping system, as illustrated in Fig. 3.1 [1].

In both low vacuum and ESEM mode, the specimen sits in a gaseous atmosphere in the ESEM chamber. Ionization of the gas by electrons emitted from the specimen results in the neutralization of the charge build-up on the specimen surface. As a result, non-conductive samples can be imaged. The ESEM mode is appropriate for the examination of hydrated, oily or outgassing samples, where it is desirable to observe the sample in its

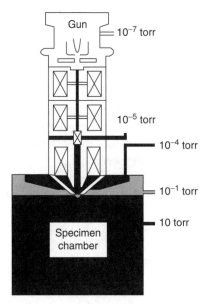

3.1 Differential pumping system.

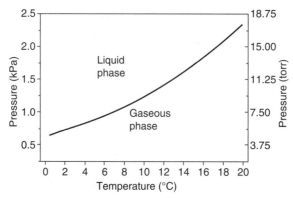

3.2 Water phase curve.

natural state. Samples are examined uncoated, with a gaseous secondary electron detector (GSED) [2], within a gaseous environment.

In the ESEM, specimens can be hydrated or dehydrated by controlling the temperature of the specimens and the chamber pressure in favour of water condensation or evaporation at different relative humidity, as presented in Fig. 3.2 [3].

The ESEM is specifically suited to dynamic experimentation at the micron scale and below. ESEM technology allows dynamic experiments at a range of pressures and temperatures and under a variety of gases/fluids. Some accessories can also be added into an ESEM to expand its observation capacity.

Application of environmental scanning electron microscope for surface characterization of uncoated textile materials

It is well known that fibre surface characteristics affect wetting, stiffness, strength, dyeing, wrinkling and other performance properties, to a large extent. Based on the understanding of fibre surface properties, novel fibres and their applications may be created or engineered.

Philips XL30 ESEM-FEG at Heriot-Watt University, UK, offers high-resolution secondary electron imaging of wet, oily, dirty, outgassing and non-conductive samples in their natural state without significant sample modification or preparation. Some new techniques to characterize textile materials have been developed.

The ESEM is able to physically examine virtually any textile materials without any special preparation or conductive coating. The ESEM images in Fig. 3.3 [4] present the different surface characteristics of glass fibre before and after gold coating. The image in Fig. 3.3(a) shows the relatively

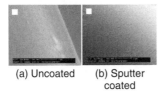

(a) Uncoated   (b) Sputter
coated

3.3 ESEM images of glass fibres: (a) uncoated, (b) sputter coated.

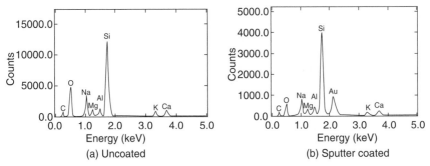

(a) Uncoated   (b) Sputter coated

3.4 EDX spectra of glass fibres: (a) uncoated, (b) sputter coated.

smooth surface of the glass fibre with some particle-like dots, but next the fibre surface is covered with the gold cluster (Fig. 3.3(b)) after sputter coating at 20 mA for 60 s. The coating has also changed the chemical composition, revealed by dispersive X-ray analysis (EDX) in the ESEM. The Philips XL30 ESEM-FEG equipped with a Phoenix energy dispersive X-ray analysis system (EDX) was used to examine the chemical compositions of the glass fibres, and an accelerating voltage of 20 kV with accounting time of 100 s was applied. Figure 3.4(a) [4] shows the EDX spectrum at an area of the uncoated glass fibre observed in Fig. 3.3(a). It can be seen that the fibre predominantly consists of Si, O, Ca, Mg, Na and K. A significant amount of Au on the fibre surface can be observed after coating (Fig. 3.3(b)) compared to the original fibre (Fig. 3.4(b)) [4].

Interface characterization of textile materials

In many applications of textile materials, interfaces are formed between two phases of either the same or different materials. The characteristics of interfaces are usually different from those of the bulk phase(s). The goal of textile interface studies is to facilitate the manufacture of technological textiles with optimized properties on the basis of a comprehensive understanding of interfacial behaviour of textile materials and their resulting influence on material processes.

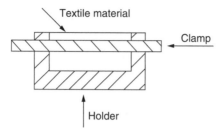

3.5 Sample holder.

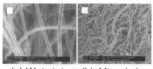

(a) Wet state   (b) After drying

3.6 ESEM of wet samples: (a) alginate sorbent in wet state, (b) alginate sorbent after drying.

The Philips XL30 ESEM-FEG is able to image not only non-conductive but also wet samples without any need for coating or preparation. Therefore it is an important tool for the interfacial studies of textile materials.

To improve the image of interfacial characteristics of textile materials, a special specimen holder was designed as illustrated in Fig. 3.5 [4]. The ESEM image in Fig. 3.6(a) [4] illustrates the absorption of wound exudate by alginate fibres. The image was taken at 20 kV at a temperature of 5°C and a pressure of 5.0 torr. Direct observation avoids artefacts or destruction, which may be caused by drying and coating the samples in a normal SEM as shown in Fig. 3.6(b) [4].

When a liquid comes in contact with the surface of a fibre, the liquid will either spread out and 'wet' the fibre surface, or it will form droplets that are 'repelled' from the surface. The wettability of a textile fibre is of importance in such systems as filters, coalescent units, sorbents, composite and biomedical materials. The wetting of fibres by a liquid is governed by the interfacial energies between the three phases of the liquid/vapour, solid/vapour and solid/liquid interfaces [5]. The angle formed at the edge of these droplets where the liquid contacts the solid surface, and the surface energy that affects this angle, form the basis of modem liquid–solid interface technology.

In the ESEM, specimens can be hydrated or dehydrated by controlling the temperature of the specimens and the chamber pressure in favour of water condensation or evaporation at different relative humidities. In the water wetting experiment, the pre-cooled fibre specimen is placed onto the

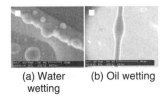

(a) Water   (b) Oil wetting
wetting

3.7 Wetting of PP fibre in ESEM: (a) water wetting, (b) oil wetting.

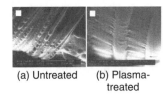

(a) Untreated   (b) Plasma-
                      treated

3.8 ESEM images of PET fibres: (a) untreated, (b) plasma-treated.
Water droplets of different shapes are visible on the fibre surfaces.

Peltier cooling stage in the ESEM chamber. The relative humidity can be adjusted by changing the pressure or the temperature of the Peltier stage within the chamber. As the relative humidity reaches 100%, water condensed onto the surface of the sample will appear. Observations on water droplets on the material can then be made at each point of interest, and the dynamic wetting process of the material can be recorded. The image in Fig. 3.7(a) [4] reveals the hydrophobic properties of the polypropylene fibre surfaces. High contact angles can be observed from the ESEM image [6].

However, the ESEM image of the same fibres shows the hydrophilic properties of the surfaces with low contact angles for oil, as presented in Fig. 3.7(b) [4]. In this oil wetting experiment, oil was added using a micro-injector [7], which can be mounted on the specimen chamber. The injection needle of the micro-injector can be placed just above the specimen. Different kinds of liquid can be used for wetting observations in the ESEM.

The fibres untreated and treated by plasma activation also show different wetting behaviours, as illustrated in Fig. 3.8 [4]. In the ESEM chamber, as relative humidity reaches 100%, the condensation of water is initiated by forming small water droplets on the fibre surfaces. It can be seen that the droplets are formed on the untreated material and fibre systems at the microscopic level *in situ*. An example of bacteria cells on PLLA fibres is presented in Fig. 3.9 [4].

Dynamic characterization

An ESEM equipped with a tensile stage can be used to examine the dynamic tensile behaviour of textile materials ranging from individual fibres to

3.9 Bacterial cells on PLLA fibres.

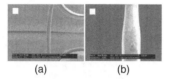

(a)   (b)

3.10 Tensile testing of single PP fibre: (a) fibre on the tensile stage, (b) necking of PP fibre.

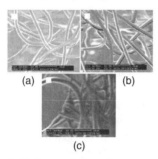

(a)   (b)

(c)

3.11 Dynamic bonding process of bicomponent fibres: (a) at room temperature, (b) at about 100°C, (c) at about 200°C.

fabrics made by different processes. The tensile stage can be placed in the ESEM chamber. The tensile process can be video-recorded, while the strain–stress curve is obtained. Figure 3.10(a) [4] illustrates polypropylene fibre on the specially designed tensile stage. The fibre was wound onto the pins, which are fixed on two movable plates in the tensile stage. In Fig. 3.10(b) [4], necking is observed as the tensile force is applied to the fibre. This observation gives insight into the neck profile and the structural features of the necked region.

Textile materials in fabric form can be easily examined by tensile stage in the ESEM [8]. ESEM technology also allows dynamic experiments at a range of pressures and temperatures and under a variety of gases/fluids. An ESEM equipped with a heating stage can be used to observe the dynamic thermal process of textile materials. Figure 3.11 [4] shows an example of thermal bonding of the PET/CoPET bicomponent fibre web (a carded web without bonding) as observed by the ESEM. The sample was placed on the heating stage. The heating stage was then fixed in the ESEM chamber. The

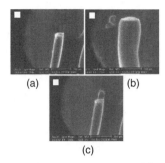

3.12 Water absorption of fibre structure: (a) experiment starts at 75% RH, (b) adsorption phase at 100% RH, (c) desorption at 75% RH.

heating rate was set at 10°C/min and the soaking time was set for 5 minutes. The heating process was video-recorded. The ESEM images show that the PET/CoPET bicomponent fibre web shows no obvious change from room temperature to 100°C, as presented in Fig. 3.11(a) and (b). It can be seen that when the temperature reaches about 200°C, the fibres are bonded together at fibre intersections, as illustrated in Fig. 3.11(c). It can be seen that fibre surface morphology has also changed due to the thermal bonding.

A series of micrographs taken during hydration and dehydration of the sodium polyacrylate fibre (shown in Fig. 3.12 [4]) also illustrates dynamic characterization. The pre-cooled sample fibre was placed onto the Peltier stage. The specimen temperature was set at 5°C, as this minimized the risk of accidental freezing. The sample was observed at 75% RH (5.0 torr) and 100% RH (6.6 torr) at 5°C. On reversal of the process, the ramping down was halted at 75% RH at 5°C. This clearly shows the effects of relative humidity on fibre structure. As can be seen from Fig. 3.12(a)–(c), the fibre diameter increased from about 43 μm to over 100 μm during adsorption at 100% RH. The swelling in the cross-section was much higher than that in the fibre axis. When the relative humidity was lowered from 100% to 75%, dehydration took place and the fibre rapidly shrank to the previous state.

Dynamic experiments can be conducted at a range of pressures and temperatures, and in a variety of gases/fluids. Different combinations of temperature, pressure and gases/fluids provide more opportunities for the study of textile materials.

3.5.2 Near infrared spectral image measurement system

For the textile industry, infrared spectroscopic methods that are based on diffuse reflectance measurements can be used for the non-destructive analysis of polymer composition of the fabric materials including their auxiliaries.

Heise *et al.* [9] used mid-infrared spectroscopy in combination with the diffuse reflectance (DR) technique for quantification of the textile finishing auxiliaries that were applied for improving various fabric characteristics. The results of a quantitative analysis of a reactive auxiliary (cyclodextrin derivative) applied on cotton fabrics up to 5% (by weight) are satisfactory, but limitations of the diffuse reflectance measurement technique still exist. Multivariate calibration based on partial least squares was employed using the specific bands of the cyclodextrin derivative within the spectral interval of 1900–1480 cm^{-1}, providing prediction results with around 5% of relative standard prediction error, based on mean sample population concentrations.

Yotsuda *et al.* [10] developed the near infrared spectral image measurement system for water absorbency of woven fabrics. This system consists of a pair of near-infrared light sources, a series of optical apparatus, a near-infrared camera, and an image processor. The developed measuring system is adequate for testing the time-dependent water absorbency of the materials by using the model samples and several woven fabrics. It could be expected that the sensed information obtained by the spectral image could control the quality of the water absorbency more precisely and effectively than the conventional test methods.

Wang *et al.* [11] studied the analysis methods of fibre by micro IR spectroscopy. In the study, 20 different kinds of fabric fibres analysed by micro IR spectroscopy were discussed.

Attenuated total reflection (ATR) is a convenient mode for single-fibre analysis by infrared microspectroscopy, particularly when transmission spectra are difficult to obtain or when surface preference sampling is desirable. Textile finishes such as spin finishes, anti-static finishes and permanent press finishes can be revealed by ATR techniques. Bicomponent fibres may be analysed by a combination of ATR techniques, transmission techniques and spectral subtraction [12].

3.5.3 Capillary electrophoresis/mass spectrometry technique

Capillary electrophoresis/mass spectrometry (CE/MS) has been used in forensic fibre comparison. CE/MS can separate extracted dye components on forensically relevant fibre samples and provide semi-quantitative estimates of dye amounts as well as qualitative information to identify the dye present (via the molecular weight and mass spectra) (https://www.facss.org).

Fibre evidence is frequently used in forensic science to associate a suspect to a victim or crime scene. The fibres are found as trace evidence in crimes of personal contact such as homicide, assault, sexual offences and hit-and-run accidents. In forensic fibre comparison, fibres are screened by visual inspection using optical microscopic techniques such as polarized light microscopy (PLM) and by spectroscopic methods such as UV/visible and fluorescence microspectrophotometry. If spectra of the known and questioned fibres are consistent, the hypothesis that the fibres originate from a common source should not be rejected. The premise of the method presented here is that additional discrimination may be achieved by extraction of the dye from the fibre, followed by trace analysis by a high-resolution separation technique.

A sensitive and selective technique such as capillary electrophoresis/mass spectrometry (CE/MS) is needed to analyse the small amount of dye (2–200 mg) present on forensically relevant fibre samples. CE/MS can separate extracted dye components and provide semi-quantitative estimates of dye amounts as well as qualitative information to identify the dye present (via the molecular weight and mass spectra). A decision tree for extraction of unknown dyes from textile fibres has been developed that employs three capillary electrophoresis methods with diode array detection for the separation and identification of dyes from the six major textile dye classes. Although this approach is destructive to the sample, only an extremely small sample is required (1–2 mm of a single 15 micron diameter fibre). Automated micro-extractions and CE offer the forensic analyst reproducible analyses (% RSDs ranging from 5 to 25%) with limits of detection in the picogram range. The combined extraction/CE-MS system is capable of achieving both highly discriminating and highly sensitive identification of dyes for improved discrimination of trace fibre evidence. These advances establish the chemical basis for discrimination of fibres.

3.5.4 Thermogravimetry analysis

Thermogravimetric analysis (TGA) has been used for measuring the degree of thermal degradation of fibres in the fabrics, especially high-performance fibres. Here are some application examples.

*Evaluating the thermal stability of high-performance
fibres by thermogravimetric analysis*

The degree of thermal degradation of eight types of high-performance fibres (HPFs) was measured under nitrogen and air atmosphere by weight loss using thermogravimetric analysis (TGA), and the characteristic degradation temperatures were obtained. The kinetics of the thermal degradation have also been analysed according to the Freeman–Carroll method and the activation energies of the HPFs were estimated. The experimental results show that para-aramids (Kevlar® 29, 49 and 129, and Twaron®2000) have similar thermal stability, but their thermal degradation temperatures and activation energies in air are different from those in nitrogen, which means that the thermostability of the fibre depends not only on its intrinsic structure but also on the atmosphere and temperature of the testing environment. Terlon® fibre shows higher degradation temperature as a copolymer of para-aramid, and its initial degradation temperature is 476.4°C in air. It can also be found that the PBO (poly(p-phenylene benzobisoxazole)) fibre has the highest thermal degradation temperature among the samples tested, but its activation energy is not the highest in both air and nitrogen atmosphere. The UHMW-PE (ultra high molecular weight polyethylene) fibre has the lowest thermal degradation temperature, and begins to degrade when the temperature reaches 321.8°C under air atmosphere [13].

Pyrolysis behaviour of rayon fibres treated with $(NH_4)_2SO_4/HN_4Cl$

The pyrolysis behaviour of rayon fibres treated with $(NH_4)_2SO_4/HN_4Cl$ was investigated by thermogravimetry (TG) analysis and the kinetics parameters were obtained by TG and DTG curves. It was found that the pyrolysis proceeded at a lower temperature and had a bigger yield of char when the rayon fibres had been treated with $(NH_4)_2SO_4/HN_4Cl$ than when they had been treated with $H_2SO_4/CO(NH_2)_2$. The effective part of $(NH_4)_2SO_4/HN_4Cl$ was produced at a wide range of temperature and in a tempered way, which resulted in a better effect of the catalyst. After treatment with $(NH_4)_2SO_4/NH_4Cl$, the order of the pyrolysis reaction of the rayon fibres increased from 1.1 to 3.2, while the activation energy decreased from 237 kJ/mol to 94 kJ/mol. Using this catalyst system, the rayon-based carbon fibres showed a tensile strength of 1.05 GPa [14].

3.5.5 Computer image processing technology

The recent development of computer technology makes it possible to apply pattern recognition techniques to many real textile industrial problems, including fabric composition testing. Shangtiao *et al.* [15] investigated some image processing techniques such as filtering, edge detection, boundary extraction and extracting the central line of the yarn to construct an image processing method to distinguish between cotton fibre and ramie fibre from the blended yarn fabric. An image-processing algorithm for classifying both fibres from the image data obtained from the blended yarn fabrics was developed and the results turned out to be successful.

With new fibres being developed continually, for relevant fabric composition testing, traditional fibre identification methods are not enough. New methods have to be investigated, such as those aimed at identification of soybean protein fabric, chitin fabric, bamboo fabric, cashmere fabric, poly-lactic acid fabric, lyocell fabric, modal fabric, milk protein fabric, and fabrics of their fibre blended with cotton, ramie, silk and wool.

For high-performance fibres (HPFs) such as para-aramid fibres (Kevlar® 29, 49 and 129, and Twaron®2000), Terlon® fibre, PBO (poly(p-phenylene benzobisoxazole)) fibre, UHMW-PE (ultra high molecular weight polyeth-ylene) fibre, basalt fibre and quartz fibre, when testing fabric composition not only the generic fibre type needs to be identified but also the chemical element discrimination and even specific distinguishing behavioural fea-tures. So fabric composition testing of high-performance fibres needs more advanced methods with more sophisticated equipment.

For newly structured, newly engineered fabrics such as multilayer fabrics, phase change fabrics and electrically conductive fabrics, more complex or complicated methods should be sought. Some applications of the achieve-ments in the above fields will be introduced in the following sections.

3.6   Applications

3.6.1 Applications in standards on fabrics

Standard test method for extractable matter in textiles

This test method covers a procedure for determining the extractable mate-rial on most fibres, yarns and fabrics. Three options are included. Option 1 uses heat and Soxhlet extraction apparatus. Option 2 uses room tempera-ture and extraction funnels. Option 3 uses either Option 1 or Option 2 extraction but provides for calculation of extractable matter from the loss in mass of the material due to the extraction rather than the extractable matter residue [16].

The solvents for use in this method are any solvents that the party or parties concerned agree on, such as halogenated hydrocarbon (HH), chlo-roform, tetrachloroethane, alcohol, etc.

3.6.2 Analysis of other vegetable textile fibres and yarns

In general, the methods and procedures described in this part deal with the identification of vegetable textile fibres of various types as well as paper yarn and woven fabrics of paper yarn. Physical characteristics, such as the decitex of yarns, the weight per unit area of fabrics, as well as the weight percent of vegetable fibres when the vegetable fibres are blended with other types of fibres, are also determined [17].

Principle

The identification of vegetable fibres in all their forms and of paper yarn is determined microscopically. The physical characteristics are visually determined then measured following prescribed methods. If required, manual or chemical identification and separation of additional fibres is performed and the resulting mass of each type of fibre is taken.

Apparatus

- Polarizing microscope comprising a light source (reflected and transmitted), a light condenser, a stage which supports the slide carrying the fibres, an ocular and objectives. A first-order red plate is desirable
- Stage movable in two directions at right angles
- Objective and ocular, capable of providing at least 100× magnification
- Low-powered stereo microscope
- Micro projector (optional), equipped with a fixed body tube, a focusable and movable stage responsive to coarse and fine adjustments, a focusable substage with condenser and iris diaphragm, and a vertical light source to give a precise magnification of up to 500× in the plane of the projected image on a flat surface
- Image analysis system (optional)
- Stage micrometer calibrated in intervals of 0.01 mm for accurate setting and control of the magnification
- Scanning electron microscope (optional)
- Suitable sectioning apparatus
- Suitable mounting media
- Cover-glass slips
- Reference samples of flax, true hemp, jute (and other textile bast fibres), sisal (and other textile fibres of the genus *Agave*), coconut, abaca (manila hemp), ramie, paper, etc.
- Analytical balance, with a range up to 200 g and accurate to 0.0001 g.
- Precision balance, with a minimum capacity of 1000 g and a readability of 0.1 g.
- Yarn reel, precision ruler or any other apparatus giving similar results
- Shirley crimp tester, or any other length-measuring apparatus giving similar results
- Ultraviolet (UV) light source
- Dissecting needles, picks or other equipment giving similar results.

3.7   Identification of new textile fibres

The differences in composition such as macrostructure and microstructure between new fibres such as lyocell fibre, modal fibre, soybean protein fibre, bamboo fibre, milk protein fibre and chitin fibre must result in discrepancies in macroscopic features, physical and chemical properties, which can be used to distinguish different fibres. Some efficient methods to identify the six new textile fibres listed above are presented.

3.7.1   Identification of lyocell fibre

The nature or generic type of a cellulosic material may be determined by a burning test and then the longitudinal feature may be observed by microscopic examination, referring to the information from fibres such as modal, bamboo, chitin, cotton and regular viscose fibres. Sodium hypochlorite may be used as a solvent to investigate the dissolution behaviour further. In addition, a tensile test may be carried out to find out the tenacity of the fibre, as the tenacity of lyocell fibre is the highest among the cellulosic fibres [18].

3.7.2   Identification of modal fibre

The nature or generic type of a cellulosic material may be determined by a burning test and then the longitudinal feature may be observed by microscopic examination, referring to the information from fibres such as modal, bamboo, chitin, cotton and regular viscose fibres. In addition, the modal fibre cross-section may be observed by microscope taking the reference as bamboo fibre and regular viscose. Subsequently, 75% sulfuric acid may be used as a solvent to observe the solubility. Tensile and wet elongation tests may be performed to determine the tenacity [18].

3.7.3   Identification of soybean protein fibre

Firstly, the nature or generic type of a protein fibre may be determined by a burning test. Then, an iodine–potassium iodide staining test may be performed to exclude wool and silk. A solution of boiling 5% sodium hydroxide may be used as a solvent to find out the dissolution behaviour: soybean protein fibre does not dissolve, milk fibres swell and become moist, and chitin fibre dissolves. Finally, boiling dimethyl formamide (DMF) may be used as a solvent to examine the protein fibre further [18].

3.7.4   Identification of bamboo fibre

Firstly, the nature or generic type of a cellulosic material may be determined by a burning test and then the longitudinal feature may be observed by microscopic examination, referring to the information from fibres such as modal, bamboo, chitin, cotton and regular viscose fibres. Subsequently, 37%

hydrochloric acid may be used as a solvent to observe the solubility at normal temperature: modal fibre dissolves quickly, while regular viscose dissolves slowly and bamboo fibre dissolves partially. Finally, a density test may be used (using a density grading column) to determine the density. The density of bamboo fibre is lower than those of regular viscose, lyocell, modal and cotton fibres [18].

3.7.5 Identification of milk protein fibre

Firstly, the nature or generic type of a protein fibre may be determined by a burning test. Then an iodine–potassium iodide staining test may be performed to exclude wool and silk. Microscopic observation may help reveal surface features. While the surface of milk fibres appears smooth and non-microporous, soybean protein fibres are not smooth and show random microporous surfaces with a spot welding effect. Subsequently, boiling dimethyl formamide (DMF) may be used as a solvent to find out the dissolution behaviour. Milk fibres moisten and swell, whereas other protein fibres such as wool, silk, chitin and soybean protein fibres exhibit no changes [18].

3.7.6 Identification of chitin fibre

Firstly, a burning test may be used to observe the fibre's burning behaviour. Chitin fibre does not melt or shrink upon being subjected to a flame, but it burns rapidly like a cellulosic fibre with the odour of burning a protein fibre. The distinguishing features of chitin compared to other fibres are that it burns rapidly to black while keeping its original shape, the ash produced being greyish in colour and easily crumbled. After initially observing its burning behaviour by a burning test, the surface may be observed using a microscope. The surface of the fibre is characterized by small openings and cracks. An iodine–potassium iodide test may reveal important information: while in a wet state, lyocell, modal, bamboo and regular viscose fibres look alike, but upon drying, iodine on the chitin fibre sublimes easily, resulting in a blackish-blue to red-brown colour. As in other protein fibres, 5% boiling sodium hydroxide may be used as a solvent to observe the solubility. Chitin fibre dissolves in this solvent. Alternatively, 88% boiling formic acid may be used: here only chitin fibre dissolves, whereas all other fibres show no dissolution [18].

3.8   Case study: identification of ancient textile fibres

Two-thousand-year-old archaeological single fibres from textile fragments excavated in the Cave of Letters in the Dead Sea region were investigated by a combined approach using microscopy (optical and SEM), X-ray micro-beam diffraction and X-ray microbeam fluorescence. In comparison with modern reference samples, most of the fibres were identified as wool, some as plant bast fibres (flax). The molecular and supermolecular structure of both keratin (wool) and cellulose (flax) were found completely intact. In many fibres, mineral crystals were intimately connected with the fibres. The fluorescence analysis of the dyed wool textiles suggests the possible use of metal-containing mordants for the fixation of organic dyes [19].

3.9   Identification of cashmere and wool fibre scale frequency using fast Fourier transform

Sometimes fine wool may be found in 'pure' cashmere garments, and wool and synthetic fibre blends are sometimes labelled as pure wool products. Therefore, both wool and special animal fibre industries require accurate classification of animal fibres. Animal fibres can be distinguished using the patterns of their cuticular scales and other techniques such as analysing their unique physical and chemical properties. The recognition of charac-teristic features of scales is still the most useful method to distinguish animal fibres such as Merino, mohair and cashmere. Natural animal fibres possess cuticle surface morphology. The cuticle is composed of flat, plate-like cells called scales. Scale patterns provide very important information about the identities of animal fibres for classification. Scale frequency is an important feature of animal fibre cuticle surface morphology. It is defined as the number of scale separations per 100 μm fibre length in the direction of the fibre axis. Robson [20] reports that using fibre scale frequency alone, 659 cashmere and wool fibres among a population of 800 'unknown' fibres can be correctly classified. When combined with fibre diameter, the accuracy for correct fibre type classification is even higher [21].

Fibres of different types possess different fibre diameter ranges and scale frequencies as shown in Table 3.4 [21]. Compared to wool fibre, cashmere is finer and has a lower scale frequency. Therefore, fibre diameter and scale frequency are often used to identify wool and cashmere.

Animal fibre scale patterns and profiles are often observed under an optical microscope or a scanning electron microscope (SEM) to determine the fibre types. Therefore, image enhancement techniques and computer aided image recognition are also used for better fibre classification. However, this process is time-consuming and tedious, making fibre classification costly. The Fourier transform provides a link between a time domain signal and a frequency domain signal. According to the Fourier theory, any signal can be expressed by a sum of sine and cosine waves of various frequencies and amplitudes. The contribution of each frequency to the total effect (i.e. the signal) is determined by the amplitude of its Fourier coefficient. Therefore,

Table 3.4 Fibre diameter and scale frequency of some animal fibres

Fibre type	Fibre diameter (μm)	Scale frequency (scales/100 μm)
Vicuna	10	11
Cashmere[1]	13–18	6–8
Cashmere[2]	15.3	7
Wool[1]	16–39	7.6
Wool[2]	17.9	9.1
Yak hair	19	9–10
Cashgora	17	6–7
Camel hair	19	9–10
Huacaya alpaca	16.6–40.1	10.5
Alpaca	23	10

[1] Chinese white cashmere, Australian fine and coarse Merino wool.
[2] Australian white cashmere, Australian fine Merino wool.

if the scale edges show a regular pattern on the rightmost and leftmost edges of a fibre projection image, the scale frequency can be readily found out by means of the fast Fourier transform (FFT) from the fibre diameter profile signal. A simple method that measures the fibre diameter profile of a single animal fibre will be introduced in the following sections. Subsequently, we will analyse the diameter profile using the FFT technique to determine whether data from the diameter profile and FFT results can be used to quantify the scale frequency of different animal fibres [21].

3.9.1 Testing for cashmere and wool fibre scale frequency

Greasy cashmere and wool fibres were sampled from sale lots, then scoured under identical conditions. They were used to measure the profiles of single-fibres. Single-fibre diameter profiles were measured by a Single Fibre Analyser (version 1: SIFAN1; version 2: SIFAN2) instrument. SIFAN1 gives more accurate measurements but fewer data points than SIFAN2. The SIFAN measures a fibre diameter profile at a set interval along the fibre length in less than a minute. The scanning intervals of 5 μm for SIFAN1 and approximately 0.4 μm for SIFAN2 were used in this work.

After scanning the profiles of single fibres by the SIFAN, their scale frequencies were measured using either an Olympus SZX12 microscope with an Olympus DP10 digital camera or a Leo 1530 field emission gun scanning electron microscope.

The fibre diameter profile digital signals were first filtered using a Butterworth bandpass digital filter, to remove components that were outside the range of 2.5 to 15.5 scales per 100 µm, as animal fibre scale frequency is in the range of 6 to 11 scales per 100 µm (Table 3.4). The order of the filter was set to 5. A Hamming window was applied to the diameter profile signal. This was to reduce the truncation effect normally encountered in data acquisition [21].

Figure 3.13 [21] shows the diameter profile of a 30 mm long wool fibre measured with the SIFAN1 at a scanning interval of 5 µm. Figure 3.13(b) [21] is a zoom view of the boxed section in Fig. 3.13(a). It can be seen from Fig. 3.13(a) that the fibre diameter is highly irregular along the fibre length and the fibre scale frequency cannot be determined by simply observing either Fig. 3.13(a) or Fig. 3.13(b).

The wool fibre was repeatedly measured five times with the same test parameter settings and the diameter profiles were then transformed into a power spectrum as shown in Fig. 3.14 [21]. From Fig. 3.14, it can be seen that the fibre scale frequency was between 5 and 10 scale edges per 100 µm and the measurements had good repeatability.

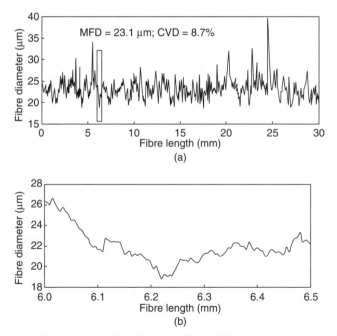

3.13 Diameter profile of a wool fibre (SIFAN1 at 5 µm interval) and the zoomed profile of a 0.5 mm segment of fibre.

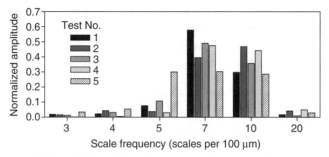

3.14 FFT analysis of five diameter profiles repeatedly measured using the same wool fibre.

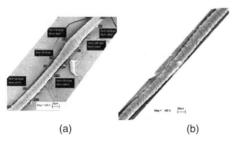

(a)   (b)

3.15 Micrographs of two sections of the fibre used for diameter profile measurement in Fig. 3.13.

In Fig. 3.15(b) [21], 380 μm long of wool fibre used reveal that the fibre scale frequency is approximately 7.6 scales per 100 μm, which is in the range of 7–10 scales per 100 μm (Fig. 3.13). This suggests that FFT analysis has great potential to determine the animal fibre scale frequency.

One of the disadvantages of the scanning electron microscope or optical microscope is the tendency to acquire fibre surface information from a few, not necessarily representative, samples. The wool fibre used in Fig. 3.13 sets an example. Its mean fibre diameter measured by SIFAN1 is 23.1 μm (Fig. 3.13) while by five measurements using the SEM images in Fig. 3.13 the diameter is 25.3 μm in Fig. 3.15(a) and 22.9 μm in Fig. 3.15(b) respectively. This suggests that due to diameter irregularity along the fibre length and limited fibre length for imaging, the mean fibre diameter measured with a microscope or SEM may not represent the fibre population, and a large sample size is needed for more accurate measurement. For this reason, the fibre diameter profile and its FFT result may have an advantage in determining fibre diameter and scale frequency as the SIFAN uses large samples (up to the whole fibre length) at a fast testing speed.

Because the resolution of the 5 μm scanning interval is too coarse, the FFT result may not cover the full scale frequency spectrum in detail.

Therefore, the scale frequency in Fig. 3.14 does not include data between 5 and 10 scales per 100 μm, which would not help to identify an animal fibre as most animal fibres have a scale frequency in the range of 6 to 11 (Table 3.4). A finer scanning interval is also necessary to analyse scale frequencies of more than 10 scales per 100 μm.

Figure 3.16 [21] shows a cashmere fibre diameter profile at 0.37 μm sampling interval and the zoomed Fig. 3.16(b) shows more details of the fibre profile compared to Fig. 3.16(a). Regardless of the fibre profile scanning intervals, the fibre scale frequency still cannot be determined by observing the diameter profile figures only.

The FFT result in Fig. 3.17 [21] suggests that the cashmere fibre has a scale frequency between 5 and 7, and the frequency of 6 scales per 100 μm is most likely the case. Optical microscope observation of the cashmere fibre revealed that the fibre has a scale frequency of 6.1, which confirms the FFT result in Fig. 3.17.

A 19 μm fine wool fibre was also analysed. The FFT result in Fig. 3.18 [21] indicates that the wool fibre has a scale frequency of 7 scales per

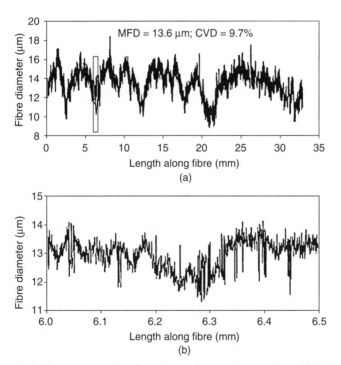

3.16 Diameter profile of an Australian cashmere fibre (SIFAN2 at 0.37 μm sampling interval) and the zoomed profile of a 0.5 mm segment of fibre.

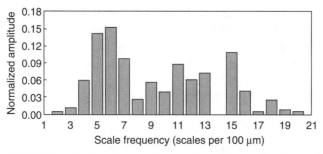

3.17 FFT analysis of the cashmere diameter profile shown in Fig. 3.16.

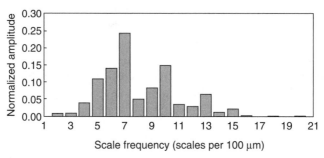

3.18 FFT analysis of the diameter profile of a 19 μm wool fibre.

Mag = 500 X   100μm

3.19 A section of the wool fibre used for the FFT result in Fig. 3.18.

100 μm. The SEM image in Fig. 3.19 reveals that the fibre has a scale frequency of 7.3 scales per 100 μm, which agrees well with the FFT result in Fig. 3.18.

Figure 3.19 [21] also reveals that the diameter of the fibre on the SEM image is approximately 18 μm. Again, due to the fact that the SIFAN can measure a longer fibre segment than the SEM, fibre diameter results measured from the SIFAN are more representative.

The instrument for the fibre diameter measurement should have a high resolution in order to accurately measure the differences due to scale protrusion, as some fibres have small fibre scale height; for example, the scale height of Huacaya alpaca is only 375 nm, that of wool 1.1 μm [22] and that of cashmere around 600 nm (Fig. 3.20) [21]. As this section presents only a few samples of wool and cashmere, we will further evaluate the method using more animal fibres to examine its robustness. The latest SIFAN,

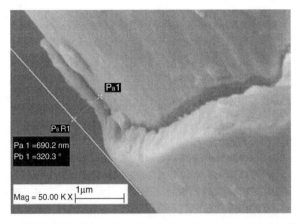

3.20 Scale height of an Australian brown cashmere fibre (fibre diameter 14 μm).

version 3, which has a claimed resolution of 0.1 μm, will be employed for fibre diameter profile scanning.

The sampling interval greatly affects the accuracy of the scale frequency measurement: a smaller sampling interval gives more detailed spectra, hence better estimation of scale frequency. The SIFAN instrument works best on a sampling interval of 5 μm or more. For high-resolution sampling (for example 300 nm or less), the reliability of the testing system needs to be investigated first to ensure accurate results.

3.10   Testing of high-visibility fabrics

'High-visibility materials for safety garments' are specified by the following three types of materials:

- Retroreflective materials (Class R)
- High daylight visibility coloured materials, which covers fluorescent and non-fluorescent colours (Class F)
- Combined retroreflective–fluorescent coloured materials (Class RF).

Safety garments are expected to be worn frequently, often under rough conditions. AWTA (Australia Wool Testing Authority) Textile Testing can perform the full range of tests for Class F colours in-house.

It is recommended that fluorescent colours be tested for colour (luminance and chromaticity) first, because of the difficulty of achieving the required shade and luminance. Dyers should note that a recipe that works on one fabric might not work on another fabric of the same fibre composition. Fabric is measured as a single layer over a very dark backing. The more open or transparent the fabric, the less of the bright fabric and the more of the dark backing is measured.

Measurement of retroreflective performance (CIL) requires specialized equipment, mainly used in non-textile fields. AWTA Textile Testing conducts the durability tests and prepares specimens for measurements, then sub-contracts the CIL measurement to a specialist laboratory. There are several standards covering this field, for example:

- AS/NZS 4602:1999 'High visibility safety garments' is a specification which refers to the materials covered in AS 1906.4:1997.
- EN471:1994 'High visibility warning garments' combines material specifications and garment design. The concepts are similar but the detail is different from the Australian Standards.

3.11   Fourier transform infared and thermal analysis of cashmere and other animal fibres

Animal fibres such as cashmere, wool and alpaca are composed of keratin of a similar structure. Cashmere is an extremely fine and luxury material, in comparison with most other animal fibres. For reducing the cost of cashmere products, fine wool is sometimes blended with cashmere in products labelled as pure cashmere. Fibre identification between cashmere and fine wool has been a tedious and difficult process. Microscopic methods have been largely employed by the textile industry and forensic services. The surface morphological differences between wool and cashmere have been observed by many researchers with the help of a scanning electron microscope (SEM). An image processing system and a hybrid artificial neural network (ANN) combined with images from an optical microscope to undertake scale feature extraction and discrimination of animal fibres have been applied [23].

The chemical constitutions of animal fibres have been analysed and the results show that the contents of amino acids, particularly the cystine component, in cashmere are different from those of wool and alpaca. The DNA amplification technique has also been explored to identify speciality animal fibres. However, DNA analysis was shown to be influenced by the amount of substances extracted from the fibres (i.e. chemical-treated fibres). Bioengineering methods (e.g. species-specific monoclonal antibodies produced by proteins extracted from different animal fibres) have also been used for anti-fraud identification. Because of the complexity among fibre varieties, satisfactory results have not been achieved so far. All of these operations also involve a time-consuming preparation process [23].

Fourier transform infrared (FTIR) microscopy, differential scanning calorimetry (DSC) and thermogravimetric analysis (TGA) techniques are relatively new in material characterization. From an FTIR spectrum, particular chemical bands can be precisely located at a certain wavelength. Raw and chlorinated or reductive chemical-treated wool have been examined by FTIR for specifying the chemical changes in treated wool. Thermodynamic properties of materials such as melting temperatures, transition enthalpies, phase transformations, etc., can be analysed by DSC. In addition, thermal stability and composition of materials can be determined by TGA. The following section will introduce the application of FTIR and DSC/TGA analysis in animal fibre identification. FTIR and thermal analysis techniques were investigated for differentiating cashmere from other animal fibres. The FTIR spectra and TGA as well as DSC curves were analysed and the differences between Chinese cashmere and other animal fibres produced in Australia were examined [23].

3.11.1 Testing for thermal properties of cashmere and other animal fibres

Both Chinese cashmere and Australian cashmere of different diameters were sampled to compare with other animal fibres, which include Chinese camel hair, Australian Merino wool and alpaca fibres. All samples were extracted by ethanol twice to remove the residual grease content. Fibre diameter profiles were measured by OFDA 100 and the results are listed in Table 3.5 [23].

Thermal properties of all samples were measured by a NETZSH STA409PC instrument (NETZSH Gerätebau GmbH, Germany) for obtaining TGA and DSC curves. Aluminium crucibles were used to hold 6–9 mg of 2 mm-long fibre snippets for measurements. Argon gas was employed for purging samples at 30 ml/min and for protection at 10 ml/min. The temperature was increased at 10°C/min after 25°C, and 500°C was set as the maximum.

A Vertex 7.0 Fourier Transform Infrared Spectroscope (FTIR) (Bruker Optics Inc., Germany) was used in the attenuated total reflectance mode. Each fibre specimen was scanned at a resolution of 4 cm^{-1} 32 times to acquire an ATR spectrum between 4000 cm^{-1} and 600 cm^{-1} wavelength. Five spectra were obtained for each sample.

3.11.2 Thermal properties by thermogravimetric analysis

Figure 3.21 [23] shows that the mass of fibres changed with an increase in temperature. For all animal fibres examined, two main phases of mass change are shown on each TGA curve. The first mass loss occurred at about 60°C and was caused by moisture evaporation from fibre structures, and the second loss resulted mainly from thermal decomposition of the fibres. Some

Table 3.5 Fibre diameter profiles and production regions

Sample no.	Label	Description	MD (μm)*	CVD (%)*
1	Aus_cashmere	Australian white cashmere	15.16	21.3
2	Aus_cashmere	Australian white cashmere	15.60	20.6
3	Aus_cashmere	Australian white cashmere	17.51	22.6
4	CN_cashmere	Chinese white cashmere	14.16	21.3
5	CN_cashmere	Chinese white cashmere	14.21	22.5
6	CN_grey cashmere	Chinese grey cashmere	20.25	29.1
7	CN_camel	Chinese brown camel hair	19.11	27.0
8	Aus_alpaca	Australian white alpaca	23.65	23.0
9	Aus_alpaca	Australian white alpaca	34.96	23.3
10	Aus_alpaca	Australian white alpaca	23.14	24.4
11	Aus_alpaca	Australian white alpaca	22.19	22.7
12	Aus_fine wool	Australian fine Merino wool	20.33	17.7
13	Aus_fine wool	Australian fine Merino wool	19.59	23.1
14	Aus_coarse wool	Australian coarse Merino wool	31.32	22.3
15	Aus_coarse wool	Australian coarse Merino wool	32.78	18.0
16	Aus_fine wool	Australian fine Merino wool	16.36	17.2

*MD: mean diameter of fibres; CVD: coefficient variation of diameter.

residual mass remained at the end of the temperature rise. A ratio of mass change within any temperature range as well as an associated onset temperature can be obtained from a TGA curve as shown in Fig. 3.21.

Based on the TGA curves, the rapid mass changes occurred between 200°C and 350°C, as a result of thermal degradation. Figure 3.22 [23] shows the average mass change within this temperature range for different fibre groups. It is clearly seen that Australian cashmere and Chinese grey cashmere have a smaller mass change than other fibres. Chinese white cashmere

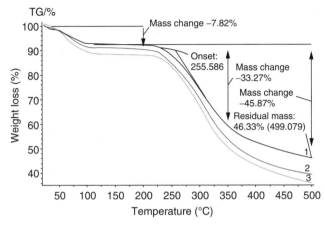

3.21 Typical TGA curves of cashmere and wool produced in China and Australia (curve 1: Australian cashmere; curve 2: Chinese white cashmere; curve 3: Australian fine wool).

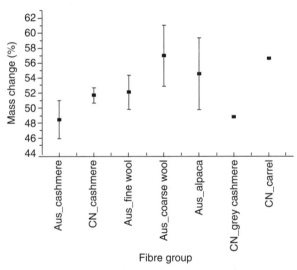

3.22 Mass changes of animal fibres in groups.

and Australian fine Merino exhibit a similar mass change, with the coarse Australian Merino wool showing the largest mass change within groups.

Fibre diameter may have some effect on the mass change. For Australian cashmere, wool and alpaca fibres, there exists a moderate linear relationship between the mass change and fibre diameter (Fig. 3.23(a)) [23]. The coarser fibres tend to have a larger mass change than finer fibres. Conversely, the residual mass of the fibres has a moderate and negative linear relationship with fibre diameter (Fig. 3.23(b)) [23].

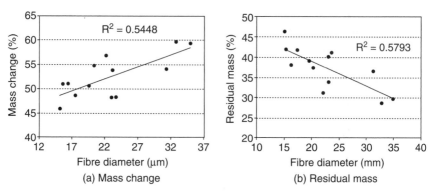

3.23 Relationships of fibre mass change and residual mass with fibre diameter.

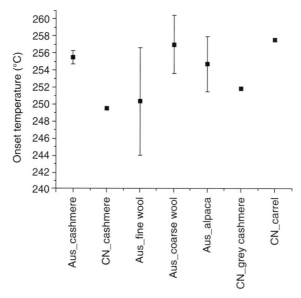

3.24 Fibre onset temperatures of mass change in groups.

Along with the associated fibre mass changes, onset temperatures within the same temperature range (200–350°C) are shown in Fig. 3.24 [23]. Chinese cashmere and Australian fine Merino wool demonstrated almost the same onset temperature, which is lower than those of the other fibres tested. Australian cashmere fibres were thermally degraded at a high temperature, as were Australian coarse wool, alpaca fibres and Chinese camel hair. However, there are considerable variations within some fibre groups, as indicated by the large error bars.

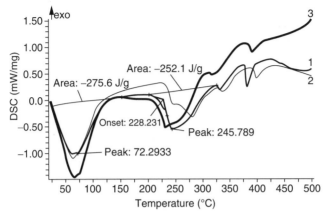

3.25 Typical DSC curves of cashmere and wool produced in China and Australia (curve 1: Australian cashmere; curve 2: Chinese white cashmere; curve 3: Australian fine wool).

3.11.3 Thermal properties by differential scanning calorimetry analysis

The heat transfer process can be viewed from differential scanning calorimetry (DSC) curves of the fibres as shown in Fig. 3.25 [23]. Corresponding to the mass change, there are two main heat transfers during the process of temperature increase. The first endothermic heat peak occurred at around 60°C due to moisture evaporation from inside the fibres. The second endothermic heat peak occurred at around 250°C, resulting in the thermal decomposition of the fibres. The onset temperature (T_o) and the peak temperature (T_p) as well as the amount of endothermic heat (ΔH) can be calculated from the DSC curve. The results for all the fibres tested are listed in Table 3.6 [23].

In Table 3.6, Australian cashmere, like Chinese camel hair, starts to absorb heat much later than the other fibres tested before the thermal decomposition of the fibres starts. The Australian alpaca fibres have the lowest T_0. Except for Chinese grey cashmere, both Australian and Chinese cashmere fibres have a very small ΔT and ΔH, which means the white cashmere fibres were thermally degraded in a shorter time and with a lower endothermic heat than others. On the other hand, Chinese camel hair is more resistant to thermal degradation.

3.11.4 Fourier transform infared analysis

The ATR spectra of fibres are shown in Fig. 3.26 [23]. Apparently, differences between Chinese cashmere and Australian cashmere as well as wool

Table 3.6 Results of DSC analysis in fibre groups

Fibre group (see Table 3.5)	T_o (°C)*	T_p (°C)*	ΔT (°C)*	ΔH (J/g)*
Aus_cashmere	230.28	260.59	30.30	247.13
CN_cashmere	229.73	263.01	33.29	209.90
Aus_fine wool	228.91	286.07	57.16	393.83
Aus_coarse wool	224.83	285.31	60.47	411.85
Aus_alpaca	220.55	281.94	61.39	529.25
CN_grey cashmere	228.31	283.68	55.38	414.00
CN_camel	231.15	295.25	64.70	693.80

* T_o: onset temperature of endothermic heat; T_p: peak temperature; ΔT: $T_p - T_o$; ΔH: endothermic heat.

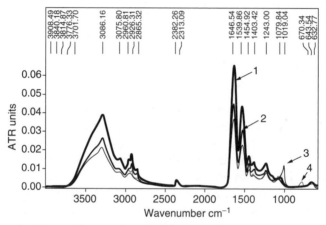

3.26 ATR spectra of cashmere and wool (curve 1: Australian cashmere; curve 2: Australian fine wool; curve 3: Chinese white cashmere; curve 4: Chinese grey cashmere).

clearly exist in a peak near the 1040 cm^{-1} wavelength which relates to the S–O stretching component of cysteic acid residues (R–SO$_3^-$). R–SO$_3^-$ is usually derived from an oxidation of disulfide bonds of cystine. Because of the mass effect, steric interactions and resonance from adjacent atoms, this peak may be shifted away from the 1040 cm^{-1} wavelength. In the present research, Chinese white cashmere had a strong absorption near 1019 cm^{-1}, while Australian cashmere and Merino wool had a weak absorption but the peak was shifted left to 1079 cm^{-1}. For Chinese grey cashmere, besides its strong absorption near 1019 cm^{-1} (same as Chinese white cashmere), there was an apparent extra peak at about 800 cm^{-1}, indicating the presence of melanin pigment inside the fibre.

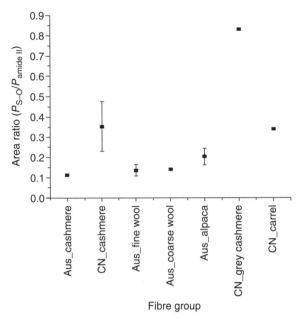

3.27 Normalized S–O stretching peak of animal fibres in groups.

A normalization of the spectrum peak near 1040 cm^{-1} ($P_{S–O}$) to the peak at 1539 cm^{-1} ($P_{amide\ II}$) for the amide II backbone group may reveal the difference between the fibres in a relative manner. Figure 3.27 [23] clearly shows that the animal fibres produced in China have a significantly larger amount of R–SO$_3^-$ group than the fibres produced in Australia. In particular, the Chinese grey cashmere has the highest amount of R–SO$_3^-$ group in all the fibres tested. In addition, Australian cashmere has a relatively lower amount of R–SO$_3^-$ group than wool and alpaca fibres. These differences may imply that the content of R–SO$_3^-$ formed in these animal fibres is closely related to the region where the animals are farmed.

3.12   References

[1] E.R. Prack, An introduction to process visualization capabilities and considerations in the environmental scanning electron microscope (ESEM), *Journal of Microscopy Research and Technique*, 1993, 25(5–6), 487–492.

[2] P. Messier, T. Vitale, Cracking in albumen photographs: an ESEM investigation, *Journal of Microscopy Research and Technique*, 1993, 25(5–6), 374–383.

[3] Q.F. Wei, R.R. Mather, A.F. Fotheringham, R.D. Yang, Observation of wetting behavior of polypropylene microfibres by environmental scanning electron microscope, *Journal of Aerosol Science*, 2002, 33(11), 1589–1593.

[4] Q.F. Wei, X.Q. Wang, R.R. Mather, A.F. Fotheringham, New approaches to characterisation of textile materials using environmental scanning electron microscope, *Fibres and Textiles in Eastern Europe*, 2004, 12(2), 79–83.

[5] R. Combes, M. Robin, G. Blavier, M. Aïdan and F. Degrève, Visualization of imbibition in porous media by environmental scanning electron microscopy: application to reservoir rocks, *Journal of Petroleum Science and Engineering*, 1998, 20(3–4), 133–139.

[6] Q.F. Wei, R.R. Mather, A.F. Fotheringham, R.D. Yang, Dynamic wetting of fibres observed in an environmental scanning electron microscope, *Textile Research Journal*, 2003, 73(6), 557–561.

[7] Q.F. Wei, R.R. Mather, A.F. Fotheringham, R. Yang, J. Buckman, ESEM Study of oil wetting behaviour of polypropylene fibres, *Oil and Gas Science and Technology – Rev. IFP*, 2003, 58(5), 593–597.

[8] R. Rizzieri, F.S. Baker, A.M. Donald, A study of the large strain deformation and failure behaviour of mixed biopolymer gels via in situ ESEM, *Polymer*, 2003, 44(19), 5927–5935.

[9] H.M. Heise, R. Kuckuk, A. Bereck, D. Riegel, Mid-infrared diffuse reflectance spectroscopy of textiles containing finishing auxiliaries, *Vibrational Spectroscopy*, 2004, 35(12), 213–218. The 2nd International Conference on Advanced Vibrational Spectroscopy (ICAVS-2).

[10] T. Yotsuda, T. Yamamoto, H. Ishizawa, T. Nishimatsu, E. Toba, Near infrared spectral imaging for water absorbency of woven fabrics, *Instrumentation and Measurement Technology Conference*, 2004, IMTC 04, Vol. 3, 2258–2261.

[11] Wang Ying, Zhang Guo-Bao, Zhao Gen-Suo, Studies on the analysis methods of fibre by micro IR spectroscopy, *Henan Science*, 2000, 18(2), 170–172.

[12] L.L. Cho, J.A. Reffner, B.M. Gatewood, D.L. Wetzel, Single fibre analysis by internal reflection infrared microspectroscopy, *Journal of Forensic Sciences*, 2001, 46(6), 5.

[13] Xiaoyan Liu, Weidong Yu, Evaluating the thermal stability of high performance fibres by TGA, *Journal of Applied Polymer Science*, 2006, 99(3), 937–944.

[14] Hui Li, Yue-Fang Wen, Yong-Gang Yang, Lang Liu, Shou-Chun Zhang, Jun-Liang Fan, A study on the pyrolysis behavior of rayon fibers treated with $(NH_4)_2SO_4/HN_4Cl$, *Hi-Tech Fiber and Application*, 2006, 31(6), 22–26.

[15] Shangtiao Zhengyi, Wang Jianmin, Hong Anfan, Xu Huixiang, On an image processing method to identify cotton fibre and ramie fibre in blended yarn fabrics, *Journal of Suzhou University (Engineering Science Edition)*, 2000, 4.

[16] ASTM Active Standard: D2257-98 (2004), Standard test method for extractable matter in textiles, *Book of Standards*, Volume 07.01.

[17] US Customs Laboratory Methods, USCL Method 53-01 (date issued 08/98), NHM-008-1998, Textiles – Methods and procedures for the analysis of other vegetable textile fibres; paper yarn and woven fabrics of paper yarn (Chapter 53).

[18] Li Zhi-hong, Ren Yu, Properties of the six new textile fibres and their identification, *Shanghai Textile Science and Technology*, 2006, 34(4), 55–58.

[19] M. Müller, B. Murphy, M. Burghammer, C. Riekel, M. Roberts, M. Papiz, D. Clarke, J. Gunneweg, E. Pantos, Identification of single archaeological textile fibres from the Cave of Letters using synchrotron radiation microbeam diffraction and microfluorescence, *Journal of Applied Physics A: Materials Science and Processing*, 2006, 83(2), 183–188.

[20] D. Robson, Animal fiber analysis using imaging techniques, Part I: Scale Pattern Data, *Textile Res. J.*, 1997, 67, 747–752.

[21] Lijing Wang, Xungai Wang, Determination of cashmere and wool fibre scale frequency by analysing diameter profiles with fast Fourier transform, *Proceedings of the 3rd International Cashmere Determination Technique Seminar – Paper Collection*, Erdos Group, Inner Mongolia, 2005, 107–117.

[22] X. Liu, L. Wang, X. Wang, The resistance to compression behavior of alpaca and wool, *Text. Res. J.*, 2004, 74(3), 265–270.

[23] Huimin Wang, Xin Liu, Xungai Wang, FTIR and thermal analysis of cashmere and other animal fibres, *Proceedings of the 3rd International Cashmere Determination Technique Seminar – Paper Collection*, Erdos Group, Inner Mongolia, 2005, 217–228.

4

Physical and mechanical testing of textiles

X WANG, X LIU and C HURREN,
Deakin University, Australia

Abstract: This chapter describes the key physical and mechanical properties of fabrics and the associated test methods. It covers fabric weight and thickness, fabric strength, fabric stretch and abrasion resistance, as well as properties related to fabric aesthetics. A brief account of future trends in this area is also provided.

Key words: fabrics, physical properties, mechanical properties, abrasion resistance, aesthetic properties.

4.1   Introduction

Fabrics made from both natural and manufactured fibres have been extensively used for clothing, decoration and industrial applications. The physical and mechanical properties of these fabrics are affected by the fibre type, yarn construction and fabric structure, as well as any treatment that may have been applied to the materials. A range of fabric performance parameters are assessed for different end-use applications.

Unlike other homogeneous materials, fabrics are heterogeneous materials. The test results differ when a fabric specimen is tested in different directions (e.g. warp or weft for wovens, course or wale for knits). While different test standards are applied to different types of fabric tests, it is important to note that the three important factors for any test are the sampling protocol, the conditions of measurement, and the instrumentation and measurement procedure.

This chapter is focused on the physical and mechanical tests of fabrics. Specifically, it covers the following tests:

- Weight and thickness
- Tensile strength
- Tear strength
- Seam strength and seam slippage
- Burst strength
- Stretch properties
- Abrasion resistance
- Drape
- Bending

- Shearing
- Compression.

While the principles of these tests have not changed much over the past 70 years, there has been considerable advance in the instrumentation used to test properties such as strength, abrasion and fabric handle. For each test and where appropriate, the different test methods and standards are introduced and compared in this chapter. The applications and future trends of these tests are briefly discussed.

4.2   Fabric weight and thickness

Weight measurement of a fabric is often a prerequisite for subsequent tests of other fabric properties. If fabric weight or dimension is not kept constant or normalised then the test results will not be comparable.

The thickness of a fabric is one of its basic properties, giving information on its warmth, weight and stiffness. Thickness measurements are very sensitive to the pressure and sample size used in the measurement, which will be briefly discussed in the section on fabric handle. In practice, fabric mass per unit area is often used as an indicator of thickness.

4.2.1 Methods for testing fabric weight and thickness

Weight can be determined by a mass per unit area or a mass per unit length of fabric. Specimens of known dimensions are taken by a cutting device or a template, to obtain a consistent specimen size. The larger the specimen size, the more accurate the measurement, and most test standards require an area of $10\,000\,\text{mm}^2$ or more to be measured. The accuracy of cutting the specimen should be within 1% of the area.

Five specimens should be selected from each fabric sample. Specimen selection should avoid taking samples from the fabric selvedge or close to the ends of a fabric piece. Testing should be conducted in a conditioned atmosphere with preconditioned samples and care should be taken to avoid the loss of fibres/threads during weighing. Results are commonly reported in grams per square metre (g/m^2).

$$m_{ua} = \frac{m}{a} \hspace{4cm} 4.1$$

where m_{ua} = mass per unit area, in g/m^2; a = specimen area, in m^2; and m = mass of specimen, in g.

If mass per unit length is required then the following formula is used:

$$\bar{m}_{ul} = \bar{m}_{ua} \times \bar{w} \hspace{4cm} 4.2$$

where $\bar{m}_{ul}$ = the mean mass per unit length, in g/m, and $\bar{w}$ = the mean width, in m.

The standards used for the weight test include:

- ASTM D3776-96(2002) Standard test methods for mass per unit area (weight) of fabric
- ISO 3801-1977 Textiles – Woven fabrics – Determination of mass per unit length and mass per unit area
- AS 2001.2.8-2001 Determination of mass per unit area and mass per unit length of fabrics.

4.3   Fabric strength

The strength tests covered in this section include tensile, tear, seam and burst strength. These mechanical properties are important for all textile users including fabric processors, garment manufacturers, designers and customers.

4.3.1 Tensile strength

Measurement of tensile stress–strain properties is the most common mechanical measurement on fabrics. It is used to determine the behaviour of a sample while under an axial stretching load. From this, the breaking load and elongation can be obtained. The principle of the tensile strength test is simple: a test piece is held in two or more places and extended until it breaks. The tensile properties measured are generally considered arbitrary rather than absolute. Results depend on specimen geometry, the fibre type and arrangement, as well as the fabric structure.

Break modes

There are two common types of tensile breaks: sharp break (Fig. 4.1) and percentage break (Fig. 4.2). A sharp break is a sudden drop in load. This test is normally called pull to break. A percentage break is generally shown as a gradual reduction in the load from its maximum as further extension is applied. A percentage drop from maximum load is often used to define an end point or break point. This test is normally called pull to yield and can have all of the same setup parameters as a pull to break. Modern tensile test instruments can be set up in both of the break modes. Most test methods report both maximum load and load at break, as the breaking strength is not always the maximum strength for the material, especially for soft and elastic fabrics.

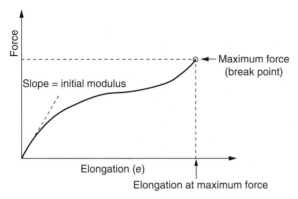

4.1 Tensile strength test curve (sharp break).

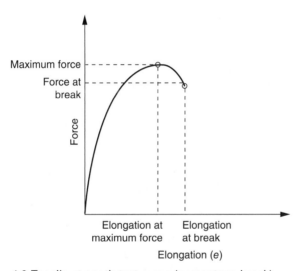

4.2 Tensile strength test curve (percentage break).

Extension

Extension is defined as the change in length of a material due to stretching. When a fabric of original length l_0 is stressed along its axis, it extends an amount dl. The strain in the sample is dl/l_0 (viz. the ratio of the extension of a material to the length of the material prior to stretching). The symbol e is normally used to represent strain, and can be referred to as elongation. Strain is a dimensionless quantity, often reported as a percentage.

Initial modulus

Young's modulus or the initial modulus (IM) is a measure of the amount of deformation that is caused by a small stress. Materials with a high modulus, often called stiff or hard materials, deform or deflect very little in the presence of a stress. Materials with a low modulus, often called soft materials, deflect significantly. In the case of fabric, initial modulus is related to the fabric handle. A higher IM means a stiffer or harsher fabric handle whereas a lower IM provides a softer fabric handle.

Tensile testing machine

Most tensile testing machines can operate in three modes:

- Constant-rate-of extension (CRE)
- Constant-rate-of traverse (CRT)
- Constant-rate-of-load (CRL).

The most commonly used mode is the CRE mode and is often required by the test standards. The main factors that need to be considered are the size and accuracy of the load cell (0.5–25 kN), the distance of cross-head travel (0.1–2 m) and the rate of cross-head travel (0.1–500 mm/min). Common tensile results include maximum load, deflection at maximum load, load at break, and deflection at break. Other data can be calculated from these results, such as work at maximum load, stiffness, work at break, stress, strain and Young's modulus. Most modern machines utilise a computer program to capture the data and calculate any additional results.

Tensile strength at break is not necessarily the best indicator of fitness for purpose. In some cases (i.e. web, linoleum and rope) the work to rupture (or break) is more important. The work to rupture is the energy absorbed by the material up to the point of rupture and is measured in joules. Work to rupture may be used to indicate fabric toughness.

Methods for testing tensile strength

Three methods (see Fig. 4.3) have been commonly used to measure tensile strength:

1. *Grab test.* In the grab test, the width of the jaws is less than the width of the specimen. An example would be for a 100 mm wide specimen where the centrally mounted jaws are only 25 mm wide. This method is used for woven high-density fabrics and those fabrics with threads not easy to remove from the edges. The grab method is used whenever it is desired to determine the 'effective strength' of the fabric in use.

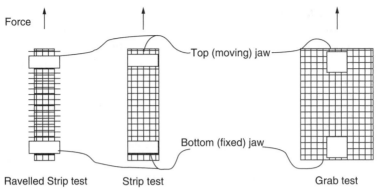

Force

Top (moving) jaw

Bottom (fixed) jaw

Ravelled Strip test   Strip test   Grab test

4.3 Sample gripping methods.

2. *Modified grab test.* The mounting geometry is the same as for the grab test; however, lateral slits are made in the specimen to sever all yarns bordering the portion to be strength tested, reducing to a minimum the 'fabric resistance' inherent in the grab method. This method is desirable for high-strength fabrics.
3. *Strip test.* There are two types of strip test: the ravelled strip test and the cut strip test. In both tests the entire width of the specimen is gripped in both the upper and lower jaws. The ravelled strip test is only used for woven fabric and specimens are prepared by removing threads from either side of the test piece until it is the correct width. The cut strip test is used for fabrics that cannot have threads removed from their sides such as knits, non-wovens, felts and coated fabrics. The test specimens are prepared by accurately cutting to size.

There is no simple relationship between grab tests and strip tests since the amount of fabric resistance depends on the fabric structure, fabric count, mobility of yarns and many other factors. The strip tests can provide information on tensile strength and elongation of fabric; however, the grab test can only give the breaking strength.

Factors affecting the tensile strength

It should be noted that many factors can affect the tensile test results. These include the number of test specimens, the gauge length used, the extension rate for the test, jaw slippage and damage to the specimen by the jaws that may cause 'jaw break'. These factors should be carefully considered when undertaking the tensile tests of fabrics.

1. *Number of test specimens.* With any test method the number of specimens tested will dictate the precision of the results. The higher the number of tests, the more precise the results.

2. *Gauge length.* A change in gauge length of a fabric will result in a change in the values obtained for maximum load, breaking load and initial modulus. The longer the gauge length, the lower the initial modulus result. The gauge length should be consistent for all tests if comparisons are to be made from test to test.

3. *Extension rate.* The extension rate (or the cross-head traverse speed) influences the elongation and break force of the fabric. Results of tests conducted at different rates of extension will not be directly comparable; however, fabrics of different elastic moduli require different test speeds.

4. *Jaws or grips.* Jaws are the part of the clamping device that grip the fabric during a test. They should be capable of holding the test piece without allowing it to slip; however, they should not over-grip, causing damage. Smooth, flat or engraved corrugated jaws can be used for clamping. Suitable packing materials can be used in the jaws (i.e. paper, leather, plastics or rubber) to avoid slip or damage during clamping. Where the test piece slips asymmetrically or slips by more than 2 mm, the results need to be discarded. To avoid slippage of smooth fabrics, capstan or self-locking jaws with an appropriate clamping face may be used.

5. *Jaw break.* Jaw break often happens before the fabric is stretched to its full potential. The test result should be discarded if the test piece breaks within 5 mm of the jaw face. In the case of a repeated jaw break, modification of the jaw material or clamping force should be considered.

Standards commonly used for tensile strength tests are as follows:

- ISO 13934-1:1999 Textiles – Tensile properties of fabrics – Part 1: Determination of maximum force and elongation at maximum force using the strip method
- ISO 13934-2:1999 Textiles – Tensile properties of fabrics – Part 2: Determination of maximum force using the grab method
- ASTM D5034-95 Standard test method for breaking strength and elongation of textile fabrics (grab test)
- ASTM D5035-95 Standard test method for breaking strength and elongation of textile fabrics (strip test)
- AS 2001.2.3.1-2001 Physical tests – Determination of maximum force and elongation at maximum force using the strip method
- AS 2001.2.3.2-2001 Physical tests – Determination of maximum force using the grab method
- AS 4878.6-2001 Determination of tensile strength and elongation at break for coated fabrics.

4.3.2 Tear strength

Tearing of a fabric can occur in a wide range of products and is involved in fatigue and abrasion processes as well as the catastrophic growth of a cut on application of a force. Tear strength is the tensile force required to start, continue or propagate a tear in a fabric under specified conditions. A tear strength test is often required for woven fabrics used for applications including army clothing, tenting, sails, umbrellas and hammocks. It may also be used for coated fabrics to evaluate brittleness and serviceability.

Methods for testing tear strength

The following methods are in use or being developed: trouser or single tear, double or tongue tear, wing tear, trapezoidal tear, ballistic pendulum (Elmendorf), puncture or snag tear, tack tear, and wounded burst tear. The test specimen shall be cut according to the design shown in Fig. 4.4, and the required dimensions are specified in relevant test standards.

The standards used worldwide for tear tests are:

- ISO 4674-1998, part 1: Determination of tear resistance
- ISO 13937-3-2000 Textiles – Tear properties of fabrics – Part 3: Determination of tear force of wing-shaped test specimens

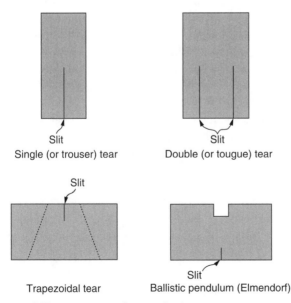

4.4 Different tear testing methods.

- ISO 13937-1-2000 Textiles – Tear properties of fabrics – Part 1: Determination of tear force using the ballistic pendulum method (Elmendorf)
- BS 3424 Method 7C, Single tear, 1973
- EN 1875-3 Determination of tear resistance – Part 3: Trapezoid tear, 1997
- ASTM D1423-83 Tear resistance of woven fabrics by falling pendulum (Elmendorf)
- ASTM D751 Tack tear, 1995
- ASTM D751 Puncture resistance, 1995
- ISO 5473 Determination of crush resistance, 1997
- AS 2001.2.10-1986 Determination of the tear resistance of woven textile fabrics by the wing-rip method
- AS 2001.2.8-2001 Determination of tear force of fabrics using the ballistic pendulum method (Elmendorf).

Two devices have been commonly used for tearing tests: the Elmendorf tearing tester and the CRE tester.

The Elmendorf tearing tester
The falling (ballistic) pendulum (Elmendorf) method is used for the determination of the average force required to continue or propagate a single-rip-type tear starting from a cut in a woven fabric by means of a falling pendulum (Elmendorf) apparatus. Part of the energy stored in the pendulum is used to produce the tearing (and any deformation of the test piece). The magnitude of this is indicated by the energy lost compared to the energy of the falling pendulum without a test piece in place. The weight attached to the pendulum can be selected based on the fabric tested and the standard used.

The basic characteristics of this test are that stresses are applied by subjecting the test piece to a sudden blow; hence the test speed (strain rate) is relatively high compared to that of a CRE machine (see below). This method is not suitable for knitted fabrics, felts or non-woven fabrics. It is applicable to treated and untreated woven fabrics, including those heavily sized, coated or resin treated.

An initial slit is made in the centre of the specimen. The principal reason for this slit is to eliminate edge tear forces and to restrict the measurement to the internal tearing force only. Cutting can be considered as the precursor to tearing.

The constant-rate-of-extension tester
The tear test can be performed on a normal tensile instrument. For the tongue method a rectangular specimen is cut in the centre of the shorter

edge to form two 'tongues' (or 'tails'). Each tongue is gripped in the clamps of a constant-rate-of-extension (CRE) machine and pulled to simulate a rip. The force to continue the tear is calculated from readings as the average force to tear.

The force registered in a tear test is irregular. The reading represents the force required for tear initiation, the subsequent reading being the force to propagate the tear. For a woven fabric, the average of the warp and weft direction tests is given as the result. The tearing force can rise rapidly; therefore the response characteristics of the apparatus are particularly important. The rate of tear is normally 100 mm/min.

The different tests in part reflect the different stress concentrations found in different products, but in many cases they are somewhat arbitrary. Consequently, the measured tear strength is not an intrinsic property of the material, and it can be difficult to correlate directly the results of laboratory tests with service performance.

The main problems encountered in carrying out tear testing are that sometimes the tear does not propagate in the direction of the jaw traverse. Tearing can occur towards the sample edge. In the tongue or double slit test, the tongue may be stretched and a tensile effect occurs, or threads may get pulled out rather than break. Under these conditions an alternative specimen shape may be chosen or a larger test piece taken and the procedure repeated. Non-woven and knitted substrates are often tested using larger samples than those initially specified in the method.

Factors affecting the tear strength

In a normal pull to break tensile test the force measured is the force to produce failure in a nominally flawless test piece. In a tear test, the force is not applied evenly but concentrated on a deliberate flaw or sharp discontinuity. In this case the force to produce a continuously new surface is measured. The force to start or maintain tearing will depend on the geometry of the test piece and the nature of the discontinuity.

The main factors that affect tear strength are yarn properties and fabric structure. The mechanism of fabric tearing is different from linear tensile failure and relates to the ability of individual yarns to slide, pack together or 'jam' into a bundle, increasing the tearing force. Thus an open fabric structure contributes to more yarn sliding and jamming, and higher tear strength. An increase in yarn density in a woven fabric will decrease the tear strength of a fabric as yarns are broken individually as they have more restriction, preventing yarn slide.

A tightly mounted fabric is easier to tear than a slackly mounted fabric because the tear force propagates from yarn to yarn as the linear force in the yarn restricts yarn slide. Staple yarn has a lower tear strength compared to filament yarn. In a trapezoid tear test, an increase in ends and picks increases tear strength. Tear resistance can also be affected considerably by the speed of the test.

4.3.3 Seam strength

The quality and performance of a sewn garment depend on seam strength and seam slippage along with appearance and other mechanical properties. Failure of the seams of the garment by breaking of the sewing thread or by seam slippage affects serviceability. The strength of the seam or its ability to resist seam opening is an important fabric property and is needed to determine seam efficiency and the optimum sewing conditions. These can include seam type, stitch type, number of stitches per unit length of seam, sewing thread size and needle size.

Seam strength relates to the force required to break the stitching thread at the line of stitching. It is often used to test the strength of a sewing thread or test joins in strong industrial fabrics.

Seam slippage is defined as the tendency for a seam to open due to the application of a force perpendicular to the seam direction. It is a measure of the yarn slippage in a fabric at the seam. Sometimes it refers to breakage of the thread used to stitch the seam. The seam slippage test is also referred to as the seam opening test. Seam slippage may occur in a garment or household item for different reasons, including:

- a low number of warp or weft threads in relation to particular yarn and fabric construction characteristics
- seam allowance too small
- high force requirements placed on the seam due to use
- improper seam selection or construction
- insufficient elasticity of the seam.

Methods for testing seam strength and seam slippage

The CRE machine is normally used and the test specimen is held the same way as in a conventional grab test. The sewn seams may be taken from sewn articles such as garments or may be prepared from fabric samples.

Seam strength
There are two geometries used for the seam strength test, transverse and longitudinal, and these are shown in Fig. 4.5. The transverse direction (Method A) is applicable to relatively inextensible fabrics, such as woven

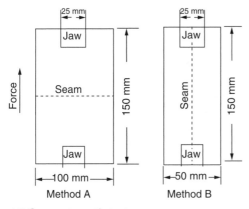

4.5 Seam strength tests.

and stable warp knit structures. The longitudinal direction (Method B) is applicable to extensible fabrics, such as knitted, elastic and highly resilient fabrics. Sample preparation is different for tests in the transverse direction compared to the longitudinal direction as shown in Fig. 4.5. A straight rather than curved seam line is required for a test in the longitudinal direction. The seam line of the seamed samples must be parallel to either the warp or weft yarns.

The test specimen is mounted centrally between the upper and lower jaws with the seam perpendicular or parallel to the jaws depending on the test method. The sample is then stretched at a constant rate until rupture occurs. In a traverse test this is when the seam ruptures, and in the longitudinal test this is when the first sign of seam rupture occurs. In the case of stitched seams, this implies the first stitch breakage. The maximum force applied to the specimen is recorded for both methods as the seam breaking force. If the fabric ruptures prior to the seam rupturing, then a statement to this effect should be made in the test report. If the specimen slips in the jaws or breaks in or at the jaws, the test result for that specimen must be discarded.

Seam slippage
Specimens for seam slippage tests are prepared according to the following steps:

1. Cut the fabric sample to rectangular specimens 175 ± 100 mm for both warp and weft directions.
2. Fold the specimens in half by placing the two shorter edges together and sew a lockstitch seam parallel to and at a distance of 12 ± 1 mm from the fold.
3. Cut the specimen along the fold after sewing.

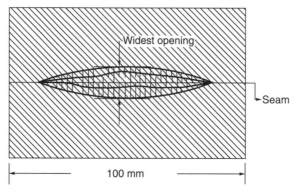

4.6 Measurement of seam opening.

Preparation of sewn seam test specimens from fabric samples requires prior specification of sewing details. These can be taken directly from the standard or can be set by the parties interested in the test results. These details often vary with the fabric end-use and include seam allowance (seam width), stitch type, stitch frequency, needle size and tread parameters.

The machine setup for the seam slippage test is similar to that for the seam strength test, except that a cross-head speed of 50 mm/min is usually used. The specimen is mounted centrally in the width of each set of jaws with the seam midway between and parallel to the horizontal edges of the jaws. The load is then increased until the selected load is reached. The jaw movement is stopped at that point, and the width of the seam opening at its widest place is measured to the nearest 0.5 mm within 10 seconds, in the direction of the applied force (Fig. 4.6). Then the force on the specimen is reduced to 2.5 N and after an interval of 2 minutes the seam opening at its widest place is re-measured. The measuring device can be a small transparent rule or a divider.

An alternative method is to increase the load until a seam opening of 6 mm is reached, at which point the load is recorded for each specimen. This method is applied to a single seam on woven fabrics. If a sample from a commercial garment has multiple seams, then an opening of 3 mm is used.

The standards commonly used for seam tests are as follows:

- ASTM D1683 Standard test method for failure in sewn seams of woven fabrics, 1990
- ASTM D751 Seam strength, 1995
- BS 3320:1988 Method for determination of slippage resistance of yarns in woven fabrics: Seam method

- AS 2001.2.20-2004 Determination of seam breaking force
- AS 2001.2.22-2006 Physical tests – Determination of yarn slippage in woven fabric at a standard stitched seam.

4.3.4 Burst strength

Burst strength testing is the application of a perpendicular force to a fabric until it ruptures. The force is normally applied using either a ball or a hydraulically expanded diaphragm. The fabric is clamped in place around the device that applies the force by a circular ring. The material is stressed in all directions at the same time regardless of the fabric construction. Ball burst testing is used as an alternative to tensile testing for materials that are not easily prepared for tensile testing or have poor reproducibility when tensile tested. These fabrics include knits, lace, non-wovens and felts.

There are fabrics which are simultaneously stressed in all directions during service, such as parachute fabrics, filters, sacks and net. A fabric is more likely to fail by bursting in service than it is to break by a straight tensile fracture, as this is the type of stress that is present at the elbows and knees of clothing. Results obtained from tensile and burst testing are not directly comparable.

When a fabric fails during a bursting strength test, it does so across the direction which has the lowest breaking extension. When a burst test is undertaken, all directions in the fabric undergo the same extension, so the fabric direction with the lowest extension at break is the one that will fail first. This is not necessarily the direction with the lowest strength. Elongation cannot be determined from a burst strength test.

Methods for testing burst strength

Ball burst method

The ball burst method uses a CRE machine to apply the perpendicular force. The attachment for the CRE machine comprises two parts: a lower fixed clamping device of fixed aperture diameter and an upper moving ball that impacts on the fabric surface. The clamping device has an upper and a lower clamp with concentric grooves and crowns that intermesh with the test piece to provide grip. Test specimens can be cut into square or circular pieces, but must be of sufficient size to protrude outside the annular rings around the complete circumference of the lower clamp. The face of the rings should be perpendicular to the direction of the application of the force. The centre portion pushes against a polished steel ball at a constant rate until it ruptures. The burst strength is then calculated from the force of rupture F and the internal cross-sectional area A of the test piece, F/A. Current

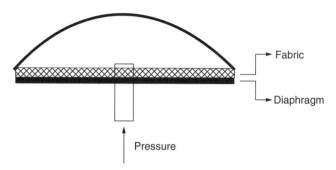

4.7 Simple principle of hydraulic diaphragm method.

research shows that a larger ball diameter of 38 mm would improve repro-ducibility; however, most standards still use a 25 mm diameter ball.

Hydraulic diaphragm method
The hydraulic diaphragm test method uses a diaphragm inflated by hydrau-lic pressure to apply the perpendicular force to the fabric. The aperture size is normally different from that used for ball burst tests. The diaphragm, normally made of rubber, is mounted below the clamped test piece (Fig. 4.7). The clamping device should provide distributed pressure suffi-cient to prevent specimen slippage during a test. During a test hydraulic fluid is introduced behind the rubber diaphragm at a known rate and the burst pressure (M) at rupture is measured using a pressure gauge. The upper clamp and sample is then removed and the tare pressure (T) to distend the diaphragm is recorded. The tare pressure (T) is subtracted from the burst pressure at rupture (M) to give the actual burst pressure (B) of the test piece, viz. $B = M - T$. The burst pressure is expressed in kilopascals. From this method, bursting distension can be measured in millimetres, immedi-ately prior to rupture, from the height change of the centre of the upper surface above the starting plane.

The above two methods are set out for determining the bursting pressure of both wet and dry fabrics. The two methods may not give the same results, as the mechanism of force application is slightly different between the two test apparatus.

The relevant standards commonly used are listed below:

- ISO 3303-1995 Determination of bursting strength
- ISO 2960 Textiles – Determination of bursting strength and bursting distension – Diaphragm method
- BS 4768 Method for determination of the bursting strength and bursting distension of fabrics

- BS 3424 Methods of test for coated fabrics – Wounded burst test
- ASTM D3787 Standard test method for bursting strength of knitted goods – constant-rate-of-traverse (CRT) ball burst test
- AS 2001.2.19-1998 Determination of bursting force of textile fabrics – Ball burst method
- AS 2001.2.4-1990 Physical tests – Determination of bursting pressure of textile fabrics – Hydraulic diaphragm method.

4.4   Fabric stretch properties

These properties are particularly important for elastic fabric and stretch fabric. Elastic or elastomeric fabric is made from an elastomer either alone or in combination with other textile material. Elastomers include polymers such as rubber, polybutadiene, polyisobutylene and polyure-thanes. Because the glass transition temperature of these polymers is below room temperature, these materials are soft or rubbery at room temperature and can easily return to their original shape after stretching. Due to the nature of these materials they do not always return to their original shape after prolonged deformation. Tests should measure size change (kickback) after long periods of extension. The tension to stretch an elastic material and the percentage stretch achievable are also important variables to be measured.

Stretch fabric is usually accomplished by incorporating a small percent-age of elastomeric fibres or filaments into a conventional woven or knitted textile fabric. Stretch fabric can also be achieved without elastromeric fibres by fabric construction or yarn selection. There are two types of stretch fabrics: comfort stretch (5–30%) and power stretch (30–50%) (Lyle, 1977). Comfort stretch fabrics are designed for low loads, and power stretch for considerably higher loads. Stretch is important in sportswear such as swim-wear or other active sports clothing, which is required to be a close fit to the body. The stretch requirements of a fabric can be gauged from the typical values of stretch that are encountered during the actions of sitting, bending or flexing of knees and elbows.

Both elastic fabric and stretch fabric require good elasticity; consequently fabric tends to recover its original size and shape immediately after removal of the force causing deformation. The three main factors of interest when testing a fabric with recoverable elongation are elongation at load, force for elongation, and recovery after load.

- Elongation at load is the amount that a fabric stretches in length from its original length after a fixed load is applied. This is commonly used to define the level of stretch within the fabric. Woven fabrics have much less stretch than knitted fabrics.

- Force for elongation defines the amount of force required to extend a fabric a certain distance in elongation. It can be called power or tension of the fabric at elongation and is important for comfort factors in garment design.
- Recovery after load is the amount a fabric returns to its original dimensions after the elongation load is released.

Recovery is possibly the most important factor as it defines whether a fabric is stretch or not. Fabrics without elastic properties are often tested for stretch and recovery to quantify the effect of stretching the fabric in use. A 100% cotton single jersey fabric will generally stretch significantly when a load is applied; however, its recovery after stretch is poor. The addition of an elastomeric fibre will increase the level of recovery, which can then define this fabric as a stretch fabric.

Recovery is often measured after a long period of load. Elastomers can break down when loaded for a long time. This is observed as a loss in fabric recovery or tension at load. This type of test is often used for elastic tapes or fabrics where the tension is an integral part of the garment design. An example is underwear where the product is rendered useless if the elastic waistband no longer holds the garment in place.

There are two main ways by which fabrics are measured for stretch and recovery. These are dynamic and static measurements. In dynamic measurement the fabric is applied with a fixed load or a fixed extension at a controlled rate of extension. Dynamic measurements can be cycled through a series of extensions before the results are taken. The CRE machine is an example of a machine used for dynamic testing. Dynamic tests generally measure tension at elongation as well as elongation and relaxation.

A static test is conducted by clamping one end of a fabric on a flat plane. The other end is then displaced by applying a fixed load or by stretching to a set elongation. Static tests generally only provide elongation and load information. However, they are commonly used to measure recovery after a long period of loading.

4.4.1 Methods for testing fabric stretch

Test for elongation of elastic fabric

When a CRE machine is used for testing tension and elongation of an elastic fabric, a straight wide or narrow elastic fabric, or a loop specimen, is prepared. The specified loads and cross-head speeds are applied to cycle (loading and unloading) the fabric for a required number. For low elongation fabrics (below 100%), use of a slower cross-head speed should be agreed. Three properties should be examined: elongation (percentage stretch), tension (power) and recovery.

Test for fabric stretch

There are a number of tests devised for stretch fabrics by various organisations, all following similar procedures but differing widely in many of the important details, such as load applied, number of stretch cycles before the actual measurements, time held at the fixed load, and time allowed for recovery. Here are some comparisons between the BS 4952:1992 and ASTM D3107-1980 standards. Two quantities are generally measured:

• The extension at a given load, which is a measure of how easily the fabric stretches
• Growth or residual extension, which measures how well the fabric recovers from stretching to this load.

In the British standard (BS4952:1992), five specimens from warp and weft directions each are tested. Two different dimensions of specimen in clamps are required for woven and knitted fabrics respectively: a width of 75 mm for both woven and knit, and a gauge length of 75 mm for knit and 200 mm for woven (as L_1). The fabric is to be stretched at a specified force (30 N for knit and 60 N for woven) at a rate of 100 mm/min, and the load is maintained for 10 s; the extension (cross-head movement) is then recorded as L_2. The sample is removed from the clamps and allowed to relax on a flat, smooth surface and its length is re-measured after 1 min as L_3. If a longer period of relaxation is required, the length is re-measured as L_4 after 30 min. The stretch and recovery results can be calculated as follows:

Stretch:

$$\text{Mean extension percent, } E = 100L_2/L_1 \qquad\qquad 4.3$$

Recovery:

$$\text{Mean residual extension after 1 min, } R_1 = 100(L_3 - L_1)/L_1 \qquad 4.4$$

$$\text{Mean residual extension after 30 min, } R_{30} = 100(L_4 - L_1)/L_1 \qquad 4.5$$

In the American standard (ASTM D3107-1980), apart from the different sample size for woven (width of 51 mm and gauge length of 500 mm), a weight of 1.8 kg also hangs at the bottom clamp. Both edges of the clamps are marked as the lower and top bench marks, and the original distance between the two marks is recorded as length A. The fabric specimen is stretched by cycling three times from 0 to 1.8 kg load with 5 s interval, then the full load is applied at the fourth time and the extension (B) is then measured. Afterwards, the weight and bottom clamp are removed and the distance between the two marks is measured after 30 s as C. The percentage fabric stretch and immediate fabric growth are calculated as follows:

$$\text{Fabric stretch percent} = \frac{B-A}{A} \times 100 \qquad\qquad 4.6$$

$$\text{Fabric growth percent} = \frac{C-A}{A} \times 100 \qquad\qquad 4.7$$

Standards commonly used for the stretch test are as follows:

- ASTM D2594-2004 Standard test method for stretch properties of knitted fabrics having low power
- ASTM D3107-1980 Standard test method for stretch properties of fabrics woven from stretch yarns
- ASTM D6614-2007 Standard test method for stretch properties of textile fabrics – CRE method
- BS 4952-1992 Methods of test for elastic fabrics.

4.5   Fabric abrasion resistance

Abrasion is defined as the wearing away of any part of the fabric by rubbing against another surface. Fabrics are subjected to abrasion during their life-times and this may result in wear, deterioration, damage and a loss of performance. However, the abrasion resistance is only one of several factors contributing to wear performance or durability. Abrasion can occur in many ways and can include fabric to fabric rubbing when sitting, fabric to ground abrasion during crawling, and sand being rubbed into upholstery fabric, and it is difficult to correlate conditions of abrasion of a textile in wear or use with laboratory tests. This may explain the reason why there are many different types of abrasion testing machines, abradants, testing conditions, testing procedures, methods of evaluation of abrasion resistance and interpretation of results.

The methods used may be described by the equipment, the test head movement or testing device setup. These include (a) inflated diaphragm; (b) flexing and abrasion (i.e. the Stoll Flex Tester); (c) oscillatory cylinder; (d) rotary platform; (e) uniform abrasion; and (f) impeller tumble. Presentations of the fabric to the abradant include in plane (or flat), flex, tumble or edge abrasion or a combination of more than one of these factors.

There are two general approaches for assessment of abrasion resistance: (1) to abrade the sample until a predetermined end-point is reached, such as the breaking of two threads or the generation of a hole, while recording the time or number of cycles to achieve this; and (2) to abrade for a set time or number of cycles and assess the fabric for change in appearance, loss of mass, loss of strength, change in thickness or other relevant property. The length of the test for the first approach is indeterminate and requires the sample to be regularly examined for failure. This need for examination is time consuming as the test may last for a long time. The second approach provides for simpler measurements; however, the change in properties such as mass loss can be slight.

4.5.1 Factors affecting abrasion resistance

A fabric's resistance to abrasion is affected by many factors, such as fibre type, the inherent mechanical properties of the fibres, the dimensions of the fibres, the structure of the yarns, the construction and thickness of the fabrics, and the type and amount of finishing material added to the fibres, yarns or fabrics.

For example, fibres with high elongation, elastic recovery and work of rupture have a good ability to withstand repeated distortion, hence a good degree of abrasion resistance. Nylon is generally considered to have the best abrasion resistance, followed by polyester, polypropylene, wool, cotton and acrylic. Longer fibres incorporated into a fabric confer better abrasion resistance than short fibres because it is harder to liberate them from the fabric structure. Flat plain weave fabrics have better abrasion resistance than other weaves because the yarns are more tightly locked in a plain weave structure and the wear is spread more evenly over all of the yarns in the fabric. Fabrics with a loose structure have a lower abrasion resistance than those with a tight structure.

The resistance to abrasion is also greatly affected by the conditions of the tests, such as the nature of the abradant, variable action of the abradant over the area of specimen abraded, the tension of the specimen, the pressure between the specimen and the abradant, and the condition of the specimen (wet or dry).

Abradants can consist of anything that will cause wear. The most common solid abradants are abrasive wheels (vitreous and resilient), abrasive papers or other fabrics, stones (aluminium oxide or silicon carbide) and metal 'knives'. The nature of abradants and the type of action will control the severity of the test. It is important that the action of the abradant should be constant throughout the test and the tension of the mounted specimen should be reproducible, as this determines the degree of mobility of the sample during abrasion. The pressure between the abradant and the sample affects the severity and rate at which abrasion occurs. Accelerated destruction of test samples through increased pressure or other factors such as heat generation may lead to false conclusions on fabric behaviour.

4.5.2 Methods for testing abrasion resistance

Three methods have been widely used over the years: the Martindale tester, the Taber abrader (rotary platform double-head abrader) and the accelerator.

The Martindale tester

The Martindale tester is designed to give a controlled amount of abrasion between fabric surfaces at comparatively low pressures in continuously changing directions. The results required determine the test and assessment method used. Assessments can include determination of specimen breakdown, mass loss or appearance change.

For the methods applying assessment of specimen breakdown or mass loss, specimens are circular of either 38 mm or 140 mm in diameter. Normally the abradant is silicon carbide paper or woven worsted wool mounted over felt. The small test specimen is sitting on the large abradant and then cycled backwards and forwards in a Lissajous motion producing even wear. A force of either 9 or 12 kPa is applied to the top of the specimen to hold it against the abradant. If assessment of appearance change needs to be carried out, then larger test pieces (140 mm in diameter) are required. The roles are reversed and the abradant is placed in the holder with the specimen as the base platform. The standard abradant should be replaced at the start of each test and after 50 000 cycles if the test is to be continued beyond this number. Behind the abradant is a standard backing felt which is replaced at longer intervals.

For assessment, the specimen is examined at suitable intervals to see whether two threads have broken, mass has changed or appearance has changed. Different fabric structures or components will require different inspection intervals. Some bias may occur if a fabric has a low abrasion resistance. Hosiery may be tested using a modified specimen holder, which stretches the knitted material, thus effectively accelerating the test. A flattened rubber ball is pushed through the sample as the holder is tightened, thus stretching it.

The Taber abrader

The rotary platform abrader (Taber abrader) applies two abrasive wheels (13 mm thick and 51 mm in diameter) under controlled pressure to a circular sample (110 mm in diameter) mounted on a rotating table or platform. The fabric is subjected to the wear action by two abrasive wheels pressing onto a rotating sample. The wheels are arranged at diametrically opposite sides of the sample so that they are rotated in the opposite direction by the rotation of the sample. These are available in different abrasive grain sizes. The load used can be 125, 250, 500 or 1000 g (or 1.23, 2.45, 4.9 or 9.81 N). The test specimen is abraded until damage (broken threads or hole) occurs or there is a visual change in the surface appearance (loss of texture, pile or surface coating). The number of cycles is recorded when the end point is reached.

The accelerator abrasion tester

The accelerator abrasion tester has an action that is quite different from most other abrasion testers. In the test a free fabric specimen is driven by a rotor inside a circular chamber lined with an abrasive cloth. The specimen suffers abrasion by rubbing against itself as well as the liner. Evaluation is made on the basis of either weight loss of the specimen or the loss in grab strength of the specimen broken at an abraded edge. For evaluation by loss in strength, two specimens measuring 100 mm × 300 mm are used for grab tests. Each specimen is numbered at both ends and then cut in half. One half is used for determining the original grab strength and the other half for determining the grab strength after abrading.

Many different standards are used worldwide for abrasion resistance tests, including:

- ASTM D3884 Standard guide for abrasion resistance of textile fabrics (rotary platform, double-head method)
- ASTM D4966-1998(R04) Standard test method for abrasion resistance of textile fabrics (Martindale abrasion tester method)
- ISO 12947-1-1998 Textiles – Determination of the abrasion resistance of fabrics by the Martindale method – Part 1: Martindale abrasion testing apparatus
- ISO 12947-2-1998 Textiles – Determination of the abrasion resistance of fabrics by the Martindale method – Part 2: Determination of specimen breakdown
- AS 2001.2.25.1-2006 Physical tests – Determination of the abrasion resistance of fabrics by the Martindale method – Martindale abrasion testing apparatus
- AS 2001.2.25.2-2006 Physical tests – Determination of the abrasion resistance of fabrics by the Martindale method – Determination of specimen breakdown
- AS 2001.2.25.3-2006 Physical tests – Determination of the abrasion resistance of fabrics by the Martindale method – Determination of mass loss
- AS 2001.2.25.4-2006 Physical tests – Determination of the abrasion resistance of fabrics by the Martindale method – Assessment of appearance change
- AS 2001.2.26-1990 Physical tests – Determination of flat abrasion resistance of textile fabrics (flexing and abrasion method)

- AS 2001.2.27-1990 Physical tests – Determination of abrasion resistance of textile fabrics (inflated diaphragm method)
- AS 2001.2.28-1992 Physical tests – Determination of abrasion resistance of textile fabrics (rotary platform, double-head method)
- AS 2001.2.30-1994 Physical tests – Determination of abrasion resistance of coated textile fabrics (oscillatory cylinder method).

4.6   Testing the aesthetic properties of fabrics

Fabric aesthetic properties include the optimised handle of fabric, good appearance in the garment and good appearance in wear. Fabric properties like thickness, compressibility, bending properties, extensibility, dimensional stability and surface properties are associated with fabric aesthetics. Generally, the aesthetic characteristics of fabrics can be measured by a mixture of subjective evaluation and objective tests.

When assessing fabric handle subjectively, the assessor usually strokes the fabric surface with one or several fingers and then squashes the fabric gently in the hand. Subjective characteristics are assessed by the sensations of smoothness or roughness, hardness or softness, stiffness or limpness. These feelings may determine whether a fabric is comfortable or uncomfortable to a wearer. However, there are many factors that influence the characters of a fabric observed through handling, for instance the type of fabric being assessed, which may be different in the material used, and differences in fabric structure made specially for apparel, upholstery or industrial uses. This subjective hand evaluation system requires years of experience and can obviously be influenced by the personal preferences of the assessor. A fabric may feel light, soft, mellow, smooth, crisp, heavy, harsh, rough, furry, fuzzy or downy soft. So there is a need to replace the subjective assessment of fabrics by experts with an objective machine-based system which will give consistent and reproducible results.

The theoretical primary hand values (PHV) of a fabric can be calculated from its mechanical properties according to the Kawabata method (Kawabata, 1980). The PHV values include *koshi* (stiffness), *shari* (crispness) and *fukurami* (fullness and softness). These hand values relate to the shear and bending properties, and consequently to the inherent fibre properties and fabric geometry. The Kawabata Evaluation System for Fabric (KES-F) consists of four specialised instruments: FB1 for tensile and shearing, FB2 for bending, FB3 for compression and FB4 for surface friction and variation. A total of 16 parameters are measured at low levels of force. The measurements are intended to mimic the fabric deformations found in use.

A set of the Fabric Assurance by Simple Testing (SiroFAST) instruments developed in Australia is used to measure the mechanical properties of wool fabrics and to predict their tailoring performance. SiroFAST gives similar information on the aesthetic characteristics of fabric as KES-F does, but in a simple manner, and is more suited to a mill environment. The SiroFAST system includes SiroFAST-1 for thickness, SiroFAST-2 for bending, SiroFAST-3 for extensibility and SiroFAST-4 for dimensional stability. The SiroFAST PressTest has also been added to complement these tests. Through the objective measurements of fabric and a data set on a chart or 'fingerprint', manufacturers can identify fabric faults, predict the consequences of those faults and identify re-finishing routes or changes in production.

The tests considered relevant to fabric hand in this chapter include drape, bending, shearing and compressibility. Different testing methods applied over many years are compared in the following sections.

4.6.1 Fabric drape

Drape is the term used to describe the way a fabric hangs under its own weight. Fabric drapability is an important factor from an aesthetic point of view. The quality of 'drape' is important to a designer as it influences a garment's appearance. The draping qualities required from a fabric will differ depending on its end use, e.g. knitted fabrics are relatively floppy and garments made from them will tend to follow the body contours. Woven fabrics are stiffer than knitted fabrics, so they are used in tailored clothing where the fabric hangs away from the body and disguises its contours. Uses such as curtains, tablecloths or women's clothing need to exhibit good drape shape and appearance. Good draping leads to the fitting of a fabric over a surface without undesirable wrinkling or tearing. Measurement of a fabric's drape assesses its ability to hang in graceful curves.

The drape coefficient (F) has been developed to describe the degree of drape and drape shape (configuration, modality). A lower F value means the fabric is softer, and its drapability is better. In other words, the higher the drape coefficient (F) the stiffer the fabric is. The drape coefficient is relevant to the drapability of fabrics but is not sufficient for characterising drape formation. Fabrics with the same drape coefficients may form different drape shapes. Hence other parameters such as number of nodes (folds) and node dimensions are also used to describe the drape quality.

The drape formation process is experimentally found to consist of three stages (Mizutani *et al.*, 2005): node generation (node appearance in the early stage), development (drapes growing from these nodes) and stabilisation (static stabilised drapes). The generation of nodes and the development process must be considered in relation to the mechanical properties of the fabrics.

When a fabric is draped, it deforms with multidirectional curvature. Draping qualities are related to fabric bending stiffness and shear properties. Factors such as fibre content, yarn structure, fabric structure and type of finish affect the drape behaviour. For example, fabric thickness (T) affects drape in different ways (Chen *et al.*, 2005): when $T < 0.4$ mm, increasing T causes a decrease in F because the weight effect imparts more influence than rigidity and flexibility at these thicknesses. This factor is reversed for $0.4 < T < 0.8$ mm, as changes in bending rigidity influence F more, causing it to rise with increased thickness. When $T > 0.8$ mm, the fabric is rigid, so the drapability is poor.

Methods for testing fabric drapability

Drape test systems currently used worldwide include the Peirce's cantilever method, the Rotrakote-CUSICK drape tester, the Fabric Research Liberating method (FRL drapemeter) (Japan), and the 3D body scanner.

The cantilever method measures fabric bending characteristics and then converts them into a measure of fabric drape. The FRL drapemeter also works on a similar principle. The cantilever method and FRL drapemeter only reflect a fabric's two-dimensional characters, and as fabric drape is actually a three-dimensional phenomenon, they are now less widely used.

The CUSICK drape tester is a simple but apt instrument which uses a parallel beam of light to cast a shadow from a circular piece of fabric, supported by a smaller circular disc. The area of shadow (A_S) is measured and compared with the area of the sample (A_D) and that of the supporting platform (A_d). The drape coefficient F is defined as

$$F = \frac{A_S - A_d}{A_D - A_d} \times 100\% \qquad 4.8$$

In the actual test, the light beam casts a shadow of the draped fabric onto a ring of highly uniform translucent paper supported on a glass screen. The surface drape pattern area on the paper ring is directly proportional to the mass of that area. So the drape coefficient (F) can be calculated in a simple way:

$$F = \frac{\text{mass of shaded area}}{\text{total mass of paper ring}} \times 100\% \qquad 4.9$$

There are three standard diameters of specimen that can be used for different types of fabrics:

- 24 cm for limp fabrics (drape coefficient below 30% with the 30 cm sample)
- 30 cm for medium fabrics
- 36 cm for stiff fabrics (drape coefficient above 85% with the 30 cm sample).

A fabric should be tested initially with a 30 cm specimen in order to see which of the above categories it falls into. When test specimens of different diameter are used, the drape coefficients measured from them are not directly comparable with one another.

The CUSICK drape tester can be fitted with a video camera and computer for instantaneous measurement of the drape coefficient. This is a trend that is adopted by most new drape measurement systems as it enables computer-aided analysis of the drape shape of fabrics and the numbers of nodes. A new apparatus (the drape elevator), designed by Japanese researchers, can also be used to evaluate drape properties continuously during the process of drape formation.

The 3D body scanner is another adaptation of the computer-aided capture of drape characteristics. A circular piece of fabric is hung over a circular disc, which allows the fabric to drape as in a CUSICK drape tester. Two scanners (one rotated 90° from the other) take around 12 seconds to capture the complete configuration (point cloud data) of the draped sample. The captured data is then processed using the Geomagic™ software to generate a 3D surface of the scanned object. The drape coefficient along with other useful drape parameters can be extracted from the processed data.

Intricate software has been developed to utilise the results obtained by electronic drape measurement in the computer design of a textile product. Drape characteristics can be simulated in a range of different designs and applications in both static and dynamic simulations.

4.6.2 Fabric bending

A bending test measures the severity of the flexing action of a material. The test can vary between bending the material sharply to bending it over a large radius and small amplitude. For thin flexible materials such as fabrics, the deformation is always intended to be at constant strain amplitude rather than stress amplitude. Resistance to bending or flexural rigidity is defined as flex stiffness. This property can influence the aesthetic appearance as well as the comfort of a fabric.

The bending length is a measure of the interaction between fabric weight and fabric stiffness in which a fabric bends under its own weight. It reflects the stiffness of a fabric when bent in one plane under the force of gravity, and is one component of drape. Thus bending length is also called drape stiffness.

The bending rigidity, which is related to the perceived stiffness, is calculated from the bending length and mass per unit area. Fabrics with low bending rigidity may exhibit seam pucker and are prone to problems in cutting out. They are difficult to handle on an automated production line. A fabric with a higher bending rigidity may be more manageable during sewing, resulting in a flat seam, but may cause problems during moulding.

The bending length is dependent on the weight of the fabric and is therefore an important component of the drape of a fabric when it is hanging under its own weight. The stiffness of a fabric in bending is very dependent on its thickness. The thicker the fabric, the stiffer it is, if all other factors remain the same. The bending modulus is independent of the dimensions of the strip tested, so that by analogy with solid materials it is a measure of 'intrinsic stiffness'.

Methods for testing fabric bending

Three methods are often used to test the stiffness of fabrics: the Cantilever test, the hanging loop test and the pure bending test conducted on a KES-FB2 bending tester. These methods are more suitable for testing woven fabrics than for testing knitted ones.

For the Cantilever test (Fig. 4.8), the Shirley Stiffness tester or the Gurley Stiffness tester is commonly used. The tester is based on the cantilever principle. In the test a rectangular strip (25 mm wide × 200 mm long) supported on a horizontal platform is clamped at one end and the rest of the strip is allowed to overhang and bend under its own weight. The bending length (C) is read from a calibrated scale in millimetres when the tip of the specimen reaches a plane inclined at 41.5 degrees. The higher the bending length is, the stiffer the fabric. The bending modulus (q) and the flexural rigidity (G) can be calculated from the bending length, the mass per unit area and fabric thickness:

$$\text{Flexural rigidity } G = 9.8MC^3 \times 10^{-6} (\mu Nm) \qquad 4.10$$

where C is bending length and M is mass per unit area.

$$\text{Bending modulus } q = \frac{12G \times 10^3}{t^3} (N/m^2) \qquad 4.11$$

where G is the flexural rigidity and t is the cloth thickness in mm.

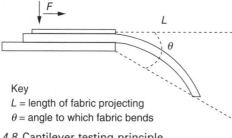

Key
L = length of fabric projecting
θ = angle to which fabric bends

4.8 Cantilever testing principle.

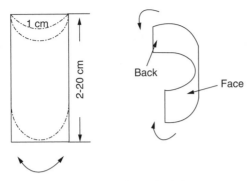

4.9 Pure bending testing principles.

If some fabrics are too flexible or limp, the hanging loop method may be used. Different shapes of hanging loops are used: ring loop, pear loop and heart loop. One end of the fabric is brought against the other end by bending through angles of 180° (pear), 360° (ring) and 540° (heart) and joined together. The length of this loop is measured when it is hanging vertically under its own weight. This hanging length is inversely related to the bending stiffness.

The KES-FB2 bending tester is a different approach used for determining stiffness and hysteresis of fabric specimens under pure bending. The precise bending momentum of the specimen can be detected. A standard size specimen 20 cm × 20 cm is mounted on two clamps (one is fixed and the other is free to move), which have a space of 1 cm between them (Fig. 4.9). The sample is then bent at a constant bending deformation rate of 0.5 cm^{-1}/s through a controlled curve pattern at a fixed torque by moving one of the clamps. The bending moment vs curvature curve can be obtained from the tester.

The digital pneumatic stiffness tester determines fabric stiffness using the ASTM circular bend test method. A plunger of 25.4 mm (1 in) diameter pushes the fabric through a 38 mm (1.5 in) diameter orifice for a distance of 57 mm (2.25 in) in 1.7 seconds and the maximum force is recorded. The machine is provided with a pneumatic cylinder, controls and a battery-operated digital force gauge of 50 kgf, 500 N or 100 lb (switchable) with peak-hold facility.

Standards commonly used for the bending stiffness test are as follows:

- ASTM D1388-2007 Standard test method for stiffness of fabrics
- BSI BS 3356-1991 Determination of bending length and flexural rigidity of fabrics (AMD 6337)
- AS 2001.2.9-1977 Determination of stiffness of cloth.

4.6.3 Fabric shearing

Shear deformation is very common during wear as the fabric needs to be stretched or sheared to various degrees as the body moves. The ability of a fabric to deform by shearing enables fabric to undergo more complex deformations than two-dimensional bending. Shearing enables a fabric to conform to complex shapes, such as the contours of the body in clothing applications. As a shearing force or moment is applied to a fabric, in-plane rotation of the yarns at the cross-over of the weave occurs along with yarn slippage at the interlacing points of warp and weft yarns, causing angle change. The shear mechanism is one of the important properties influencing the draping, pliability and handle of woven fabrics. It also affects their bending and tensile properties in various directions.

The shear behaviour of a woven fabric can be characterised by two shear parameters: shear rigidity and shear hysteresis. Shear rigidity determines fabric stiffness or softness. Fabric with low values of shear rigidity distorts easily, giving rise to difficulties in laying up, marking and cutting. A high value of shear rigidity means that a fabric is difficult to mould. Shear hysteresis is the energy loss when the direction of shear is reversed within a shear deformation cycle. This is due to the fact that when a fabric is sheared, most of the force expended is used in overcoming the frictional forces that exist at the intersection of warp and weft. Shear hysteresis can be related to various handle characteristics such as crispness, scroopiness, and how noisy the fabric is when handled. There is a strong linear relationship between shear rigidity and shear hysteresis.

The shear deformation depends upon the frictional and elastic forces within a fabric, so the values of shear properties are greatly affected by the fabric structure and finishing process. For example, the values of shear rigidity and shear hysteresis increase with the increase in the weft density of woven fabrics. The finishing process releases residual bending stress existing in the yarns, thus it can reduce the shear rigidity of the finished fabric.

Methods for testing fabric shearing

Several simplified methods for testing the shear of fabrics have been developed by workers in this area, i.e. KES-FB1 (Japan) and SiroFAST-3 (Australia) as illustrated in Fig. 4.10. Method (a) in Fig. 4.10 is based on the test principle employed by the KES-F system. A rectangular fabric sample is subjected to a pair of equal and opposite stresses F which are acting parallel to its edges. The fabric deforms to a slant position, though its area remains constant. This is in-plane shear. Figure 4.11 shows a typical shear stress vs shear strain curve. The shear strain is defined as the tangent of the angle θ of shear. That is:

Shear strain = $\tan \theta$

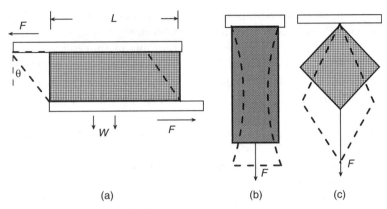

(a)   (b)   (c)

4.10 Shear testing methods.

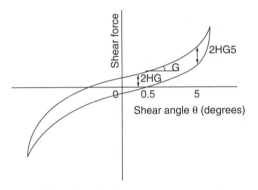

4.11 Shear hysteresis.

The shear rigidity (G) is the slope of the shear stress–strain curve:

$$G = F / \tan\theta$$

However, the shear deformation is not always a simple shear at constant area. Fabrics subjected to compressive forces in the plane of the material tend to buckle at very low values. In order to delay the onset of buckling, a vertical force W is applied to the fabric by using a weighted bottom clamp. The horizontal force F which is required to move the bottom clamp laterally is measured together with the shear angle θ. Then:

$$\text{Effective shear force} = F - W\tan\theta$$

The shear stress is defined as the shearing force divided by the sample width (L). A height:width ratio of 1:10 is considered to be the limit for practical measurements. The shear hysteresis parameters 2HG and 2HG5 are used in this method (2HG = hysteresis of shear forces at 0.5°, 2HG5 = hysteresis of shear forces at 5°).

Although the principle of method (a) in Fig. 4.10 is relatively simple, in practice it can be rather complicated to perform. Because the bias extension to a fabric is actually equivalent to shear, the test for it is easy to carry out on a CRE machine. Methods (b) (based on the test principle of SiroFAST-3) and (c) (a CRE machine can be used) are therefore the most appropriate for industrial use. The uniaxial tension is applied to a bias-cut specimen. Shear rigidity can be calculated from the extension of a fabric in the bias direction. For example, using a 5 gf/cm (or 4.9 N/m) tension (the same as that required in a SiroFAST-3 tester), if the extension on the bias (45°), EB5, is measured in %, then the shear rigidity (G) in N/m is simply calculated as $G = 123/EB5$.

It has been found experimentally that there were inconsistencies between the fabric properties measured in simple shear and by bias extension, due to a number of factors including the geometry of the test specimen, the assumption in the analysis that the threads were inextensible, and the variation that takes place in the normal stress during bias extension.

4.6.4 Fabric compression

The compression test for fabric is used to determine the fabric thickness at selected loads, and reflects the 'fullness' of a fabric. When measuring compression properties of fabric, it must be appreciated that all fabrics contain air as well as fibres and yarns. When a fabric is compressed, three distinct stages in the deformation of a fabric have been identified (Saville, 1999):

1. Individual fibres protruding from the surface will become bent and/or compressed. The resistance to compression in this region comes from the fibre bending stiffness.
2. The yarns come into close contact and are flattened and straightened, at which point the inter-yarn and inter-fibre friction as well as the yarns' bending stiffness provide the resistance to compression until the fibres are all in contact with one another.
3. The yarns are compacted, and the individual fibres are squashed against each other. The resistance is controlled mainly by the transverse properties (or lateral compression) of the fibres themselves.

These stages of compression involve elastic deformation, frictional forces and also elastic recovery of the fibres from bending and lateral compression. So the compression property contains information about the handle of the fabric. The greater the radius of curvature of the transition between the first and third stages, the softer is the fabric in compression.

Methods for testing fabric compression

The SiroFAST-1 compression meter and the KES-FB3 compression instrument are commonly used for measuring fabric thickness and compressibility. Older methods may rely on making direct observations of the fabric cross-section using a microscope, but these have mostly been abandoned because of the difficulty of identifying the edges of the fabric sample.

In the SiroFAST system, compression property tests include fabric thickness, fabric surface thickness and released surface thickness as shown in Fig. 4.12. The fabric is considered to consist of an incompressible core and a compressible surface. The fabric thickness is measured on a 10 cm² area at two different pressures, firstly at 2 gf/cm² (equivalent to 0.195 kPa or 19.6 mN/cm²) and then at 100 gf/cm² (equivalent to 9.807 kPa or 981 mN/cm²). The difference between these two values gives a measure of the thickness of the surface layer. The fabric thickness measurements are repeated after steaming on an open Hoffman press for 30 s in order to determine the stability of the fabric finish.

From the KES-FB3 tester, the compression energy, compressibility, resilience and thickness of a specimen can be obtained. A circular compressing board of 2 cm² attached with a sensor is used to apply the force on the fabric specimen (Fig. 4.13). The applicable compression force is 0.1 gf/cm²

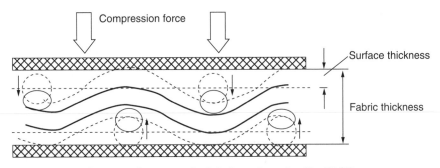

4.12 Compression test principle based on the SiroFAST system.

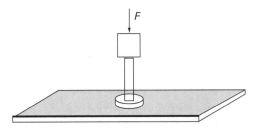

4.13 Compression test on the KES-F system.

(minimum) to 2.5 kgf/cm² (maximum) and the machine is running at different compression deformation rates from 0.1 mm/s to 10 mm/s.

4.7   Applications and future trends

Fabric objective measurements provide a scientific means to quantify the quality and performance characteristics of fabrics. This forms the basis for fabric specification, product development, process control, product failure analysis and quality assurance. It also facilitates communication between consumers, manufacturers, designers and researchers in the whole textile chain.

The tests and results can be used to simulate and predict fabric performance in use. For example, fabrics with a low tensile strength and low tear strength may be susceptible to mechanical damage. This can occur when tension is put on the fabric during wear or cleaning. Sharp objects in contact with a fabric may also cause rips, tears or holes. In some cases, mechanical damage may be attributed to the basic construction of the fabric itself. Early identification of the problems in fabrics allows remedial action to be taken before the cost of rejects becomes an issue.

One fabric may have very different performance properties due to the interactions of durability factors. If the fabric durability is affected by environmental influences, such as ultraviolet radiation and atmospheric temperature, not only is the fabric aesthetic property changed, but also the breaking strength may change because of the deterioration of fibres. For example, fibre deterioration in curtain, drapery and sportswear fabrics results from exposure to either direct or indirect rays of the sun. Hence, one test method may or may not predict how a fabric may perform in consumer use. It is often necessary to test a combination of fabric properties.

On the other hand, the interrelationships of all mechanical properties are complex and are affected by many factors such as fabric geometry, setting, finishing, coating, laminating and so on; they can influence performance properties in very different ways. For instance, if other factors such as fibre type and fabric finish are held constant, a tight fabric construction generally contributes to high tensile strength but also lower tear strength and vice versa. A moderate structure, not too tight or too loose, could be expected to yield best abrasion resistance. Thereby, the early tests for the fabric properties enable the best processing route to be selected from the outset, to produce the optimal performance for an intended application.

Developments in modern fabric testing instrumentation have followed two broad routes: simplicity and versatility. This trend will continue, but with increased objectivity and intelligence built into the instruments. Generally, simplicity is preferred in the industry while research organisations prefer versatile test instruments, which are often complex. The challenge to researchers and instrument developers has been to quantify a complex fabric attribute with a simple parameter. A good example is fabric handle, which is affected by many factors. A simple approach to measuring fabric handle involves extracting a fabric specimen through a fixed diameter nozzle using a CRE machine (Alley, 1978; Alley and McHatton, 1976). A quantity termed 'handle modulus' is calculated from the force–displacement data, the geometric considerations of the nozzle, fabric coefficient of friction and fabric effective thickness. Studies have shown that results from the nozzle measurement are in fairly good agreement with those from other more complicated hand evaluation systems, such as the KES-F system and physical tests related to fabric hand. Other simple techniques include the ring or slot test (Grover *et al.*, 1993) and the pulling force measurement by pulling a fabric through a set of parallel pins (Zhang *et al.*, 2006). These methods consider the combined effect of fabric surface properties and bending stiffness. More developments are likely in this direction, with increased intelligence and sophistication (Pan, 2006).

4.8   Sources of further information and advice

In addition to the references, the following materials and websites provide good sources of further information on fabric testing:

1. *Annual Book of ASTM Standards.*
2. *Fabric Assurance by Simple Testing – Instruction Manual*, 1989, CSIRO Division of Wool Technology, Sydney, Australia.
3. www.iso.org
4. www.astm.org
5. www.sdlatlas.com
6. www.kestato.co.jp
7. www.bodyscan.human.cornell.edu
8. Brown, R., ed. (1999), *Handbook of Polymer Testing, Physical Methods*, Marcel Dekker, New York.
9. Hu, J. L. (2004), *Structure and Mechanics of Woven Fabrics*, Woodhead, Cambridge, UK.
10. Jeong, Y. J. and Phillips, D. G. (1998), 'A study of fabric-drape behaviour with image analysis, Part II: The effects of fabric structure and mechanical properties on fabric drape', *Journal of the Textile Institute*, 89(1): 70–79.
11. Kawabata, S., Postle, R. and Niwa, M., eds (1985), 'Objective measurement: applications to product design and process control', *Proceedings of the Third Japan–Australia Joint Symposium on Objective Measurement: Application to Product Design and Process Control*, Kyoto, Japan, The Textile Machinery Society of Japan.

12. Peirce, F. T. (1930), 'The "handle" of cloth as a measurable quantity', *Journal of the Textile Institute*, 21: T377–416.
13. Saville, B. P. (1999), *Physical Testing of Textiles*, Woodhead, Cambridge, UK.
14. Sen, A. K. (2001), *Coated Textiles*, Technomic Publishing, Lancaster, PA.
15. Tao, X. M., ed. (2002), *Smart Fibres, Fabrics and Clothing*, Woodhead, Cambridge, UK.
16. Warner, S. B. (1995), *Fiber Science*, Prentice-Hall, Englewood Cliffs, NJ.

4.9   References

1. Lyle, D. S. (1977), *Performance of Textiles*, John Wiley & Sons, New York.
2. Kawabata, S. (1980), *The Standardization and Analysis of Hand Evaluation*, 2nd edition, The Textile Machinery Society of Japan, Osaka, Japan.
3. Mizutani, C., Amano, T. and Sakaguchi, Y. (2005), 'A new apparatus for the study of fabric drape', *Textile Research Journal*, 75(1): 81–87.
4. Chen, H., Shen, Y., Liu, X. and Chen, Y. W. (2005), 'Linear regression of fabric's thickness versus drape behaviour', *Journal of Zhejiang Sci-Tech University*, 22(2): 110–113.
5. Saville, B. P. (1999), *Physical Testing of Textiles*, Woodhead, Cambridge, UK.
6. Alley, V. L. (1978), Nozzle extraction process and handlemeter for measuring handle, US Patent 4,103,550, 1 August.
7. Alley, V. L. and McHatton, A. D. (1976), A proposed quantitative measure of fabric handle and the relative characterization of some aerospace materials by handle moduli, *Ninth Air Force Geophysics Laboratory Scientific Balloon Symposium*, Portsmouth, NH.
8. Grover, G., Sultan, M. A. and Spivak, S. M. (1993), 'A screen technique for fabric handle', *Journal of the Textile Institute*, 84: 1–9.
9. Zhang, P., Liu, X., Wang, L. and Wang, X. (2006), 'An experimental study on fabric softness evaluation', *International Journal of Clothing Science and Technology*, 18(2): 83–95.
10. Pan, N. (2006), 'Quantification and evaluation of human tactile sense towards fabrics', *International Journal of Design and Nature*, 1(1): 1–13.

5

Fabric chemical testing

Q FAN, University of Massachusetts, USA

Abstract: The fundamental structure of fabrics is fibers. The identification of fibers is often the first task when an unknown fabric has to be analyzed. The chemical identification of fibers based on the solubility, stain and burning tests is presented. The determination of common fabric chemical qualities, such as pH and metals that are mostly concerned by textile wet processors is discussed together with fabric whiteness and absorbency that are important for finished textiles. The analysis of auxiliaries such as sizing agents, surfactants and softeners is provided for quality control purpose. The identification of dye classes and finishing agents often required in a forensic investigation, confirmation of dye and finish application, and environmental, health and safety concerns is given. The testing for damage to textiles is also mentioned.

Key words: chemical testing, chemical analysis, dyes, finishing agents, formaldehyde, auxiliaries, quality.

5.1 Introduction: definition and role of tests

Fabric chemical testing is a broad topic and can range from a simple pH test to a complicated colorant identification using HPLC. As the name indicates, chemical testing is basically the test of the fabric samples either using chemicals or for their chemical properties, and very often both. Sometimes, a Bunsen burner (oxidation reaction) is sufficient to conduct the fiber identification, while the dye analysis may need to use an HPLC (adsorption and desorption processes). It is also very important that the professionals involved in the chemical testing have fundamental knowledge of fibers, dyes, auxiliaries and finishing agents as well as the basic understanding of yarns, fabrics, and their formations. With the right lab facilities and well-trained personnel, fabric chemical testing can be performed successfully.

As a rule of thumb, the fabric samples need to be prepared properly before the testing according to the test method to be used. The sample preparation usually means the samples should be conditioned at regulated temperature and humidity. It also requires that the sampling process should be conducted following certain protocols in order to make sure the sample to be tested can represent the batch/shipment/bulk. During the chemical testing, personnel protection is one thing that will never be overstressed, due to the nature of the chemicals used in the testing. Overlooking the possible hazardous situation is a guarantee of mishaps. Therefore, wearing of goggles and gloves is a must for handling any chemicals. Preparation of chemical stocks and standard solutions should be performed cautiously. If volatile organic chemicals (VOC) are to be involved in the chemical testing, a ventilated fume hood should be used. At the end of testing, the test results should be reported based on statistical analysis using industrial standard formats to facilitate the communication between the testing lab and the client.

The testing performed on the fabric samples can meet one or a combination of the following requirements:[1]

- To provide a service to the customer
- To conform to a standard
- To assess the effect of textile processes
- To develop a new product
- To establish a specification
- To investigate problems a product has.

Many standard testing methods (AATCC, ASTM, EN, JIS, and ISO) are available nowadays. Whenever possible, the standard testing methods should be used unless customers have different requirements, or the lab/sample conditions necessitate the modifications. If the test is not performed exactly as the standard test, any modifications should be presented in the testing report.

This chapter presents the basic description of some standard methods for fabric chemical testing. For detailed procedures in each testing, readers should access the reference materials listed at the end of this chapter.

5.1.1   Instrumentations

The basic facilities a fabric chemical testing lab should have include the following:

- Balance, capacity up to 300 grams with accuracy 0.1 mg
- A pH meter, ideally with three-point calibration function and readability 0.1 unit
- Microscope, optical magnification up to ×1500
- Standard light booth, ideally with daylight, horizon light, cool white light and UV light, or lights to be useful for the localized applications
- Spectrophotometer, for visible range between 380 and 780 nm, ideally with UV range (200 to 400 nm) measurement and dual modes for both transmission and reflection measurements
- Stopwatch, with readability of 1/100 second
- A lab-sized atmospheric dyeing machine

- Oven with temperature control
- Crockmeter
- Gray scales
- Access to water, gas, electricity, and drainage.

A well-equipped testing lab can also have the following:

- Soxhlet extractor
- Auto-titrator
- HPLC, high performance liquid chromatography apparatus
- FTIR, Fourier transform infrared spectrometer
- DSC, differential scanning calorimeter.

A temperature and humidity conditioned environment is a must for textile physical testings, and a plus for chemical testing. Many fabric chemical testings can be performed under non-environmentally controlled conditions. However, it is sometimes more practical to use an environmentally controlled container to have fabric samples conditioned and then conduct testing in a non-controlled environment. This is especially true for some small in-house testing facilities. Consequently, the test results would deviate from the real values and should be reported accordingly, if environmentally controlled conditions are listed in the relevant standard testing methods.

5.1.2   Factors affecting tests

The general factors that could affect the testing results include sample size/weight, sample contamination, chemical concentrations/assays, instrument calibration, personnel error, and environmental variables such as temperature and humidity. The effects of these factors are generally detectable if an error source is identified. The most important managerial action is to have a quality control and quality assurance system in place, which can significantly minimize the effects of the influential factors. It is worth noting that many of these factors can be eliminated if the quality control and quality assurance system is maintained and executed properly.

5.2   Chemical identification of fibers

5.2.1   Fiber classification

Textile fibers can be easily classified based on their sources, natural fibers and synthetic fibers. The common natural fibers can be divided into two groups, cellulose and protein. In the cellulose group, cotton and rayon (viscose) are the most common ones. The other cellulosic fibers that have become more and more popular are linen, ramie, and bamboo. The protein group has silk and wool. Sometimes, cashmere, angora, and mohair are used in knitting. The common synthetic fibers are polyester, nylon, and acrylic.

The comprehensive fiber identifications and both qualitative and quantitative analysis can be found in AATCC Test Methods 20 and 20A.[2] The methods include physical approaches: visual and microscopic examination, density, drying twist, and melting point; chemical pathways: burning, solubility and stain; and instrumental ways: FTIR, DSC and SEM, etc. In this section, only the qualitative chemical identifications of common fibers are introduced.

5.2.2 Chemical identification

Solubility test

Separate a few fibers from the fabric. Cut them into short lengths of about 5–7 mm (¼ inch). Put four or five cut fibers into a test tube with 10 ml of solvent as listed in Table 5.1 or Table 5.2. Observe their dissolution in each solvent. If necessary, raise the temperature of the solvent by immersing the test tube in a temperature-controlled water bath. For the common natural fibers, using the inorganic solutions shown in Table 5.1 under the set conditions, a positive identification can be made.[2] For the common synthetic fibers, using the organic solvents shown in Table 5.2 under the set conditions, a positive identification can be made.[2]

Table 5.1 Solubility of common natural fibers in different inorganic solutions and conditions

	5% sodium hypochlorite, 20°C, 20 min	59.5% sulfuric acid, 20°C, 20 min	70% sulfuric acid, 38°C, 20 min
Cotton	Insoluble	Insoluble	Soluble
Rayon	Insoluble	Soluble	Soluble
Silk	Soluble	Soluble	Soluble
Wool	Soluble	Insoluble	Insoluble

Table 5.2 Solubility of common synthetic fibers in different organic solvents and conditions

	85% formic acid, 20°C, 5 min	Dimethyl formamide, 90°C, 10 min	*m*-cresol, 139°C, 5 min (use a glycerol bath)
PET	Insoluble	Insoluble	Soluble
Nylon 6	Soluble	Soluble	Soluble
Nylon 66	Soluble	Insoluble	Soluble
Acrylic	Insoluble	Soluble	Plastic mass formed

Table 5.3 Burning behavior of common fibers

	Melts near flame	Shrinks from flame	Burns in flame	Continues to burn	Appearance of ash	Smell from flame
Cellulose	No	No	Yes	Yes	Light grayish	Burning paper
Silk	Yes	Yes	Yes	Slowly	Soft black bead	Burning hair
Wool	Yes	Yes	Yes	Slowly	Irregular black	Burning hair
PET	Yes	Yes	Yes	Slowly	Hard black	Burning wax
Nylon	Yes	Yes	Yes	Yes	Hard gray	Characteristic odor
Acrylic	Yes	Yes	Yes	Yes	Hard black	Characteristic odor

Stain test

For uncolored fabrics, staining or dyeing can provide a quick identification of unknown fibers in a sample including the blend samples. DuPont Fabric Dyestain #4 (available from Testfabrics, Inc., West Pittiston, PA) is a common stain widely used in textile testing labs, which is a mixture of C.I. Acid Blue 298, Acid Red 182, Direct Blue 218, Disperse Orange 25, Disperse Yellow 3, and Direct Yellow 11.[3]

The unknown sample strip, measuring 2.5 × 10 cm (1 × 4 inches), is boiled with a multifiber standard (also available from Testfabrics) in a 1% DuPont Fabric Dyestain #4 solution at pH 4.5 at a liquor ratio of 1:20 for 10 min. After rinse and drying, the color of the sample strip is checked against the multifiber standard to identify the unknown fiber. Depending very much on the number of fiber types on the multifiber standard, many commercially available textile fibers can be positively identified.

Burning test

The burning test is the simplest test for fiber identification. A Bunsen burner is all it needs. However, caution should be exercised in order to avoid confusion caused by fiber blends and fiber finishes. Generally, the observations shown in Table 5.3 can help distinguish a few types of fiber.[2]

5.3   Fabric quality

5.3.1   pH value[4]

Treat 10 grams of fabric in 250 ml boiled distilled water at boiling temperature for 10 minutes, cool down, remove the sample, then measure the pH of the solution.

5.3.2 Whiteness[5]

Before measuring the whiteness using a spectrophotometer, it is necessary to make sure the fabric has no fluorescent whitening agents (FWA). A UV light in the light booth can be used to detect the presence of FWA on the samples in a dark room. The fluorescence shown indicates the existence of FWA, which will affect the whiteness measurement. A specialized spectrophotometer should be used to compensate or control the FWA effect.

A sample with an adequate opacity, usually folded a few times to achieve this, is measured by a reflectance spectrophotometer using D_{65} illuminant and 10° observer, and a whiteness index is reported by the software associated with the spectrophotometer. Usually, a sample needs to be measured at least four times to get an averaged result. The higher the whiteness index, the whiter the sample.

5.3.3 Absorbency[6]

The absorbency of the fabrics is determined by the time the sample takes to absorb a fixed amount (usually a drop) of distilled or deionized water. Sample conditioning is extremely important as the residual moisture on the sample to be measured can affect the absorbency results considerably. The sample under a set tension (mounted on an embroidery hoop) is placed vertically at the prescribed distance from the end of a burette with water. The time starting from the moment the water drop touches the surface of the sample to the moment the water drop is completely absorbed by the sample (disappearance of the liquid reflection) is taken on a stopwatch. One sample should be measured at least five times to get an averaged result. The longer the time, the lower the absorbency. A more critical examination of fabric absorbency can be achieved using a 50% or 65% sucrose solution, because the sucrose solution is more viscous and is absorbed more slowly than water, thus distinguishing between samples with very small differences in their water absorbency.

5.3.4 Metals and NO_x detection[7]

Calcium

Ash 5 g fabric in a crucible, move the ash to 10% hydrochloric acid, and add ammonium chloride and ammonia ($d = 0.88$) until the solution is alkaline. Filter off any precipitate. Acidify the ammoniacal filtrate with acetic acid. Precipitate calcium out as calcium oxalate using oxalic acid. Spot the precipitate onto a platinum wire and place it into a flame. A red flame color indicates the presence of calcium.

For quantitative determination, wash the filtered calcium oxalate with a minimum amount of distilled water and move it into warm 20% sulfuric acid. Titrate the solution using $0.1N$ potassium permanganate solution. Each 1 ml $0.1N$ potassium permanganate solution consumed in the titration is equivalent to 2.004 mg calcium. The total calcium percentage in the sample is calculated using the following equation:

$$\text{Calcium \%} = \frac{\text{calcium from titration, mg}}{\text{sample weight, mg}} \times 100\%$$

Iron

Place a small piece of sample on a watch glass and add 1–2 drops of 5% nitric acid. Alternatively, dissolve the ash of a small piece of colored sample in 10% nitric acid. Add 2–4 drops of 10% potassium thiocyanate after allowing the sample to stand for 3 minutes. A pink color indicates a low content of iron, while a dark red color indicates a high content. The color intensity is proportional to the content of iron in the sample. Copper may interfere to generate a yellow-brown color.[8]

Copper

Spot the ash of the sample with 5–10 drops of 10% nitric acid. Add diluted ammonia until the ash is alkaline. A blue color shows the presence of copper. The test may be performed on fabric directly using 0.1% sodium diethyl dithiocarbamate solution after the above-mentioned procedures. A brown color shows the presence of copper. However, iron may interfere to generate a brown color too.

NO_x

The presence of NO_x can cause yellowing on bleached fabrics, including those with FWAs. The detection of NO_x can be performed using Saltzman's reagent. Add 1–2 drops of 1% w/v sulfamic acid (H_3NSO_3) in 30% acetic acid to the test sample. After 1 minute, add 1–2 drops of 0.05% w/v N-(1-naphthyl)ethylenediamine hydrochloride. A deep to vivid purplish red color indicates the presence of NO_x.

5.4   Auxiliaries testing – sizing agents[9]

5.4.1   Starch and its derivatives

Starches are natural polymers called polysaccharides that have multiple anhydroglucose units. The chemical formula for starch and its derivatives is $(C_6H_{10}O_5)_n$. Starch has chemically two moieties, an amylose part which is anhydroglucopyranose units joined by α-D-1,4 glycosidic bonds, and an amylopectin part which is anhydroglucopyranose units linked by α-D-1,6 glycosidic bonds as shown in Fig. 5.1. Amylose has a linear chain and amylopectin a branched chain. Starch can be decomposed by pyrolysis and acidic hydrolysis. British gums and dextrins are the products derived from starch. They have lower molecular weight and better water solubility.

Qualitative testing of starch and its derivatives can be conducted using an aqueous solution containing 1.4 g/L I_2 and 2.4 g/L KI.[10] The existence of starch (specifically amylose) will turn the yellow-orange color of the solution to a dark blue color. When tested with I_2/KI solution, British gums and completely degraded dextrins give a reddish brown color, partially degraded dextrins a violet color, and white dextrin a blue color.[11]

5.4.2 Cellulose derivatives

Cellulose is another natural polymer belonging to polysaccharides. The chemical formula for cellulose is also $(C_6H_{10}O_5)_n$. It has a chemical structure very similar to amylose. Instead of α-D-1,4 glycosidic bonds, cellulose has only β-D-1,4 glycosidic bonds as shown in Fig. 5.2. Due to these β-D-1,4 glycosidic bonds, the molecular chain of cellulose can extend quite linearly, making it a good fiber-forming polymer. In order to use cellulose for sizing purposes, it should be modified to shorten the molecular structure. The two most used cellulose derivatives are carboxymethyl cellulose (CMC) and hydroxyethyl cellulose (HEC).

Carboxymethyl cellulose (CMC) is manufactured with alkali cellulose and sodium chloroacetate. The hydrogen atoms of hydroxyl groups on C2

5.1 Chemical structure of amylopectin.

5.2 Chemical structure of cellulose.

and C6 are partially substituted with —CH$_2$COONa or —CH$_2$COOH depending on reaction conditions. The degree of substitution (DS) is usually between 0.2 and 1.5 (0.2 to 1.5 carboxymethyl groups (—CH$_2$COOH) per anhydroglucose unit). CMC with DS 1.2 and below is water soluble. The final product always contains sodium salt. A foaming test can distinguish sodium CMC from other cellulose ethers, alginates and natural gums.[12] Sodium CMC solution, after vigorous agitation, would not produce any foam layer. Uranyl nitrate can be used to detect the existence of CMC. A 4% uranyl nitrate is used to precipitate CMC between pH 3.5 and 4.[13] A precipitate with off-white color indicates the presence of CMC. A 0.5% methylene blue methanol solution may also be used to detect CMC on fabric.[7] After rinsing in distilled water and drying, the methylene blue treated sample may show a blue/purple color which confirms the existence of CMC or acrylic sizes. A separate extraction of the fabric with toluene can exclude the acrylic sizes from the test.

When the hydrogen atom of the hydroxyl group on C6 of cellulose is partially substituted with the hydroxyethyl (—CH$_2$CH$_2$OH) group in a reaction with ethylene oxide under alkaline condition, hydroxyethyl cellulose (HEC) is produced. So far there are no known testing methods for HEC detection. However, if one wants to distinguish CMC from HEC, an ion tolerance test can be conducted. CMC is anionic and can be precipitated from an aqueous solution with a cationic surfactant. Since HEC is non-ionic, its aqueous solution is compatible with cationic surfactants. Based on the same ionic tolerance principle, a high salt concentration can precipitate CMC, not HEC.

5.4.3 Alginates

Alginates are linear copolymers of randomly arranged β-D-1,4 mannuronic acid (M) and α-L-1,4 guluronic acid (G) blocks as represented in Fig. 5.3. Their chemical structure is similar to that of cellulose except that they have a carboxylic group on the C5 position instead of a methylol group in the case of cellulose. Alginates have good water solubility. Divalent and higher

5.3 Chemical structure of alginate.

valent metal ions, strong acids and bases can precipitate alginates out of its aqueous solutions.

In order to distinguish alginates from other thickening agents, precipitation methods can be tried.[12] 2.5% $CaCl_2$ can cause 0.5% sodium alginate solution to precipitate. Aqueous solutions of gum arabic, sodium carboxymethyl cellulose, carrageenan, gelatin, gum ghatti, karaya gum, carob bean gum, methyl cellulose and tragacanth gum would not be affected. Saturated $(NH_4)_2SO_4$ would not precipitate 0.5% sodium alginate, but agar, sodium carboxymethyl cellulose, carrageenan, de-esterified pectin, gelatin, carob bean gum, methyl cellulose and starch would be affected. The existence of sodium alginate can be tested with acid $Fe_2(SO_4)_3$.[12] Five minutes after a sample is in contact with a ferric sulphate solution, a cherry red color appears and this gradually changes to a deep purple color. This confirms the presence of sodium alginate in the sample.

5.5   Synthetic sizing agents

5.5.1 Polyvinyl alcohol

Polyvinyl alcohol (PVA) is the hydrolysis product of polyvinyl acetate. Depending on the hydrolysis conditions, there are fully hydrolyzed PVA and partially hydrolyzed PVA as shown in Fig. 5.4(a) and (b) respectively. Fully hydrolyzed PVA usually has a degree of hydrolysis (DS) of 98% to 99.8%, and can dissolve in water only at >80°C. The solubility of partially hydrolyzed PVA with a DS between 85% and 90% is dependent upon its molecular weight. The one with high molecular weight needs to have high temperature for dissolving.

Specific detection of PVA on fabrics can be achieved using potassium dichromate ($K_2Cr_2O_7$).[14] Two solutions are used. Solution A consists of 11.88 g $K_2Cr_2O_7$ and 25 ml concentrated H_2SO_4 in 50 ml distilled water. Solution B contains 30 g NaOH in 70 ml distilled water. After solutions A and B are applied to a white fabric sample sequentially, a brown color developing after rubbing the spotted area for a complete reaction indicates the existence of PVA. A yellow-green color can be triggered by unsized

$$—CH_2—CH—CH_2—CH—CH_2—CH—$$
$$\quad\ \ |\qquad\quad |\qquad\quad |$$
$$\quad\ \ OH\qquad OH\qquad OH$$

$$—CH_2—CH—CH_2—CH—CH_2—CH—$$
$$\quad\ \ |\qquad\qquad |\qquad\qquad |$$
$$\quad\ \ OH\qquad\ \ O\qquad\quad OH$$
$$\qquad\qquad\qquad |$$
$$\qquad\qquad\quad COCH_3$$

(a) Fully hydrolyzed   (b) Partially hydrolyzed

5.4 Chemical structures of PVA.

$$CH_2\!=\!CH$$
$$O\!=\!C\!-\!OH$$

5.5 Chemical structure of acrylic acid.

goods, potato starch, styrene–maleic anhydride copolymer, alginates, guar, gelatin or CMC.

5.5.2 Acrylics

Acrylic is a generic term for a large group of homopolymers and copolymers derived from acrylic acid shown in Fig. 5.5. Since the hydrogen atoms of the carboxylic group and the vinyl group can be substituted with many different chemical groups, a huge variety of polyacrylic acid and polyacrylates are currently available for many different applications. Most of them are used as an emulsion.

The analysis of acrylics is almost impossible without using sophisticated instruments. There are no known simple methods for the tests to be done with wet chemistry, which could be because (1) the analysis with wet chemistry is too complicated; (2) the analysis has to deal with too many different types of polymers; (3) the analysis involves the use of many toxic organic solvents. If the analysis is needed, it is recommended to use either FTIR or GC to get the results quickly and accurately. FTIR analysis of acrylics can show very distinctive absorption peaks between 1100 and 1150 cm^{-1} for the alkyl C—O—C stretching band and at 1750 cm^{-1} for the C=O stretching band. If the acrylic polymer system contains some vinyl monomers, a broad peak at 3020 cm^{-1} and a strong peak at 1660 cm^{-1} indicate the stretching bands of C—H and C=C in CH=CH structure respectively.[15] With GC analysis, acrylic acid, ethyl methacrylate, n-butyl acrylate, 2-ethylhexyl acrylate, isobutyl acrylate, methyl methacrylate, ethyl acrylate and methyl acrylate can all be detected and quantitatively determined.[16] ASTM Test Method D3362[17] is a standardized method for the purity analysis of acrylate. It is certain that more types of acrylates can be successfully analyzed nowadays with modern instruments.

5.6   Surfactants

The detection of surfactants on fabrics requires extraction of the surfactants from the fabric sample. First, agitate a fabric sample in water at a low liquor ratio (<1:20) overnight at room temperature, then remove the sample and concentrate the solution containing surfactants for testing using the following methods at room temperature. Any colors, especially the ionic dyes, removed during the agitation may interfere. The following methods can also be used to detect the ionic type of the surfactants concerned.

5.6.1 Anionic surfactants

Acidic methylene blue test[18]

Methylene blue is a cationic dye soluble in water and insoluble in chloroform. It can form with anionic surfactants a blue compound which is soluble in chloroform.

Acidic methylene blue solution: Slowly add 12 g H_2SO_4 to 50 ml water; after cooling down, add 0.03 g methylene blue and 50 g Na_2SO_4 anhydrate; dilute the whole solution to 1 liter.

Test: Add 5 ml of 1% sample surfactant solution into a mixture of 10 ml methylene blue solution and 5 ml chloroform in a test tube; shake vigorously, then allow it to stand until two layers are formed. If the chloroform layer (bottom layer) shows blue, add another 2–3 ml of the surfactant solution. Shake well and leave for layers to form. The chloroform would show dark blue and the water layer would be almost colorless. This is a positive result for the existence of anionic surfactant in the sample solution. This test is suitable for alkylsulfate and alkylbenzolsulfonate surfactants. Soap cannot be tested because it would precipitate in the strong acidic medium.

Basic methylene blue test[19]

Add one drop of 5% sample solution to a mixture of 5 ml 0.1% methylene blue solution, 1 ml $1N$ NaOH solution and 5 ml chloroform. Shake well and observe the color of the chloroform layer. If a blue-purple color is shown, there exists an anionic surfactant in the sample. This test is suitable for any type of anionic surfactants.

Thymol blue test[20]

Thymol blue solution: Add three drops of 0.1% thymol blue to every 5 ml of $0.005N$ HCl solution.

Test: Add 5 ml neutralized sample solution to 5 ml thymol blue solution. Shake well and observe the color of the mixture. A reddish-purple color is the evidence of existence of anionic surfactants in the sample solution.

Precipitation test[21]

A few drops of sample solution are added into 5 ml of 5% *p*-toluidine hydrochloride aqueous solution. If a white precipitate appears, there is anionic surfactant in the sample solution.

5.6.2 Cationic surfactants

Methylene blue test

Cationic surfactants can also be tested using methylene blue solution. First add two drops of a known anionic surfactant solution to a mixture of 5 ml methylene blue solution and 5 ml chloroform, shake well and leave it to stand for the chloroform layer to show blue. Then add a few drops of the sample solution, shake well and leave it for layers to form. If the blue color in the chloroform layer becomes lighter or colorless, the existence of cationic surfactants in the sample solution can be confirmed.

Bromophenol blue test[22]

Bromophenol blue solution: Add 20 ml of 0.1% bromophenol blue in 96% ethanol to a mixture of 75 ml $0.2N$ sodium acetate and 925 ml $0.2N$ acetic acid. Adjust the pH of the solution to 3.6–3.9.

Test: Add 2–5 drops of a neutralized sample solution to 10 ml of bromophenol blue solution. Shake well and observe the color of the mixture. If a blue color is shown, the existence of a cationic surfactant is confirmed.

Alternatively, add one drop of 5% sample solution to a mixture of 5 ml chloroform, 5 ml 0.1% bromophenol blue dilute ethanol solution and 1 ml $6N$ HCl. Shake well and observe the color of the chloroform layer. If a yellow color appears, there exists a cationic surfactant in the sample.

Precipitation test[20]

A diluted aqueous solution of either sodium salicylate, sodium benzoate, or sodium succinate can precipitate cationic surfactants.

5.6.3 Non-ionic surfactants

Methylene blue test

This test is conducted as on page 136. If the aqueous layer is emulsified to a milk-like state, or both layers have the same color, the existence of non-ionic surfactants can be confirmed.

Cloud point test[23]

The solubility of polyoxyethylene surfactants is dependent on their hydrogen bonding with water. At a high temperature, the hydrogen bonds of the surfactants would be dissociated, leading to lower solubility of the surfactant. Therefore, the solution of the surfactant becomes cloudy at the high temperature. Based on this principle, the polyoxyethylene surfactants can be detected.

A 1% sample solution is gradually heated with a thermometer in the solution to monitor its temperature. When the solution becomes cloudy, stop heating. Let the solution cool down slowly. The cloud point is reached when the solution turns to clear.

Amphoteric surfactants[19]

Amphoteric surfactants contain both anions and cations. They should show positive results when tested with either the basic methylene blue test for anionic surfactants or the alternative bromophenol blue test for cationic surfactants.

A saturated bromine aqueous solution can also be used to determine the type of amphoteric surfactants. Add 5 ml of 1% sample solution to 1.5 ml saturated bromine aqueous solution. Observe the color of the precipitate. Heat the mixture and observe the change of the precipitate. If the precipitate is yellow to yellow-orange and is dissolved to form a yellow solution after heating, the sample is imidazoline or alanine types of amphoteric surfactants. If the precipitate is white to yellow and insoluble after heating, the sample is the other types of amphoteric surfactants.

5.7   Fluorescent whitening agents

Fluorescent whitening agents (FWAs) are a special type of chemicals that can significantly increase the apparent whiteness of treated fabrics. They absorb UV radiation and re-emit the absorbed energy in the blue visible light range, which makes the treated fabrics appear whiter complementary with the yellow color. The detection of the presence of FWAs on fabrics is simply a visual examination under UV light in a dark room. An FWA-treated fabric fluoresces under the UV light.

5.8   Colorants testing

Before conducting any testing listed in this section, it is necessary to positively identify the fiber first. Care should be taken to separate blend fibers carefully. A burning test of the fiber samples can often help distinguish from main fiber types. If a cellulosic sample is resin finished, treat the sample with 1% hydrochloric acid at 75°C for 20 minutes, then wash with hot water, 0.1% ammonia solution, and cold water in succession; finally dry.

5.8.1 Azoic, vat and sulfur dyes[1]

Boil the sample in sodium formaldehyde sulfoxylate (Formosul, C. I. Reducing Agent 2) solution at pH 11 for a few minutes. If vat and sulfur dye are present, discoloration happens. However, the original color is usually restored by exposing the rinsed sample to air. The color of azoic dyes changes permanently and cannot be restored.

Sulfur dyes can be confirmed[24,25] by boiling the sample in 16% hydrochloric acid for 30 seconds in a test tube. Cool and add 3 mg pure zinc dust and warm for 2–3 minutes. A filter paper wetted with a solution of 5% lead acetate and 25% glycerol shows a brown or black stain when placed on the test tube, and vapors give smell of hydrogen sulfide. Other dyes containing sulfur element may give a positive reaction.

5.8.2   Ionic dyes[25]

Boil 0.5 g cellulosic fabric sample for 2 minutes in a 5% sodium hydroxide solution (stripping). Add a small piece of bleached cotton fabric to the warm stripping solution. If the bleached cotton fabric is deeply stained, direct dye is confirmed. Reactive dye is confirmed when little or no staining happens to the bleached cotton fabric.

Boil a protein fiber/fabric or nylon fiber/fabric in dimethylformamide for 2–4 minutes. Acid dyes bleed off except some supermilling acid dyes which are difficult to distinguish from reactive dyes. Metal-complex acid dyes may bleed slightly, and mordant acid dyes do not bleed at all. Acid dyes can be stripped in 2% ammonia solution and can redye white wool in acidified solution except the acid dyes with high wet fastness.

Boil an acrylic sample in 85% formic acid for 2 minutes (stripping) and dilute with an equal amount of water after removal of the stripped sample. The stripped basic dye can be transferred to an undyed acrylic (Orlan) sample by boiling.

5.8.3   Disperse dyes[24,25]

Boil a nylon/polyester/acetate sample in ethanol for 2 minutes. Some dye can be stripped and transferred to undyed secondary acetate in the same solution.

Treat an acrylic sample in liquid paraffin at 160°C for 5–10 minutes in a glycerol bath. Disperse dyes can be partially stripped and then partially transferred to undyed secondary acetate.

5.8.4   Pigments[24,25]

Pigments can be identified using a microscope when the fiber sample is immersed in ethyl salicylate as an optical solvent. If the fibers are mass-pigmented, pigment particles are visible within the fiber. If the fibers are binder-bonded to the pigments in full depths, the particles are visible on the surface of the fiber. For pale depths, the test method listed in Section 5.8.5 can help.

Table 5.4 Dye test using *N*-methyl-2-pyrrolidone (NMP)

	Water layer	Toluene layer
Acid and direct dyes	Strongly colored	Colorless
Basic dyes	Colored	Colored
Disperse dyes	Colorless to slightly tinted	Colored similarly to the original dyeing
Azoic dyes	Colorless to slightly tinted	Colored darker than the water layer
Vat dyes	Colorless to slightly tinted	Colorless to slightly tinted

Azoic and vat types of pigments can be tested using the test methods mentioned above for the corresponding types of dyes. Phthalocyanine pigments show a bright green tone when spotted with concentrated nitric acid.

5.8.5   *N*-methyl-2-pyrrolidone extraction test[26]

Boil a 2.5 × 2.5 cm (1 × 1 inch) sample in 5 ml of 25% *N*-methyl-2-pyrrolidone (NMP) in a test tube immersed in a water bath for 20–40 seconds. Take the test tube out of the water bath, remove the sample, and add 5 ml toluene and 25 ml of water. Shake well, record the color of each layer after standing for 30 minutes or until settling, and check against the results shown in Table 5.4.

5.9   Finishes testing

5.9.1   Formaldehyde

Spot test[27]

Place one drop of 1% phenylhydrazine in 40% sulfuric acid on a white or pale colored fabric sample and let it sit for 30 seconds. Apply a drop of 10% ferric chloride solution. Within 30 seconds, a pink, red, or brown color shown indicates the presence of formaldehyde.

Free formaldehyde detection using cold sulfite[28]

Weigh a 1.5 g fabric sample accurately, cut it into small pieces, place them into a conical flask with 50 ml water at <5°C, and add 1 ml 2*N* sodium sulfite solution and four drops of thymolphthalein indicator. If the mixture shows no color, add 2*N* sodium carbonate slowly until a blue color is shown. The mixture is then neutralized to colorless with 0.3N hydrochloric acid and stood in an ice bath for 7 minutes. Add 5 ml ice-cold 1N acetic acid and a few drops of starch indicator. Titrate the excess sulfite using 0.01N iodine solution to the blue end point. Then:

$$\text{Free formaldehyde \%} = \frac{\text{volume of iodine} \times N \text{ of iodine} \times 1.5}{\text{weight of sample, g}} \times 100\%$$

Free and released formaldehyde in a sealed jar[29]

Suspend 1 g fabric sample in a sealed jar with 50 ml water. The jar is kept in an oven at 49°C for 20 hours or alternatively at 65°C for 4 hours. The free and released formaldehyde is absorbed in water and reacted with NASH reagent, which consists of 150 g of ammonia acetate, 3 ml of acetic acid, and 2 ml of acetylacetone and deionized water totaling to 1000 ml, at 58°C for 6 minutes. The amount of formaldehyde (μg/ml) is determined colorimetrically using the absorbance measured at 412 nm against a prepared calibration curve.

5.9.2 Silicones

Silicones are used widely as softeners and lubricants in textile finishes. The most common silicone is polydimethylsiloxane as shown in Fig. 5.6.

It was reported[30] that using a carefully selected fluorescent dye, the silicone finish on the dye-specific fiber can be detected by microscopy. To achieve the best results, Rhodamine B was selected for cellulose, polyester, and polypropylene fibers and Phloxin B for nylon fibers. Auramine can be used in some cases, but the fluorescence from the delustering agent such as TiO_2 used in the synthetic and regenerated fibers can interfere.

Silicones can be easily detected using FTIR ATR analysis. The representative IR spectrum is shown in Fig. 5.7.[31] Alternatively, silicone can be extracted using pentane in a Soxhlet extractor. The pentane is then evaporated and the residue taken up in carbon disulfide is ready for IR analysis.[32] The IR absorption peak at 1262 cm^{-1}, which represents the CH_3 symmetrical deformation of Si–CH_3, is the most characteristic absorption for organosilicon compounds. This band is intense and sharp and always occurs at 1262 ± 5 cm^{-1}. Therefore, it can also be used for quantitative analysis of silicones.[33]

5.6 Chemical structure of polydimethylsiloxane.

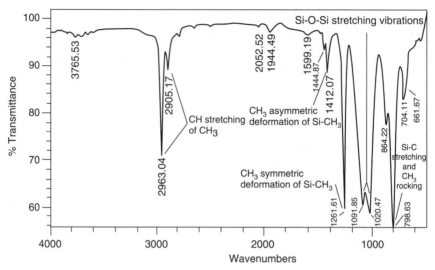

5.7 IR spectrum and peak assignments of polydimethylsiloxane.

5.9.3 Halogens[34]

Boil the fabric sample in deionized water to extract the halogen. Concentrate the water solution. Add a few drops of 1% silver nitrate into the concentrated extraction solution. The presence of a halogen is indicated by a white turbidity.

5.9.4 Hypochlorite[34]

Agitate a fabric sample in water at a low liquor ratio (<1:20) overnight at room temperature, then remove the sample. Test the extraction solution with 4% alkaline solution of thallous sulfate. A brown precipitate indicates the presence of hypochlorite.

5.9.5 Silicate[34]

Agitate a fabric sample in water at a low liquor ratio (<1:20) overnight at room temperature, then remove the sample. A drop of the extraction solution and a drop of ammonium molybdate solution (5 g ammonium molybdate in 100 ml of cold water poured into 35 ml of nitric acid) are placed on a filter paper and warmed gently over a hot plate. A drop of benzidine solution (0.5 g of benzidine dissolved in 10 ml of acetic acid and diluted with water to 100 ml) is added and the paper suspended over ammonia show a blue color indicating the presence of silicate.

Alternatively,[7] ash 5 g of a fabric sample in a crucible. After cooling, mix with five times the amount of 50/50 w/w mixture of sodium carbonate and potassium carbonate and heat until a clear melt is formed. Cooling down again, dissolve the solid in distilled water, mix with solid ammonium molybdate, and acidify with 20% nitric acid. An intense yellow color or a yellow precipitate indicates the presence of silicate.

5.10 Degradation testing

5.10.1 Oxidative damage to cellulose

Cellulose can be oxidized in the bleaching operation, generating many carboxylic groups that can bind basic dyes under acidic conditions. Two undyed fabric samples are dyed with methylene blue for 1 minute at boil at neutral and acidic pH, respectively. The oxidatively damaged cellulose shows a darker color with the acidic pH than that at neutral.[35] It was reported[36] that a test dye, Oxycarmine, can be four times as sensitive as methylene blue, and is specific for oxycellulose.

The other qualitative staining tests[7] for the evaluation of damage to cellulose due to the formation of oxycellulose and hydrocellulose are the Fehling's solution (brownish red color), Nessler's reagent (yellow then gray color), Schiff's reagent (red color), alkaline silver nitrate (gray to black color), and stannous chloride and gold chloride (purple color).

5.10.2 Alkali damage to wool[34]

The cystine disulfide link ($-CH_2-S-S-CH_2-$) in wool can be attacked by alkali and converted to a lanthionine sulfide link ($-CH_2-S-CH_2-$). Due to the variation in the sources and fineness of the wool, the absolute content of the cystine group in wool samples is variable. Therefore the damage analysis of wool fibers based on chemical principles requires the undamaged wool sample as a comparison reference. Otherwise, the test cannot give quantitative results.

A common test for estimating the alkaline damage of wool is the urea/bisulfite solubility test. Urea can cleave the hydrogen bonds and bisulfite the bisulfide bonds. The testing solution contains 50 g urea, 3 g sodium metabisulfite and 2 ml of $5N$ sodium hydroxide per 100 ml. The alkaline-damaged wool shows only <10% weight loss in the test due to the reduced amount of cystine disulfide links after the alkaline attack. The undamaged wool can have 40–50% weight loss, while the acid-damaged wool can have a weight loss of more than 80%.[35]

Alternatively,[37] Kiton Red G dye (C. I. Acid Red 1) can be used to test the damaged wool. Mix 1 g of wool fiber for 10 minutes with 100 ml of 0.1%

Kiton Red G acidified with 5 ml of $0.1N$ HCl. Undamaged wool shows mainly as uncolored, while damaged or chlorinated wool shows a red color, of which the intensity is proportional to the degree of the damage or chlorination up to 50%. A well-chlorinated sample can show 25–35% damage.

5.10.3 Damage to elastane (spandex)

The most common damage to elastane fibers is due to heat at temperatures above 170°C and chlorine attack. The result of such damage is loss of elasticity as well as strength reduction. Therefore the damage analysis of elastane fibers can be performed using a universal material tester, such as the Instron tester. The following methods can be adopted to evaluate damage to textile materials with elastane components:

- Stress (strength)–strain (elongation) curve of the elastane fiber: ASTM D2653-07, Standard Test Method for Tensile Properties of Elastomeric Yarns (CRE Type Tensile Testing Machines)
- Elastic properties test of the elastic yarns: ASTM D2731-07, Standard Test Method for Elastic Properties of Elastomeric Yarns (CRE Type Tensile Testing Machines)
- Stretch properties of fabrics: ASTM D3107-07, Standard Test Methods for Stretch Properties of Fabrics Woven from Stretch Yarns.[38]

5.11   Conclusions

Fabric chemical testing is a task requiring a good understanding of chemistry as well as of textiles. Because many textile chemicals are applied to substrates based on the affinities between the textile material and the chemical, conducting a fiber identification first can often provide a good direction towards the following dye/chemical analysis. However, it is not the intention of this chapter that the test methods presented can be applied directly without further reading the details listed in the references. Instead, this chapter serves only as an introduction to the test methods, and to give guidance to anyone who wants to perform the test what could be done to start with. It is also worth mentioning that this chapter in no way delivers an exhaustive chemical test list for textiles, though some of the tests are equally applicable to fabrics, yarns, and fibers. Therefore, readers are greatly encouraged to access the further reading and references to find out more (and current) developments in the field of textile chemical testing.

With the advent of more and more affordable computer technologies, many traditional wet chemical testing methods are being replaced by instrumental analyses. Some of the testing methods presented may well be conducted using the newly developed instruments. In addition, the principles and chemistry of testing textiles can also be different from those presented. The latest research and development in nanotechnologies for textiles has challenged current testing methods with regard to how textiles processed with nanotechnology can be effectively tested in terms of quality, performance, and safety. The most critical issues would be the environmental, health and safety (EHS) concerns. The US government has just released a research strategy on the topic,[39] stating that 'EHS research is focused in particular on understanding general mechanisms of biological interaction with nanomaterials and on developing broadly useful tools and tests for characterizing and measuring nanomaterials in various environments, including in the body'. Therefore, it is envisioned that related testing methods will come to the market following the strengthened research efforts in this area. Readers concerned with textile ecology should read the Oeko-Tex Standard 100[40] revised in January 2008 to understand how ecological issues are addressed in textile chemical testing. Oeko-Tex Standard 100 tests for harmful substances such as formaldehyde, heavy metals, pesticides, chlorinated phenols, allergic dyes, flame retardants, biologically active products, volatile organic compounds, etc., as well as for skin-friendly pH (4 to 7.5) on four classes of textiles: baby; direct skin contact; non-direct skin contact; and decoration. After a textile product has successfully passed all Oeko-Tex Standard 100 tests, the certified lab which conducted the conformity tests can issue the Oeko-Tex 100 label that can then be placed on the product, showing the eco-friendly nature of the textile goods.

5.12   Sources of further information and advice

For more detailed information and testing methods as well as for further developments of new and updated tests and standards, readers should access the following:

- AATCC, American Association of Textile Chemists and Colorists, *AATCC Technical Manual*, updated annually, http://www.aatcc.org/
- ASTM, American Society for Testing and Materials, *Annual Book of ASTM Standards, Section Seven, Textiles*, http://www.astm.org
- BSI, British Standards Institution, http://www.bsi-global.com/en/ Standards-and-Publications/
- C.I., Color Index, jointly published by the Society of Dyers and Colourists (SDC, UK) and the American Association of Textile Chemists and Colorists (AATCC, USA), http://www.colour-index.org/
- EN, European Standards, http://www.cen.eu/catweb/cwen.htm
- ISO, International Organization for Standardization, TC 38: *Textiles*, and ICS 59: *Textile and Leather Technology*, http://www.iso.org/
- JIS, Japanese Industrial Standards, http://www.webstore.jsa.or.jp/ webstore/JIS/SearchEn.jsp?lang=en

- Oeko-Tex Standard 100, http://www.oeko-tex.com/en/main.html
- Textile journals: *AATCC Review, Coloration Technology, Melliand Textilberichte, Textile Research Journal, Journal of the Textile Institute, Dyes and Pigments*, etc.
- Textbooks on analytical chemistry, inorganic chemistry, organic chemistry, polymer chemistry, instrumental analysis, etc.

5.13   References

1. J. Park and J. Shore, *Dyeing Laboratory Manual*, Roaches International Ltd, Leek, UK, 1999
2. *AATCC Technical Manual*, AATCC, Research Triangle Park, NC, 2008
3. Q. Fan, *TES 303 Dyeing, Printing & Finishing Lab Manual*, internally used teaching materials, University of Massachusetts, Dartmouth, MA, 2001
4. AATCC Test Method 81-2006, *AATCC Technical Manual*, AATCC, Research Triangle Park, NC, 2008
5. AATCC Test Method 110-2005, *AATCC Technical Manual*, AATCC, Research Triangle Park, NC, 2008
6. AATCC Test Method 79-2007, *AATCC Technical Manual*, AATCC, Research Triangle Park, NC, 2008
7. *A Bleachers' Handbook*, Solvay Interox, 1998 (service booklet without editor and place of publication)
8. J. H. Hoffman, Qualitative spot tests, in *Analytical Methods for a Textile Laboratory*, 3rd edition, AATCC, Research Triangle Park, NC, 1984
9. Q. Fan, Analysis of chemicals used in fiber finishing, in *Chemical Testing of Textiles*, Woodhead Publishing, Cambridge, UK, 2005
10. H. B. Goldstein, *American Dyestuff Report*, **36**, 629, 1947
11. S. Budavari, M. O'Neil *et al.*, *The Merck Index*, 12th edition (CD-ROM), Merck & Co., 2000
12. Joint FAO/WHO Expert Committee on Food Additives, *Compendium of Food Additive Specifications. Addendum 5* (FAO Food and Nutrition Paper – 52 Add. 5), Food and Agriculture Organization of the United Nations, Rome, 1997
13. J. H. Hoffman, Qualitative spot tests, in *Analytical Methods for a Textile Laboratory*, 3rd edition, AATCC, Research Triangle Park, NC, p. 200, 1984
14. J. H. Hoffman, Qualitative spot tests, in *Analytical Methods for a Textile Laboratory*, 3rd edition, AATCC, Research Triangle Park, NC, p. 197, 1984
15. A. J. Gordon and R. A. Ford, *The Chemist's Companion*, John Wiley & Sons, New York, 1972
16. F. J. Welcher, *Standard Methods of Chemical Analysis*, 6th edition, Vol. 2, Part B, Van Nostrand Reinhold, New York, 1963
17. ASTM D3362, Standard test method for purity of acrylate esters by gas chromatography, *Annual Book of ASTM Standards*, Vol. 06.04, 2000
18. A. S. Weatherburn, Determination of the ionic type of synthetic surface active compounds, Canadian Textile Journal, **71**(16), 45, 1954
19. *Analysis of Surfactants*, edited by Institute of Surfactants Analysis, Saiwai Shobo, Japan, 1980
20. M. J. Rosen and H. A. Goldsmith, *Systematic Analysis of Surface-active Agents*, 2nd edition, John Wiley & Sons, New York, 1972

21. T. U. Marron and J. Shifferli, Direct volumetric determination of the organic sulfonate content of synthetic detergents, *Industrial and Engineering Chemistry, Analytical edition*, **18**, 49, 1946

22. C. Kortland and H. F. Dammers, Qualitative and quantitative analysis of mixtures of surface-active agents with special reference to synthetic detergents, *Journal of the American Oil Chemists' Society*, **32**, 58, 1955

23. ASTM Method D2024-65 (2003), Standard test method for cloud point of nonionic surfactant, *Annual Book of ASTM Standards*, Vol. 15.04, 2003

24. C. H. Giles, M. B. Ahmad *et al.*, Identification of the application class of a colorant on a fibre, *Journal of the Society of Dyers and Colourists*, **78**(3), 125, 1962

25. R. V. R. Subramanian and K. S. Taraporewala, Identification of application classes of dyes on man-made fibres, *Journal of the Society of Dyers and Colourists*, **88**(11), 394, 1972

26. D. Haigh, The identification of dyes on secondary cellulose acetate, *Journal of the Society of Dyers and Colorists*, **80**(9), 479, 1964

27. W. J. van Loo Jr. *et al.*, A rapid spot test for the identification of aminoplasts on textiles, *American Dyestuff Reporter*, **45**, 397, 1956

28. S. H. Yoon, Determination of formaldehyde, in *Analytical Methods for a Textile Laboratory*, 3rd edition, AATCC, Research Triangle Park, NC, p. 253, 1984

29. AATCC Test Method 112-2003, *AATCC Technical Manual*, AATCC, Research Triangle Park, NC, 2008

30. W. Schindler and P. Drescher, Fluorescence marking of applied silicones to control their distribution, *Melliand Textilberichte*, **80**(1/2), E20, 1999

31. http://www.wcaslab.com/GIF/FTIR2.GIF, accessed 15 February 2008

32. A. L. Smith, *The Analytical Chemistry of Silicones*, John Wiley & Sons, New York, p. 85, 1991

33. S. Varaprath, D. H. Stutts and G. E. Kozerski, A primer on the analytical aspects of silicones at trace levels – challenges and artifacts – A review, *Silicon Chemistry*, **3**(1/2), 79–102, 2006

34. J. H. Hoffman, Qualitative spot tests, in *Analytical Methods for a Textile Laboratory*, 3rd edition, AATCC, Research Triangle Park, NC, p. 179, 1984

35. R. S. Merkel, Analyzing damage, in *Analytical Methods for a Textile Laboratory*, 3rd edition, AATCC, Research Triangle Park, NC, p. 51, 1984

36. W. Schindler and E. Finnimore, Chemical analysis of damage to textiles, in *Chemical Testing of Textiles*, Woodhead Publishing, Cambridge, UK, p. 177, 2005

37. S. R. Trotman, H. S. Bell and H. Saunderson, Properties of chlorinated wool and the determination of damage in chlorinated knitted woolen goods, *Journal of the Society of Chemical Industry*, **53T**, 267, 1934

38. *ASTM Book of Standards*, Vol. 07.01, ASTM International, West Conshohocken, PA, http://www.astm.org

39. Subcommittee on Nanoscale Science, Engineering, and Technology, Committee on Technology, National Science and Technology Council, *Strategy for Nanotechnology-related Environmental, Health, and Safety Research*, http://www.nano.gov/NNI_EHS_Research_Strategy.pdf, Arlington, VA, 13 February 2008

40. International Association for Research and Testing in the Field of Textile Ecology, Oeko Tex Standard 100, http://www.oeko-tex.com/xdesk/ximages/470/16459_100def2007.pdf, Zürich, Switzerland, 8 January 2008

6

Fabric appearance testing

X BINJIE and J HU, The Hong Kong
Polytechnic University, China

Abstract: Fabric appearance is always considered to be one of the most important aspects of fabric quality. Testing for fabric appearance is the process of inspecting, measuring and evaluating characteristics and properties of a fabric surface; the purpose of testing and evaluation is to assess the performance of a fabric or predicate its performance in conjunction with its end use. It can be performed according to well-defined test methods by industry-wide organizations. There are many aspects of fabric appearance that can contribute to a woven or knit fabric's overall value and consumer satisfaction. In this chapter, an attempt has been made to explain the testing methods for assessing various fabric appearance attributes.

Key words: fabric appearance, pilling evaluation, wrinkling testing, seam puckering, dimensional stability, light reflectance, objective evaluation of fabric appearance.

6.1   Introduction

Fabric appearance is always considered to be one of the most important aspects of fabric quality. However, the definition of fabric appearance is quite complicated; it is a general term including the visible properties of fabric material universally, which is related to many factors, such as structure, material properties, surface morphology and reflectance properties. Usually, fabric appearance can be described separately in terms of different attributes: pilling, wrinkling, seam puckering, fuzziness, dimensional change and luster. These kinds of typical attributes are always happening and observed during the daily wearing or washing of fabric products.

Fabric appearance testing is the process of inspecting, measuring and evaluating characteristics and properties of a fabric surface; the purpose of testing and evaluation is to assess the performance of a fabric or predicate its performance in conjunction with its end use. Fabric appearance testing can be performed according to well-defined test methods by industry-wide organizations, such as AATCC, ASYM, ISO, etc. The testing methods can be divided into two main types: subjective and objective. The differences between these two types of methods rely on whether they are based on visual assessment or on digital assessment with the aid of testing machines. Subjective methods have the advantages of simplicity and low cost of equipment; however, they tend to be time consuming, inconsistent and labor-intensive. Although objective methods could avoid these disadvantages, they are not widely used in practice in the industry.

In some cases, woven fabrics and knitted fabrics might have different standards for the evaluation of appearance in view of their different material properties and internal structures. There are many aspects of fabric appearance that can contribute to a knit fabric's overall value and consumer satisfaction. For woven fabrics, wrinkling and seam puckering evaluation seems to be considered much more than other factors; for knitted fabrics, usually there are four key factors: facing-up, wrinkling/surface smoothness, cockling/loop distortion in knits, and spirality. In this chapter, we will give a brief introduction to these testing methods for assessing fabric appearance in terms of those attributes.

6.2 Appearance testing: pilling testing and evaluation

Pilling is a phenomenon of fiber movement or slipping out of yarns, which is usually happening on the fabric surface during abrasion and wear. The development of pilling could be divided into four stages: fuzz formation, entanglement, growth, and wear-off. The formation of fuzz and pills suspended on the fabric surface could affect the fabric aesthetics and its ultimate acceptance by customers. Many researchers are investigating how to improve the pilling resistance ability of cloth, including the optimization of fiber manufacture, yarn manufacture and fabric manufacture. In this case, the standards and testing methods for evaluating the pilling grade of cloth are very important in guiding the technology of pilling resistance optimization. The photographs in Fig. 6.1 show standard pilling images of woven fabrics from grade 1 to grade 5 (see Table 6.1).

Facing-up [1] can be defined as the generation of unwanted surface fibers leading to a change in the appearance of the garment, as illustrated in Fig. 6.2. Facing-up is one kind of surface change which is less serious than pilling. Facing-up is normally associated with knitted yarns made from worsted and possessing a clean surface. Fabrics and knit made from wool spun yarns generally are finished to create a fuzzy or hairy surface and during wear/abrasion the reverse effect is sometimes an issue. Facing-up can occur all over a garment or in localized areas. It is caused by the gradual withdrawal of fibers from the surface layer of yarns (migration) and can in many cases lead to pilling. The migration of surface fibers from the body of the fabric to the surface is due to frictional forces applied to the fibers on contact with other surfaces, which might also include the same fabric. Consequently, facing-up tends to occur when surface abrasion forces are high such as during tumble drying.

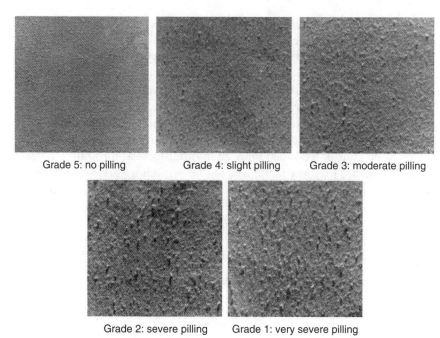

Grade 5: no pilling   Grade 4: slight pilling   Grade 3: moderate pilling

Grade 2: severe pilling   Grade 1: very severe pilling

6.1 Pilling standards of woven fabrics.

Table 6.1 Visual assessment of pilling

Grade	Description
5	No change
4	Slight surface fuzzing and/or partially formed pills
3	Moderate surface fuzzing and/or moderate pilling. Pills of varying size and density partially covering the specimen surface
2	Distinct surface fuzzing and/or distinct pilling. Pills of varying size and density covering a large proportion of the specimen surface
1	Dense surface fuzzing and/or severe pilling. Pills of varying size and density covering the whole of the specimen surface

6.2.1 Fabric pilling: standards

Fabric pilling or related surface change is commonly tested in the laboratory using specific machines by generating pilling on the fabric by simulating wear. A sample of the original fabric is fixed in the machine and wear is simulated by the action of abrasive materials. Generally, the machines are supplied with a standard reference consisting of photographs of samples

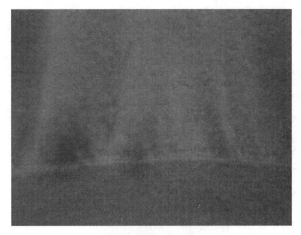

6.2 Facing-up of wool knits.

with different degrees of pilling. The abraded fabric is then compared with standard photographs that have been developed by the standards institutions such as ASTM, AATCC, IWS, BIS, JIS, etc., and a degree of pilling is assigned accordingly.

6.2.2 Fabric pilling: instruments

The most popular abrasion machines (Fig. 6.3) to simulate wear conditions in the market are the ICI pilling box tester, the Martindale tester, and the random tumble pilling tester. The methods of abrasion differ in the following ways.

- *ICI pilling box tester.* Specimens are mounted on the polyurethane tubes and tumbled randomly in a cork-lined box for a certain time.
- *Martindale tester.* Flat abrasion as specified in the ASTM D4970 pilling test. The instrument subjects specimens to a rubbing motion in a straight line that widens into an ellipse and gradually changes into a straight line in the opposite direction. This pattern of rubbing is repeated until fabric threads are broken or until a shade change occurs in the fabric being tested.
- *Random tumble pilling tester.* The specimen is placed in a cylindrical chamber and tumbled around within the chamber which is lined with mildly abrasive materials to brush the specimens to free fiber ends.

The method of abrasion has a significant effect on the pilling appearance. In comparing abrasion instruments, it was noted that the test results could be affected by instrument type, as well as instrument settings. Therefore, the

ICI pilling box   Martindale tester

Random tumble pilling tester

6.3 Pilling testers.

choice of pilling tester should depend on the consideration of material properties and the end-use of products.

6.2.3 Fabric pilling: evaluation

Subjective evaluation

Once the fabric has been subjected to abrasive action of some kind, subjective or objective evaluation can be carried out to assess the degree of fabric pilling. Subjective evaluation is performed by comparing the tested specimens with visual standards, which may be actual fabrics or photographs of fabrics, showing a range of pilling resistance. The observed resistance to pilling is reported on an arbitrary scale ranging from 5 (no pilling) to 1 (very severe pilling) (see Table 6.1). The viewing condition is shown in Fig. 6.4; it is a sketch map of apparatus for fabric appearance evaluation, also adopted for other appearance evaluation.

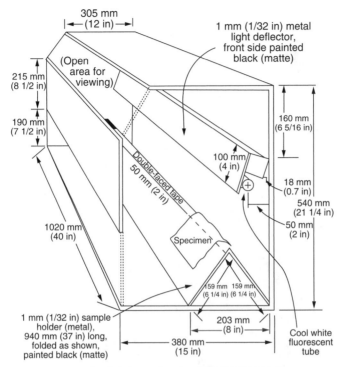

305 mm
(12 in)

1 mm (1/32 in) metal
light deflector,
front side painted
black (matte)

(Open area for viewing)

215 mm
(8 1/2 in)

190 mm
(7 1/2 in)

160 mm
(6 5/16 in)

Double-faced tape
50 mm (2 in)

100 mm
(4 in)

18 mm
(0.7 in)

540 mm
(21 1/4 in)

1020 mm
(40 in)

Specimen

50 mm
(2 in)

159 mm   159 mm
(6 1/4 in) (6 1/4 in)

1 mm (1/32 in) sample
holder (metal),
940 mm (37 in) long,
folded as shown,
painted black (matte)

203 mm
(8 in)

380 mm
(15 in)

Cool white
fluorescent
tube

6.4 Apparatus for fabric evaluation (ASTM 3514-02).

Objective evaluation

Considerable research has been undertaken on the objective evaluation of fabric pilling. These methods are mainly based on digital technologies, such as digital image analysis or the laser scanning method. The laser scanning system applies the laser triangulation technique to measure the 3-D height field of the fabric surface and to identify pills or fuzzes through the variation of its height; or a 2-D CCD/CMOS imaging system can be used to digitalize the reflectance properties under a certain illumination by using the intensity of image pixels, and some algorithms are then developed to characterize the pills or fuzzes to distinguish them from the common texture. In the latter case, the algorithms rely greatly on the design of the imaging style and the setting of the imaging condition. In comparison with laser scanning, the CCD/CMOS imaging system is simple, easy to implement and low-cost; however, it is much more sensitive to the color and texture of fabric samples. Konda *et al.* [2] first attempted to use image processing techniques to evaluate fabric pilling (Fig. 6.5); Xu [3] developed an image analysis system that aims at characterizing and rating fabric pilling appearance using Fast Fourier Transform (FFT) and other related image-processing techniques; Hsi *et al.*

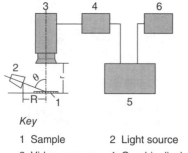

1 Sample   2 Light source
3 Video camera   4 Graphic display
5 Computer   4 Printer

6.5 Two-dimensional image analysis system for fabric pilling evaluation (from Ref. 2).

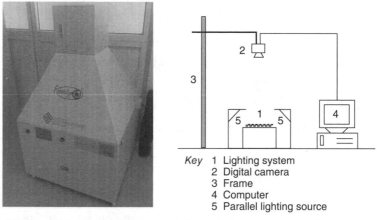

Key 1 Lighting system
     2 Digital camera
     3 Frame
     4 Computer
     5 Parallel lighting source

6.6 Two-dimensional image capturing system.

[4] developed a hardware device and software based on image analysis techniques to detect and describe fuzz on fabric surfaces; and Kang *et al.* [5] developed a non-contact 3-D measurement method for the objective evaluation of fabric pilling based on stereovision.

6.2.4 Image capturing

The 2-D image capturing system used for pilling evaluation is illustrated in Fig. 6.6 and the 3-D surface reconstruction system based on stereovision is illustrated in Fig. 6.7. The lighting box is designed to illuminate fabric samples uniformly, simulating daily lighting conditions. Each fabric sample was cut into lengths of approximately 105 mm, and put on the testing platform of the imaging system for the purpose of digitalization.

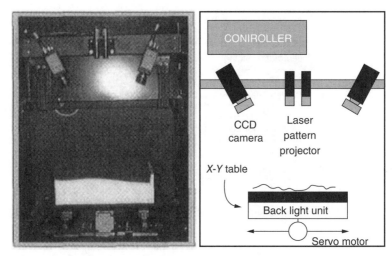

6.7 Stereovision system for fabric pilling evaluation (from Ref. 4).

The principle of the two-dimensional image capturing system is very simple. The image of the cloth surface is digitalized by a kind of photoelectric sensor called a CCD (charge coupled device). However, the principle of 3-D reconstruction is a little more complex and the hardware is also more expensive than for a 2-D imaging system; stereovision is based on the geometry of the camera and can calculate the 3-D coordinate of each surface point accurately. In this chapter, we introduce the objective evaluation method based on 2-D image analysis technology.

6.2.5 Method for pilling evaluation

After an image is acquired by hardware, two major functions are implemented by the algorithms of software, including the identification of pills/fuzzes and the measurement of their features. A general scheme of the image analysis procedure is presented in Fig. 6.8, where the processing steps are integrated in blocks.

Pill template training

The remarkably varied structure of fabrics creates a challenge for detecting pills if a wide variety of fabrics is considered. However, pills are spaced and distributed relatively far apart and are elevated compared with the yarn body in fabrics. By illuminating fabrics at an oblique angle, pills are much brighter than the ground of a fabric. For image analysis, pills can be regarded as elliptical or circular objects that contain a centered white circle

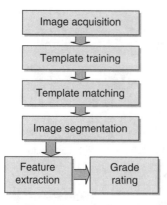

6.8 Flow diagram of pilling evaluation.

surrounded by black pixels. In this case, a two-dimensional Gaussian function should be the most suitable template for pill detection.

Template training

Actual pill images from fabric images were used for the parameter estimation of the Gaussian function; σ_x and σ_y are the most important parameters to generate the pill template. As a criterion of optimality, the mean square error (MSE) is used to describe the estimation reliability. Here,

$$\mathrm{MSE} = \sqrt{\frac{\sum\limits_{i=1}^{n}(z_e - z_i)^2}{n-1}}$$

where z_e is the estimated function and z_i is the original data. When σ_x is equal to σ_y, the shape of the pill template will be circular, otherwise the shape will be elliptical. It is found that the value of σ_x and σ_y influences the contrast of the pill template: the lower it is, the more contrast there will be.

Template matching

Template matching is the process of moving the template over the entire image and calculating the similarity between the template and the covered window on the image. Template matching is implemented through two-dimensional convolution. In convolution, the value of an output pixel is computed by multiplying elements of two matrices and summing the results. One of these matrices represents the image itself, while the other matrix is the template, which is known as a convolution kernel.

If the original image and the template are denoted $f(x, y)$ and $t(x, y)$, and the enhanced image after convolution is denoted $g(x, y)$, the convolution may be abbreviated by:

$$g(x, y) = f(x, y) \circ t(x, y) \tag{6.1}$$

where $\circ$ is used to indicate the convolution of two functions. Discrete convolution is given by

$$g(i, j) = \sum_{m=1}^{M} \sum_{n=1}^{N} t(m, n) \cdot f(i - m, j - n) \tag{6.2}$$

where M, N is the size of the template.

Image segmentation

After template matching, an enhanced fabric image is generated for pill evaluation: The bright areas in the filtered image indicate pill locations; most of the pixels of the filtered image are of low gray level and belong to the background. In this step, we intend to segment pills from the background in order to obtain a binary image with white pixels (pills) on a black background. Choosing a suitable threshold is the key problem for image segmentation.

If we examine the histogram of the filtered image after template matching, as shown in Fig. 6.9, a single peak corresponding to the background can be observed in the center of the histogram (solid line). We model the histogram curve with a 1-D Gaussian distribution (dashed line); x_0 is the position of the peak center and σ is the standard deviation of the Gaussian function. We calculate the threshold t according to the formula:

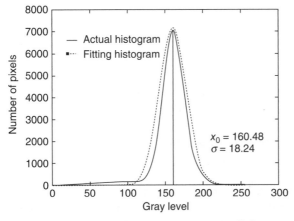

6.9 Histogram of the filtered image and its fitting curve.

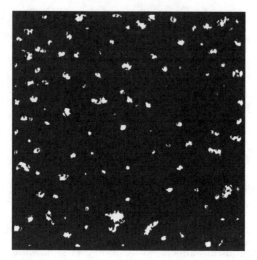

6.10 The binary image after a global binarization.

$$t = x_0 + \lambda \cdot \sigma \qquad\qquad 6.3$$

where λ is the coefficient which determines the threshold. Consider a normal population with mean x_0 and standard deviation σ; the interval $[x_0 - 3\sigma, x_0 + 3\sigma]$ contains 99.73% of the population values, so $\lambda = 3$ in this chapter. The result obtained by applying the threshold defined by the above formula is shown in Fig. 6.10.

Feature extraction

In subjective evaluation, judges tend to rate the pilling appearance of a fabric by comparing pill properties such as number, area (size), contrast and density. All these properties can be measured in the binary fabric images objectively using image analysis techniques.

Number
Pill number n is an important factor influencing pilling appearance. It can be measured by counting the number of white objects in the binary image. Note that very small objects with area less than 4 pixels should be removed, as these are considered to be noise objects.

Area (size)
The area s_i of one pill is expressed in terms of the number of pixels with 255 gray values that make up this pill, where i is the label number of the pill. Then if $\bar{s}$ is the mean area of the pills, the total area S can be calculated

as the product of the pill number n and the mean area $\bar{s}$. The size of one pill d_i is expressed in terms of its equivalent diameter in pixels, and $\bar{d}$ is the mean size of pills. Since d_i has a good correlation with s_i, both can be used to describe both big and small pills.

Contrast

Pill contrast is a measure of how much the gray level of a pill differs from the gray level of the base fabric in the fabric image. It is calculated by the following expression:

$$C_i = \frac{g_{p-i}}{g_b} \qquad\qquad 6.4$$

where g_{p-i} is the mean gray level of pill i, and g_b is the mean gray level of the background. $\bar{C}$ is the mean value of C_i.

Density

Pill density is another important factor influencing pilling appearance; a reasonable estimator of pill density has been introduced in Ref. [3]. First, a number of points were randomly generated inside the binary image. At each of these points i, a pill closest to the point was searched and the distance r_i between this pill and the point can be calculated. Then another pill nearest this point was found and the distance between these two pills x_i can also be calculated. After n random points have been counted, the population-density estimator D can be determined by the following equation:

$$D = \frac{\sqrt{2n}}{\pi\sqrt{\sum(r_i^2)\sum(x_i^2)}} \qquad\qquad 6.5$$

This estimator is insensitive to the clumping pilling, while the density estimated by the number of pills in a unit area is sensitive to it.

Grade rating

In order to make the rating result by the image analysis techniques consistent with the visual standards, five samples are selected as the initialization of system training, which are subject to different grades (from grade 1 to grade 5) according to ASTM standards. Since the pilling of textile fabrics is affected by many factors, such as type of fiber or blends, fiber dimensions, yarn and fabric construction, and fabric-finishing treatments, it is better to select the same fabric as the standard image to make the correlation between the objective evaluation and subjective evaluation. Figure 6.11 shows the images of five grades and their binary images after image segmentation. Table 6.2 shows the parameters extracted from these images.

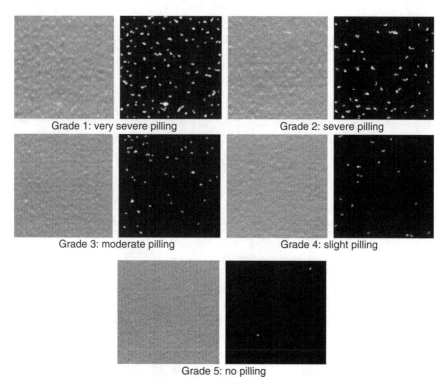

Grade 1: very severe pilling   Grade 2: severe pilling

Grade 3: moderate pilling   Grade 4: slight pilling

Grade 5: no pilling

6.11 Standard pilling images.

Table 6.2 Pill parameters of standard pilling images

Image no.	n	$\bar{s}$ (pixel)	S (pixel)	$\bar{C}$	D (1/pixel²)
1	109	93.3395	10 174	1.2926	0.000029
2	57	63.7368	3 633	1.26858	0.000015
3	46	54.2609	2 496	1.27377	0.000012
4	21	48.6667	1 022	1.20683	0.000004
5	2	45.5	91	1.29234	0.000000

From Table 6.2, it is easily found that pill number n, mean area $\bar{s}$, total area S and density D decrease with subjective grade decrease, while mean contrast $\bar{C}$ has no correlation with subjective grade. In this case, the empirical rating equations for the pilling grade are established for these four image features by using the linear regression method. Among them, the pill total area S is the highest, so for simple calculation, S can be chosen in our research to build the rating formula, which is

$$y = -7\text{E}-12x^3 + 2\text{E}-07x^2 - 0.0012x + 5.0904$$

where x is the pill total area of the image, and y is the objective pilling grade.

Having established the relationship between pilling grade and objective parameters, the objective parameters of the other 40 samples used in our experiment were measured. In order to identify the efficiency of objective evaluation, the grade values using the four rating formulas were compared with the subjective grade, and satisfactory agreement can be achieved in our experiment, as shown in Fig. 6.12. There are four objective grade values for each sample calculated by different rating formulas, for total area, pill number, mean area and density. Three have good correlation with the subjective grade (0.8468, 0.8374, 0.8421), except the objective grade from mean area (0.4715). It is an interesting finding that the parameter mean area has a good correlation with the standard grade (0.9993) but a poor correlation with the practical grade from judges (0.4715). This proves that judges are insensitive to the mean area of pills, and more sensitive to pill number, total area and density; the pill number and total area are more important factors for subjective evaluation, so is the distribution density of pills.

Although much valuable research work has been done in recent years, the objective evaluation of pilling appearance is still at the stage of academic research; although some commercial prototypes have been developed, the algorithms to eliminate the color and pattern effects and improve the system reliability should be optimized in the future.

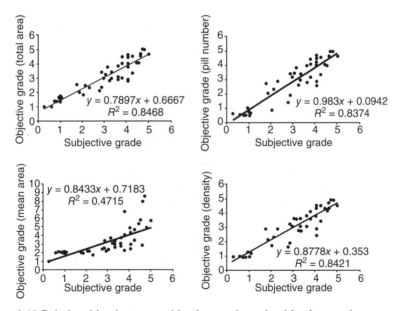

6.12 Relationships between objective grade and subjective grade.

6.3   Fabric wrinkling testing and evaluation

Wrinkles are three-dimensional versions of creases, and form when fabrics are forced to develop high levels of double curvature, which result in some degree of permanent in-plane and out-of-plane deformations. Most fabric will generate some wrinkles after laundering, dressing, and folding. These wrinkles seriously compromise the cloth's acceptability. For example, for most customers it is an unpleasant feeling to be wearing a wrinkled shirt while attending some social activities. The photographs in Fig. 6.13 show standard wrinkling images of fabrics from grade 1 to grade 5.

6.3.1 Fabric wrinkling: standards

The method most often used in the laboratory to evaluate the wrinkle recovery of a fabric is AATCC test method 128, 'Wrinkle recovery of fabrics: appearance method'. The principle of this method is to induce wrinkles in the fabric under standard atmospheric conditions in a standard wrinkle device under a predetermined load for a certain period of time. Then the specimen is rated by comparing it with AATCC wrinkle replicas. AATCC test method 66, 'Wrinkle recovery of woven fabrics: recovery angle', is used to determine the wrinkle recovery of woven fabrics. It is applicable to fabrics made from any fiber, or combination of fibers. The principle is to fold the specimen and compress it under controlled conditions of time and force to create a folded wrinkle. The test specimen is then suspended in a test instrument for a controlled recovery period, after which the recovery angle is recorded. Usually, AATCC test method 128 is used to evaluate the wrinkling appearance of fabrics.

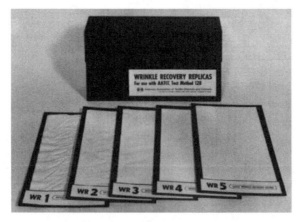

6.13 AATCC wrinkle replicas for gradings.

6.14 AATCC Wrinkle Tester.

6.3.2 Fabric wrinkling: instruments

The AATCC Wrinkle Tester illustrated in Fig. 6.14 is designed to introduce wrinkles in the test specimen under controlled conditions. This apparatus is based on a development of ENKA/AKU Research.

6.3.3 Fabric wrinkling: evaluation

Subjective evaluation

Once the fabric has been induced to be wrinkled under a certain load for a standard period of time, subjective or objective evaluation can be performed to assess the degree of fabric wrinkling. Subjective evaluation compares the tested specimens with wrinkle replicas, showing a range of wrinkling status. The observed wrinkling is reported on an arbitrary scale ranging from 5 (no wrinkling) to 1 (very severe wrinkling). The viewing conditions are shown in Fig. 6.15.

Objective evaluation

Objective evaluation of fabric wrinkling is also an interesting application of digital technology to the textile field. According to the way in which the wrinkled appearance is detected and measured, these evaluation methods could be classified into three main categories: the contact, laser scanning and image processing methods.

Contact method
A contact method-based instrument like the 'Wrinklemeter' [6] consists of a movable platform, a speed-controlled motor, a small probe linked to a

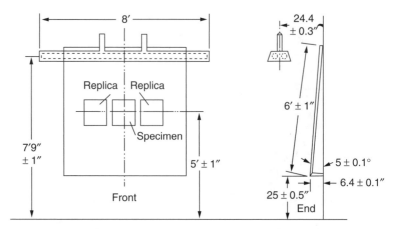

6.15 Apparatus for fabric evaluation (AATCC test method 128).

shutter, a light source, a photovoltaic cell and a signal recorder. The principle of this method is to record the surface profile of wrinkled fabric on the recorder paper, and calculate the mean wrinkle height of the fabric to characterize the degree of wrinkling. The wrinkled surface of a fabric could be regarded as a random rigid surface; for a particular cross-section of the 3-D surface profile, four parameters could be defined to describe the severity of wrinkling quantitatively: wrinkle height, wrinkle slope, density of zero points and density of extreme points [7]. However, this method is based on contact with the fabric surface; it might flatten the wrinkled surface under a certain pressure, in which case it could result in an error in the assessment of fabric wrinkle.

Laser scanning method

The laser scanning method is based on the laser triangulation or stereovision technique. It is accomplished by projecting a laser line or point on an object and then capturing its reflection with a sensor located at a known distance from the laser's source. The resulting reflectance angle can be interpreted to yield 3-D measurements of the part. Ramgulam *et al.* [8] used a laser scanning system to measure the 3-D surface profile of a wrinkled surface. The sample was placed and fixed on a platform of an *X–Y* table driven by step motors to do the laser scanning. Amirbayat and Alagha [9] applied the same 3-D laser scan method to measure the wrinkle replicate standards and extracted certain geometrical features – surface area, average surface length in *X* and *Y* directions, covering volume, average curvature and average maximum skewness – to characterize the fabric wrinkle status. In their research, the last two parameters could describe the wrinkle status reasonably, and the relationship between this method and standards could achieve 0.984. This method is not affected by color and pattern; however, the scanning speed is a little slow.

The three-dimensional scanning device developed by Park and Kang [10] to measure wrinkle degree also consists of a laser scanner to detect the magnitude of a wrinkle. On the basis of hardware, Kang and Lee [11] proposed a method based on fractal dimensions to characterize the severity of the fabric surface wrinkles. The surface contours of wrinkled fabrics or puckered seams were scanned using a laser scanning system, and the fractal dimensions of the surface were then counted using a box-counting method. Neural networks were applied by Kim [12] to judge the wrinkling grade based on 64 × 64 points of sampling data using a similar scanning system. The study revealed a linear relationship between objective evaluation and subjective evaluation results. Turner *et al.* [13] also proposed a laser-based surface profiling system that utilizes a smart camera to sense the 3-D topography of fabric specimens (see Fig. 6.16). The system incorporates methods based on anisotropic diffusion and the facet model for characterizing edge information that ultimately relate to a specimen's degree of wrinkling. It is reported that this system is as good as the current American Association of Textile Chemists and Colorists (AATCC) smoothness grading system.

Xu *et al.* [14] combined laser triangulation and image processing techniques to develop a laser profilometer for assessing fabric smoothness appearance. The profilometer consists of a laser line projector, CCD camera, rotating stage, computer and special software. They introduced the basic principle of laser triangulation, image processing techniques for extracting surface profiles, wrinkle characterization methods, and the results of a trial

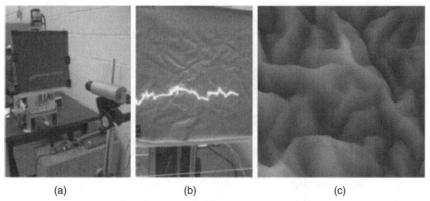

    (a)   (b)   (c)

6.16 (a) Overall view of the laser acquisition system; (b) laser line being projected onto a sample fabric; (c) sample range image (from Ref. 13).

test. The profilometer can generate results that are consistent with those of human observers, and the patterns and colors of the fabric do not affect the measurements. In addition, the profilometer is essentially insensitive to patterns of wrinkle orientations. This new evaluation method also solves a problem encountered in research on other instrumental evaluation techniques – the ability to discriminate differences in fabrics whose smoothness appearance falls between AATCC test method 124 replicas SA-3 and SA-3.5. This method is faster than Amirbayat and Alagha's laser scanning [9], but the resolution is so high that it does not give 3-D recovery for the whole fabric surface.

Image processing method
Image analysis is another objective evaluation method for the evaluation of fabric wrinkling; the low cost of image digitalization makes it easily acceptable by the textile industry. There has been much valuable research work on this kind of method. For example, Xu and Reed [15] digitalized the standard wrinkle images using a panel scanner and defined two variables – surface ratio and shade ratio – to quantify wrinkled appearance, then studied the correlation between these two parameters and subjective grades using a non-linear regression method to generate the objective evaluation judging grade. Na and Pourdeyhimi [16] analyzed the AATCC replicate standards using a combination of texture and profile analysis techniques. Their research shows that wrinkling can be reliably measured using gray level and surface statistics, co-occurrence analysis and power spectral density of image profiles. However, previous methods cannot work well when fabrics have different colors and irregular patterns, because 2-D gray images cannot describe the 3-D status of fabric surfaces and can confuse some color or pattern shades with surface height information. Fazekas *et al.* [17] designed some special illumination for fabric image capture (bi-illumination, Y-illumination and X-illumination). After preprocessing (intensity adjustment, image filter, histogram stretching), depth information for the fabric surface was calculated and good correlation between the experimental results and subjective evaluation could be achieved. Chen *et al.* [18] also digitalized fabric wrinkle images using a panel scanner, and applied a box-counting method to calculate the fractal dimension of gray level surface and the fractal dimension of wrinkle block area, both of which could be used for wrinkle evaluation.

Hu *et al.* [19] proposed a new method for measuring fabric wrinkle based on integrating photometric stereo and image analysis techniques. The photometric stereo method was used for the first time to extract 3-D surface profiles of fabrics and successfully eliminate the influence of color and pattern. Image analysis techniques were developed further to extract wrinkling features of these reconstructed 3-D images, and there was good correlation between objective parameters and subjective AATCC grades. Yang and Huang [20] proposed a similar approach to reconstruct the fabric 3-D surface profile from multi-illuminated images based on the photometric stereo method; they measured the wrinkling degree using four index values to indicate the variation of surface height.

Wrinkle characterization

No matter which method is used to digitalize the wrinkled surface, certain features are essential to characterize the wrinkling status of a fabric surface. According to the type of feature definition, wrinkle features can be divided into two major categories: geometrical features and textural features.

Geometrical features
Common geometrical features can be defined in terms of height, area, volume and its variance or statistical parameters:

- Wrinkle height and its variance is a measure of the height of wrinkles.
- Mean length of the paths over the wrinkled surface along the X and Y directions.
- Surface area and volume under the surface.
- Mean principal curvature.
- Mean maximum twist.
- Wrinkle roughness is a measure of the size of wrinkles, with no consideration of their shape, and is characterized by four different quantitative measures [14].
  - Arithmetic average roughness:

$$R_a = \frac{1}{n}\sum |Z_i - m|$$ 6.6

  - Root mean square roughness:

$$R_a = \sqrt{\frac{1}{n}\sum (Z_i - m)^2}$$ 6.7

In these two equations, Z_i is the height of the profile at the ith point, n is the number of points selected, and m is the height of the mean line which fits in the middle of the profile. Both these measures compute the average height of the wrinkles from the mean middle line.
  - Ten-point height R_z: the average distance between the five height peaks and the five lowest valleys on the curve.
  - Bearing length ratio t_p: a measure obtained by establishing a reference line parallel to the mean line at a predetermined height between the highest peak and the lowest valley of the profile. The line intersects the profile, generating one or more subtended lengths; t_p is the

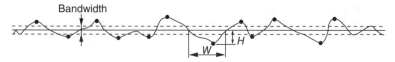

6.17 Illustration of bearing length ratio.

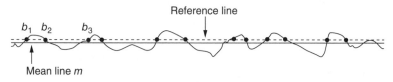

6.18 Illustration of sharpness and peak-and-valley count.

ratio of the sum of the subtended length to the sampling length of
the curve as illustrated in Fig. 6.17.

- Wrinkle sharpness k represents the shape of the wrinkle, describing the
  point of the wrinkle which forms a definite peak as illustrated in
  Fig. 6.18. The ratio of the height to the width of the wrinkle is used to
  quantify sharpness.
- Wrinkle density can be quantified by the peak and valley count, which
  is the number of peaks and valleys along the selected bandwidth symmetrical to the mean line of the profile. The selection of bandwidth is
  important to avoid tiny peaks and valleys which may correspond to
  noise signals.

Textural features

Usually the co-occurrence matrix and fractal dimension are used to describe
the textural properties. The co-occurrence matrix $M(d, \theta)$ consists of probability $P_\delta(i, j)$, in which the pixel of gray level i appears separated a distance
$\delta = (d, \theta)$ from the pixel of gray level j, where the parameters d and θ are
the distance and positional angle between the gray level pair i, j. The four
parameters and the fractal dimension D are defined as follows [21]:

- Angular second moment (ASM), which is the parameter for the evenness distribution degree of the texture:

$$\text{ASM} = \sum_{i=0}^{n-1}\sum_{j=0}^{n-1}\{P_\theta(i, j)\}^2, i = 1, 2, \ldots, n-1; j = 1, 2, \ldots, n-1 \qquad 6.8$$

- Contrast (CON), which is a measure of the gray level contrast:

$$\text{CON} = \sum_{k=0}^{n-1} k^2 P_{x-y}(k), |i - j| = k, k = 1, 2, \ldots, n-1 \qquad 6.9$$

- Correlation (COR), which is a measure of gray-tone linear dependencies:

$$\text{COR} = \frac{\sum_{i=0}^{n-1}\sum_{j=0}^{n-1} i \cdot j \cdot P_\delta(i, j) - u_x u_y}{\sigma_x \sigma_y} \qquad 6.10$$

- Entropy (ENT), which is a measure of the complex degree of the texture:

$$\text{ENT} = -\sum_{i=0}^{n-1}\sum_{j=0}^{n-1} P_\delta(i, j) \cdot \log\{P_\delta(i, j)\} \qquad 6.11$$

- Fractal dimension D:

$$D = \frac{\log(c) - \log(N(r))}{\log(r)} \qquad 6.12$$

where $P_x(i) = \sum_{i=0}^{n-1} P_\delta(i, j)$, $P_y(j) = \sum_{j=0}^{n-1} P_\delta(i, j)$, $P_{x-y}(k) = \sum_{i=0}^{n-1}\sum_{j=0}^{n-1} P_\delta(i, j)$, u_x, u_y, σ_x

and σ_y are the means and standard deviations of $P_x(i)$, $P_y(j)$, c is a positive constant, r is the side length of the cube and $N(r)$ is the number of cubes which cover the images.

Grading model

Generally speaking, the grading model establishes the relationship between the final wrinkling degree and those essential features defined previously to characterize the wrinkling status objectively. Linear regression, non-linear regression and neural networks are the three main methods to model this kind of relationship, as illustrated in Fig. 6.19.

Let $X = [x_1, x_2, \ldots, x_n]$ represent the vector of wrinkling features and G represent the final wrinkle degree; then after feature extraction using previous digital technology, including both contact and non-contact methods, one set of standard wrinkle replicas or specimens is needed to estimate the parameters of the grading model; this step is called training of the grading model. Linear regression, non-linear regression and neural network or another related non-linear fuzzy-logic method could be selected to model the relationship between X and G. Xu *et al.* [14] used the logarithmic equations corresponding to each wrinkling feature to establish the grading model in the form $G = a(\log x_i - b)$. This method is simple and effective, and can determine and correlate the essential features very easily.

Artificial neural network models or simple neural nets are known by different names such as connectionist models, parallel distributed models, and neuromorphic systems. They can be viewed as circuits of highly interconnected parallel processing units called neurons. The network is a directed

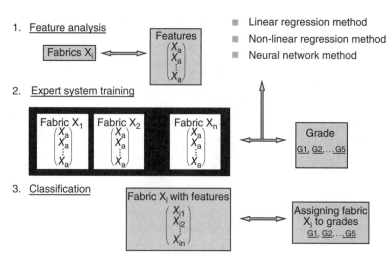

6.19 General flowchart of objective evaluation system.

graph consisting of nodes and edges. An artificial neural network (or neuro-computing) is based on the mathematical modeling that simplifies biological neurons and their connections. Neural networks have been widely used in the objective evaluation of fabric wrinkling [12, 21].

6.4   Seam puckering evaluation

Seam puckering consists of ridge, wrinkle or corrugation of the material, or a number of small wrinkles running across and into one another, which appear on sewing together two pieces of fabric. Seam pucker is regarded as a matter of primary concern in garment manufacture. When the sewing parameters are not appropriate to the material selected, puckering occurs along the garment seams, and the aesthetic appeal of the garment deteriorates. AATCC 88B from the American Association of Textile Chemists and Colorists establishes a set of photographic replicas for the subjective evaluation of seam puckering, as shown in Fig. 6.20.

There are two standard photographic replicas for seam puckering evaluation, one for single needle seams and the other for double needle seams. Seamed fabric specimens are subjected to standard home laundering practices. A choice is provided of hand or machine washing, alternative machine wash cycles and temperatures and alternative drying procedures. The test specimen is mounted on the viewing board with the appropriate photographic standard placed alongside. Evaluation is performed using a standard lighting and viewing area by rating the appearance of specimens in comparison with appropriate reference standards.

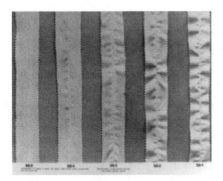
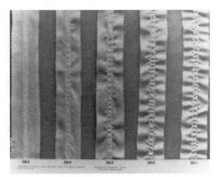

(a) For single needle seams   (b) For double needle seams

6.20 AATCC sample for seam gradings 5 down to 1.

6.4.1 Objective evaluation methods

The evaluation of seam puckering is quite similar to wrinkling evaluation in terms of surface profiling and digitalization. Techniques for the objective characterization of seam pucker may be classified into two categories: contact and non-contact methods, in which the non-contact tests could be further divided into laser scanning methods and image analysis methods. Since the non-contact type of testing has the advantage of accuracy, speed and reproducibility, our introduction will focus on it.

6.4.2 Laser scanning method

The principle of laser scanning was introduced in Section 6.3. Many researchers have also applied this method to evaluate seam puckering appearance. Among them, Kawabata *et al.* [22] used the laser scanning method to measure seam pucker and analyzed the sensory evaluation of seam pucker using the Weber–Fechner law. From the height signal, they calculated a surface roughness parameter and found that sensory evaluation of seam pucker follows the Weber–Fechner law, which states that a sensory value is proportional to the logarithm of the magnitude of the quality of the physical stimulation. Based on the above theory, they developed an equation for the objective evaluation of seam pucker which shows an almost linear relationship between the subjective seam pucker grade and physical quantity.

Park and Kang [10] also applied the laser scanning method to profile the seam puckering surface and evaluate it using artificial intelligence. Their scanning system (a displacement meter), consisting of a laser diode, could reconstruct the surface profile of the seams with little influence from color and texture. FFT was applied to calculate the specific features for the neural network modeling in order to simulate the judging behavior of the human experts by comparing AATCC rating standards with testing samples. Their research shows that it is possible to predict and optimize the quality of seam pucker from material properties and processing parameters.

Fan and his co-workers [23, 24] developed an objective method for the evaluation of seams on a 3-D garment surface through the application of laser scanning technology. A commercial 3-D laser scanning system, consisting of a laser scanning head, robot arm, computer and some special software for data acquisition, was used to scan garment seams. They used 2-D filters to obtain 3-D pucker profiles by removing the high frequency components in the seam profiles that might have been contributed to by the individual threads of the fabric or noise, as well as the lower frequency components that might have been contributed to by the smooth garment surface. They defined four geometrical parameters to characterize the seam puckering profile: average displacement, its variance, the skewness of the height distribution, and the kurtosis of the height distribution. It was found that the logarithm of the average displacement from the mean magnitude and its variance were linearly related to the severity of seam puckering. The addition of the logarithm of the skewness and kurtosis of the height distribution hardly improved the correlation, so that the first two parameters were selected as the objective measures of seam puckering.

In their research, Fan and Liu [24] used 10 men's shirts, made from different fabrics of similar weight and density, one white polyester/cotton and the other red and white cotton check, as samples. They found a relationship between the logarithm of variance and the subjective grade of seam pucker close to four regions on the sample garment – yoke seam, pocket seam, placket seam and armhole seam. Although this proved that their method is effective on the objective evaluation of seam puckering, further efforts are needed to reduce the cost of the system and make it more robust for industrial use.

6.4.3 Image analysis method

Some researchers have also investigated the objective evaluation of seam puckering based on image analysis methods. For example, Stylios and Sotomi [25] developed a so-called Pucker Vision System, consisting of a CCD camera to capture the surface image of seam pucker instead of human eyes, and a software program used to evaluate the image features. The system was designed to capture the images of two groups of seam stripes produced from the same fabric, one the unstitched seams and the other sewn with puckers. Using the mean reflection of the unstitched seams as a reference, the system assessed the configuration of the pucker by identifying the pucker wavelength and pucker amplitude to develop a pucker severity index. The consistency of the light source and the influence of the pattern and color of the fabric were the major limitations of the system.

Richard [26] developed a computer-based seam pucker measurement system to quantify seam surface irregularities using digital image analysis. A video camera was used to capture seams in the immediate vicinity of the seam formation area. The measurement of the pucker index on a scale of 1 to 5 was very rapid and the results were incorporated into a fabric sewability report, together with the measurement of the dynamic force of the sewing process.

In our previous research [19], a special lighting box was designed to illuminate the testing sample from four different directions, and one digital camera was located on top of this lighting box to capture one image group, which includes the four images with the same imaging geometry, though with different illumination. The four-directional illumination and imaging geometry are illustrated in Figs 6.21 and 6.22. In our research, 29 samples

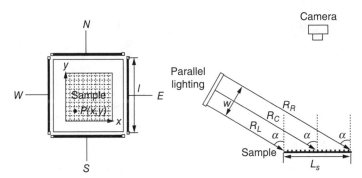

6.21 Lighting system.

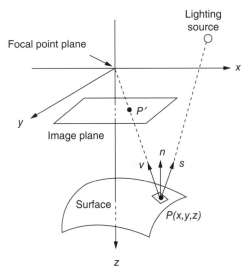

6.22 Surface model and observation system.

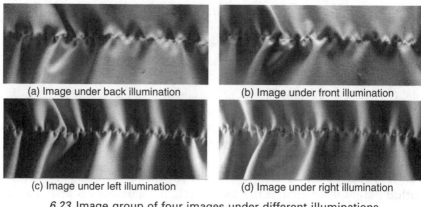

(a) Image under back illumination   (b) Image under front illumination

(c) Image under left illumination   (d) Image under right illumination

6.23 Image group of four images under different illuminations (Grade 1).

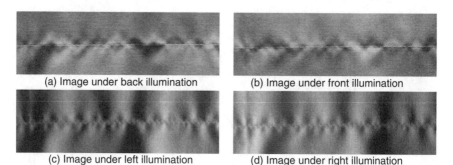

(a) Image under back illumination   (b) Image under front illumination

(c) Image under left illumination   (d) Image under right illumination

6.24 Image group of four images under different illuminations (Grade 2.5).

were sewn from gray cotton woven fabrics with different setting parameters to achieve different seam puckering grades. Each fabric specimen was cut into 380 mm × 120 mm sections, and given varying grades of seam puckering by adjusting sewing conditions as illustrated in Fig. 6.23. Four images of each sample were captured at a resolution of 640 × 480 pixels under four different illuminating directions. The region of the image was set at 300 × 300 pixels for all the samples, and the sample surface to be captured was 150 mm × 150 mm, thus making the image resolution 50.8 dpi. Each pixel was assigned a gray-level value from 0 for black to 255 for white.

From the comparison of Fig. 6.23 (Grade 1), Fig. 6.24 (Grade 2.5) and Fig. 6.25 (Grade 5), it is found that the seam puckering appearance with different puckering severity demonstrates different changes under different illumination. The seam puckering appearance with a rough surface profile has a quite different image intensity at the same position under different illuminations; however, the seam puckering appearance with a smooth and flat surface profile has a similar image intensity at the same position even

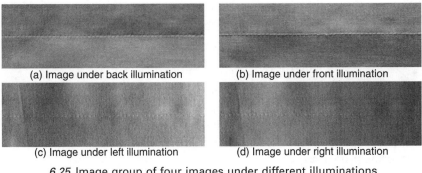

(a) Image under back illumination   (b) Image under front illumination

(c) Image under left illumination   (d) Image under right illumination

6.25 Image group of four images under different illuminations (Grade 5).

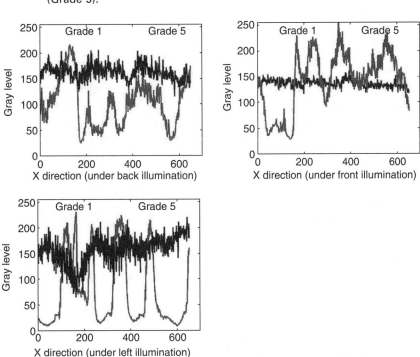

6.26 Comparison of three image profiles under different illuminations (Grade 5 and Grade 1).

under different illuminations as illustrated in Fig. 6.26. In this case, this phenomenon could be used to characterize the seam puckering.

Seam puckering characterization

In this chapter, three line profiles along the seam direction were selected to characterize the severity of seam puckering; one is the central line of seam puckering, the other two are symmetrical to this central line at a certain distance from it. The fractal dimension of these three line profiles can then be calculated according to the following equation:

$$D = \frac{\log(c) - \log(N(r))}{\log(r)} \qquad 6.13$$

where c is a positive constant, r is the side length of the cube and $N(r)$ is the number of blocks which cover the profile. Since four images were captured under four different illuminations, in this case a feature vector $\vec{V} = [D_{Li}, D_{Ri}, D_{Fi}, D_{Bi}], i = 1, 2, 3$, could be defined to characterize the seam puckering properties.

System training and grade classification

We used the five standard seam puckering specimens from grade 1 to grade 5 for system training, and the other 19 specimens for testing. The system training procedure is that of calculating the feature vector of these five standard specimens and establishing a feature space according to the values of their features. To account for the distance between one specimen and five grade clusters, we used a simplified Bayes distance as the metric to determine 'closeness'. Under the conditions that the features are independent and Gaussian, the Bayes distance provides the maximum likelihood (ML) classification. Under these assumptions, the likelihood function for a feature vector $\vec{v}$ belonging in fabric appearance grade l is

$$p\left(\vec{V} = \vec{v} \mid \vec{L} = \vec{l}\right) = \prod_{i=1}^{n} \frac{1}{\sqrt{2\pi}\sigma_{i,l}} e^{-(v_i - u_{i,l})^2 / 2\sigma^2_{i,l}} \qquad 6.14$$

where v_i represent the elements of the feature vector, and n corresponds to the total number of features. The grade classification is achieved by choosing the grade l, which minimizes the simplified Bayes distance function:

$$d_l = \sum_{i=1}^{n} \left[2\ln\sigma_{i,l} + \left(\frac{v_i - u_{i,l}}{\sigma_{i,l}}\right)^2 \right] \qquad 6.15$$

In this chapter, we calculated 12 fractal dimensions under different illuminations to make up a fractal vector, and trained the system using five standard specimens from grade 1 to grade 5, each grade class having only one standard seam puckering sample as specimen, so we simplified the Bayes distance function as below:

$$d_l = \sum_{i=1}^{n} (v_i - u_{i,l})^2 \qquad 6.16$$

Having calculated the fractal vectors of the 19 test specimens and the distances between the five standard clusters and the 19 test specimens, we can decide which grade class each test specimen should belong to. The prelimi-

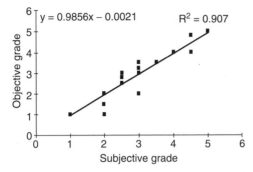

6.27 Correlation between subjective and objective evaluation.

nary experimental results show that a good relationship between objective evaluation and subjective evaluation could be achieved based on this method, as illustrated in Fig. 6.27, which integrates the suitable choice of both the hardware design and the feature definition.

6.5   Fabric dimensional stability testing

Dimensional stability refers to a fabric's ability to resist a change in its dimensions. A fabric or garment may exhibit shrinkage in some dimensions or growth in other dimensions under conditions of washing, drying, steaming and pressing. Fabrics are often given shrinkage-resistant finishes to minimize dimensional change. Relaxation shrinkage, progressive shrinkage and thermal shrinkage are three major types of shrinkage that may occur when fabric is subjected to heat and/or moisture. Relaxation shrinkage results from the relaxation of stresses imposed during weaving or knitting of the fabric. Progressive shrinkage is dimensional change that continues through successive washings. Thermal shrinkage is limited to fabrics composed of thermoplastic fibers; upon imposition of heat to these fabrics, the polymer molecules in the fibers move and assume a random, non-linear form, shortening the fibers and causing the fabric to shrink and alter its shape.

For knitted fabrics, cockling has been defined as 'an irregular surface effect caused by loop distortion' [1]. In general it appears as localized groups of distorted knitted loops which have twisted out of the symmetrical configuration. The fault is usually found in the plain knit structure, which is relatively unstable, and especially in yarns spun from animal fibers such as wool or mohair. Wool knitwear made from worsted yarns is more prone to cockling/loop distortion because, unlike wool knitwear, no milling is carried out during finishing which helps to conceal faults under surface fuzz. There are three types of cockling: rib/plain interface cockling, panel-edge cockling,

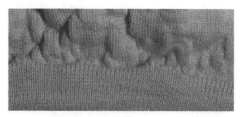

(a) Rib/plain interface cockling

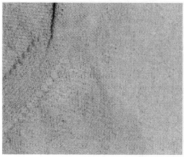 

(b) Panel-edge cockling   (c) Random overall cockling

6.28 Cockling of knit fabrics.

and random all-over cockling, as illustrated in Fig. 6.28. Rib/plain cockling is caused by a difference in the relaxed widths of the two structures. Panel-edge cockling is caused by a difference in the relaxed dimensions of neighboring structures. Panel edge loops are stretched in length when fashioning takes place and this tends to result in the contraction of the adjacent plain knit fabric, allowing cockling to take place. Random all-over cockling can almost always be assigned to using unsuitable yarn. Yarns which cause random all-over loop distortion also tend to cause rib/plain and panel edge cockling as well. Cockling could be considered as one type of three-dimensional change, similar to seam puckering. The testing method could be referred to as dimensional change and seam puckering evaluation. In some cases, manufacturer and buyer could make a specific standard according to their actual requirements.

Spirality is defined as angular displacement of filling yarns or knitted courses from a line perpendicular to the edge or side of a fabric or garment (see Fig. 6.29). Spirality in knitwear is caused by using an 'unbalanced' yarn, i.e. a single yarn or a two-fold yarn where the incorrect ratio of singles to folding twist has been used. When spirality occurs in a fabric there is little or nothing that can be done in finishing to alleviate it and usually the only option is to use replacement yarn having a balanced twist. However, sometimes spirality can be prevented by steam setting an unbalanced yarn when knitted in a single bed structure, and also some yarns which cause spirality

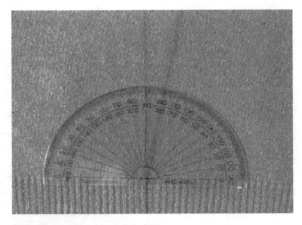

6.29 Spirality of knit fabrics.

can usually be knitted in a rib structure, since the skewing tendencies on the front and back beds should cancel out.

6.5.1 Fabric dimensional stability: standards

AATCC Test Method 135 details procedures for determining the dimensional stability of woven and knitted fabrics; this method provides the standard testing method for the measurement of dimensional change in home laundering. AATCC Test Method 96 specifies the use of a wash wheel to simulate commercial equipment for determining the stability of woven or knitted fabric in commercial laundering. AATCC Test Method 160 describes a procedure for applying dimensional restoration to textile items after a standard laundering procedure to simulate the stretching force or pressing they receive before or during use.

6.5.2 Fabric dimensional stability: evaluation

The basic procedure for testing fabric for dimensional change is measurement of length and width benchmarks before and after a selected refurbishing process. The benchmarks drawn with indelible ink are placed 25 cm apart, or 50 cm for better precision, in each direction. After a standard laundering or drycleaning method is applied, the marks are re-measured and dimensional change (DC) is calculated from:

$$DC(\%) = \frac{100(B-A)}{A}$$ 6.17

where A is the original dimension and B is the dimension after treatment. Length and width changes are calculated separately. Shrinkage is reported

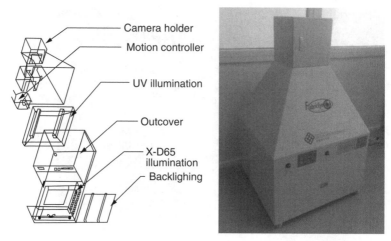

Camera holder
Motion controller
UV illumination
Outcover
X-D65 illumination
Backlighing

6.30 Image analysis system for the measurement of dimensional stability.

as a negative number, while growth is reported as a positive percentage. Specimens are conditioned accordingly before each measurement.

A digital imaging system may be used as a measuring device in place of the prescribed manual measurement devices if it is established that its accuracy is equivalent to that of the manual devices. Recently, a new image analysis system was developed to measure the dimensional change and spirality of woven/knitted fabric by HKRITA and PolyU. This digital imaging system, as illustrated in Fig. 6.30, consists of a lighting box with standard controlled illumination from four directions, a high resolution digital camera, and a set of software to control the digital camera and calculate the dimensional change and spirality of fabrics before and after home laundering or other industry finishing process. All the testing samples are marked with four cross-shaped markers using a specially designed marking template according to the standards, which cannot be washed during the finishing or laundering process, as illustrated in Fig. 6.31.

Image capturing module

Samples were put on the testing platform after the marking; a digital camera was controlled by the software to capture the image of the fabrics (Fig. 6.32) under the standard illumination (D65), then the fabric images were transferred from the digital camera and recorded in the computer.

Feature analysis module

The major function of this module is to identify the markers and calculate their movement before and after the finishing or laundering process. An

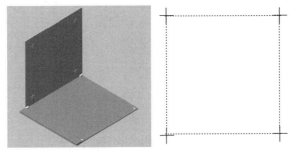

(a) Sample marker template   (b) Markers on fabric surface

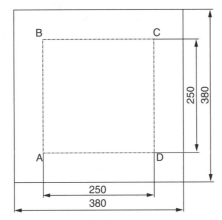

(c) Standard distance of markers (mm)

6.31 Benchmarking on fabric surface.

6.32 Fabric images captured.

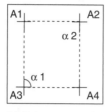 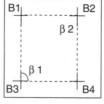

(a) Before processing   (b) After processing

6.33 Dimensional changes before and after processing.

automatic detector based on image analysis technology was developed to determine the image coordinates of these four markers, and three parameters (D_h, D_v, S) are defined to characterize the dimensional change and spirality as illustrated in Fig. 6.33.

D_h is the dimensional change along the horizontal direction:

$$D_h = \frac{\frac{d(B1, B2)+d(B3, B4)}{2} - \frac{d(A1, A2)+d(A3, A4)}{2}}{\frac{d(A1, A2)+d(A3, A4)}{2}} \times 100\% \quad 6.18$$

D_v is the dimensional change along the vertical direction:

$$D_v = \frac{\frac{d(B1, B3)+d(B2, B4)}{2} - \frac{d(A1, A3)+d(A2, A4)}{2}}{\frac{d(A1, A3)+d(A2, A4)}{2}} \times 100\% \quad 6.19$$

S is the spirality in terms of the changes of skewness angle:

$$S = \frac{\frac{\alpha_1+\alpha_2}{2} - \frac{\beta_1+\beta_2}{2}}{\frac{\alpha_1+\alpha_2}{2}} \times 100\% \quad 6.20$$

The advantages of this digital imaging system for the purpose of dimensional change and spirality measurement lie in its speed, accuracy and use of digital technology. This system can also be used to evaluate other surface changes, such as hairiness evaluation.

6.6   Light reflectance of a fabric

Light reflectance of a fabric mainly refers to the color and luster properties. Color is one of the most important factors in consumer acceptance of fabric materials. The color appearance of an object depends on the source of illumination, the object's interaction with light and the response of sensors in the observer's eye to the light reflected from the object. Current technology offers the possibility of measuring color very precisely. Two theories – the tristimulus theory and the opponent theory – form the basis of instrumental color measurement.

The CIELAB color system is widely used in color measurement of textiles. The CIELAB values L, a and b are mathematically derived from the tristimulus color values X, Y and Z. CIELAB provides a method for quantifying color difference into a single term. The overall color difference between two specimens can be designated by the term ΔE.

6.6.1 Color measurement instruments

The instruments can be categorized according to their sophistication and precision in color measurement, how the measurements incorporate the components of color, the specimen viewing geometry and instrument portability. All include a light source, a port or opening onto which the specimen is placed, and a detector. The number of color sensors used in an instrument affects its accuracy in color measurement. There are three types of instruments: colorimeters, spectro-colorimeters, and spectrophotometers; the principles of colorimeters and spectrophotometers are illustrated in Fig. 6.34. Colorimeters evaluate the reflected color of a specimen using three sensors that respond to red, green and blue. Colorimeters are useful in color measurement, but they cannot detect mesmerism. They should not be used for setting standards; they are the least accurate and least expensive of the three types. Spectro-colorimeters are more accurate because they have more color sensors that sense particular intervals of the spectrum. The accuracy depends on the number of sensors, which can be between three and 16. Spectro-colorimeters also offer more flexibility in the choice of illuminants than is possible in colorimeters. Spectrophotometers are the most accurate of the three levels of instruments and can safely be used to

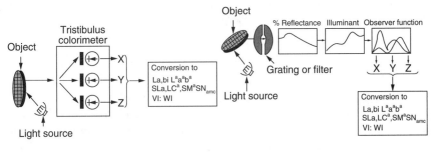

(a) Colorimeter   (b) Spectrophotometer

6.34 Sketch maps of colorimeter and spectrophotometer.

set color standards. Spectrophotometers sense reflected color of a specimen at several individual wavelengths within the visible spectrum, at either 10 nm intervals or 20 nm intervals throughout the visible spectrum. The instruments are sometimes referred to as 16 point or 31 point instruments according to the number of measurement points across the visible spectrum, 400–700 nm.

6.6.2   Color measuring procedure

A proper procedure to perform the measurement of colors is important for industrial applications. A normal measurement procedure involves four major steps: instrument setup, calibration, verification, and sample preparation and measurement [27].

Instrument setup

The spectrophotometer has to be set up in the proper mode prior to color measurement, according to the instrument manufacturer's recommendations. For example, in the case of a sphere-type reflectance spectrophotometer, the possible list of parameters to be set up includes the spectral range, the polychromatic/monochromatic mode, the specular component in/out, the sample port size, and the selection of filters.

Instrument calibration

After the instrument has been properly set up, it has to be calibrated for its photometric scale with respect to the 100% line and the 0% line using a white and black standard template, respectively, in the case of a reflectance spectrophotometer. For a transmission spectrophotometer, a clear solution for the solution sample (or air for the color filter sample) is used to set up the 100% reference line, and the 0% reference line is normally established by blocking the illumination beam with an opaque sample. The calibration procedure establishes a set of correction factors at each wavelength and is applied to the subsequent spectral measurements to obtain absolute spectral data.

Instrument verification

The performance of the spectrophotometer in terms of precision and accuracy can be checked by the measurement of color standards with calibrated spectral data. Precision refers to how repeatable the measurements are for the same sample over a period of time, while accuracy refers to how close the measured reading of a sample is to its absolute true reading.

Sample preparation and measurement

The importance of the sample itself in providing reliable color measurement data should not he overlooked. There are a number of factors that may affect the measurement precision and accuracy. The following outlines the major items to be observed during the measurement of textile samples.

Sample temperature and moisture content
The temperature and moisture content of a textile sample could change its color appearance significantly and hence its measurements. It is therefore important to condition all textile samples in a room or chamber with controlled humidity and temperature for a suitable period prior to color measurement.

Sample format
A good technique of sample presentation for measurement is to ensure an identical format of presenting all the samples to be intercompared at the instrument sample port for color measurement. The textile usually is folded to complete opacity to avoid background influence during color measurement. Thus it is important that all samples to be intercompared are folded to the same number of layers. Yarn samples should be wound onto a rigid card uniformly with identical layers. Loose fibers should be placed into a transparent cup holder with identical thickness under identical pressure. Color measuring instruments are generally designed for measurement of flat samples to be placed at the sample port. If the sample extends inside the port or is displaced away from the port, different measured readings may result. If the textile sample flatness is difficult to achieve, due to surface texture, measurement behind glass will help. However, the measurement results must be corrected for effects of the cover glass, such as the Fresnel reflection.

Thermochromic and photochromic property
Some textiles have colorants that are sensitive to heat and light. Color change on exposure to heat and light is called thermochromism and photochromism respectively. Both the thermochromic effect and the photochromic effect can be eliminated or reduced by minimizing the time of sample exposure to the illuminating source during measurement.

In general, it is always good practice to average multiple measurements at different locations and orientations of the test sample to achieve repeatable measurement.

6.6.3 Color evaluation applications

The primary applications of color measuring systems in the textile industry include color fastness assessment, whiteness evaluation, yellow evaluation, luster evaluation, etc.

Color fastness assessment

Most methods in color fastness assessment of textile materials involve treating the dyed material in a standardized manner and then comparing the treated textile and the original untreated textile visually against a gray scale that carries a series of pairs of color chips with increasing color difference magnitude. There are two kinds of gray scales, one for the staining test and the other for change of shade. The obvious disadvantage of determining the color fastness rating by means of visual assessment is the poor reproducibility from observer to observer. Methods for instrumental assessment of staining and change of shade have been developed by various professional bodies: Textiles – Tests for colour fastness (ISO-105 standards) and Textiles – Tests for color fastness – Instrumental assessment of the degree of staining of adjacent fabrics (GB/T 6410-86).

Whiteness evaluation

White is a color of freshness, purity and cleanliness. It has been used as an indicator of qualities such as freedom from contamination. The determination of the degree of whiteness has been an interesting subject for many years. In principle, it can be measured by the amount of departure from the perfect white position in a three-dimensional color space. However, agreement on the perfect white has not been reached because of some problems. The major problem is that strong preferences in the concept of whiteness are governed by trade, nationality, habit and product. This problem is further enhanced by the introduction of fluorescent whitening agents, the conditions of observation, and the measurement accuracy. The principles of deriving a whiteness formula were described by Ganz [28]. In 1981, the CIE recommended field trials of a new whiteness formula, and the CIE whiteness formula was adopted by the AATCC in 1989 as AATCC Method 110-1989.

Yellow evaluation

The preferential absorption of white light in the short-wavelength region (380–440 nm) by the material usually causes an appearance of yellowness. Interest has developed in determining the degree of yellowness as it is considered to be associated with soiling, scorching and product degradation by exposure to light, atmospheric gases, and other chemicals. Some yellow-ness standards have been developed over the years (ASTM D1925).

Luster evaluation

Some textile fabrics are finished with a lustrous appearance by means of a calendering process. The luster is the gloss appearance associated with the contrast between the specularly reflecting area of fabric and the surround-ing diffusely reflecting area. Hunter [29] has developed a formula to express this relationship:

$$\text{Luster} = 100(1 - R_d/R_s)$$ 6.21

where R_d is the diffuse reflectance factor and R_s is the specular reflectance factor.

6.7   Conclusions

The testing methods and related standards for fabric appearance evaluation are reviewed in this chapter, including pilling, wrinkling, seam puckering, dimensional change and light reflectance. Some basic technologies related to these testing methods are also introduced briefly. Traditionally, fabric appearance evaluation relies to a large extent on subjective evaluation methods based on the perception of human eyes, but with the development of digital technologies, more and more testing standards are being replaced by new objective testing methods which have the advantages of speed, accuracy and repeatability. In this chapter, actual applications of digital evaluation methods on the evaluation of pilling, wrinkling, seam puckering and dimensional change are also reported. These methods could contribute to upgrading quality control in the textile industry. The aim of this chapter is to provide a simple but general background introduction to fabric appear-ance testing and evaluation.

6.8   References

1. 'Fabric appearance', Technical paper from Australian Wool Innovation, 2007.
2. Konda A, Liang C X, Takadera M, Okoshi Y, Toriumi K, Evaluation of pilling by computer image analysis, *J. Textile Mach. Soc. Japan*, 1988, **36**(3), 96–107.
3. Xu B, Instrumental evaluations of fabric pilling, *J. Text. Inst.*, 1997, **88**, 488–500.
4. Hsi C H, Bresee R R, Annis P A, Characterizing fuzz on fabrics using image analysis, *Textile Res. J.*, 2000, **70**(10), 859–865.
5. Kang T J, Cho D H, Kim S M, Objective evaluation of fabric pilling using ste-reovision, *Textile Res. J.*, 2004, **74**(11), 1013–1017.
6. Shiloh M, The effect of fabric structure on wrinkling, studies in modern fabrics, *Text. Inst.*, 1970, **61**, 14–29.
7. Fan J, Yu W, Hunter L, *Clothing Appearance and Fit: Science and Technology*, Woodhead, Cambridge, UK, 2004.

8. Ramgulam R B, Amirbayat J, Porat I, Measurement of fabric roughness by a noncontact method, *J. Text. Inst.*, 1993, **84**(1), 99–106.

9. Amirbayat J, Alagha M J, Objective assessment of wrinkle recovery by means of laser triangulation, *J. Text. Inst.*, 1996, **87**(2), 349–354.

10. Park C K, Kang T J, Objective rating of seam pucker using neural networks, *Textile Res. J.*, 1997, **67**(7), 494–502.

11. Kang T J, Lee J Y, Objective evaluation of fabric wrinkles and seam puckers using fractal geometry, *Textile Res. J.*, 2000, **70**(6), 469–475.

12. Kim E H, Objective evaluation of wrinkle recovery, *Textile Res. J.*, 1999, **69**(11), 860–865.

13. Turner C, Sari-Sarraf H, Zhu A, Hequet E, Lee S, Preliminary validation of a fabric smoothness system, *J. Electronic Imaging*, 2004, **13**(3), 418–427.

14. Xu B, Cuminato D F, Keyes N M, Evaluating fabric smoothness appearance with a laser profilometer, *Textile Res. J.*, 1998, **68**(12), 900–906.

15. Xu B, Reed J A, Instrumental evaluation of fabric wrinkle recovery, *Textile Res. J.*, 1995, **86**(1), 129–135.

16. Na Y, Pourdeyhimi B, Assessing wrinkling using image analysis and replicate standards, *Textile Res. J.*, 1995, **65**(3), 149–157.

17. Fazekas Z, Komilves J, Renyi I, Surjan L, Towards objective visual assessment of fabric features, image processing and its applications, *Conference Publication No. 465*, ©IEE 1999, 411–416.

18. Chen J, Wu Zh, Yan H, The fractal characterization of fabric wrinkle image, *Journal of China Textile University (Eng. Ed.)*, 1999, **16**(3), 86–89.

19. Hu J, Xin B, Yan H, Measuring and modeling 3D wrinkles in fabrics, *Textile Res. J.*, 2002, **72**(10), 863–869.

20. Yang X B, Huang X B, Evaluating fabric wrinkle degree with a photometric stereo method, *Textile Res. J.*, 2003, **73**(5), 451–454.

21. Mori T, Komiyama J, Evaluating wrinkled fabrics with image analysis and neural networks, *Textile Res. J.*, 2002, **72**(5), 417–422.

22. Kawabata S, Mori M, Niwa M, An experiment on human sensory measurement and its objective measurement; case of the measurement of seam pucker level, *Int. J. Clothing Sci. Techn.*, 1997, **9**(2–3), 203–206.

23. Fan J, Lu D, Macalpine J M K, Hui C L P, Objective evaluation of pucker in three-dimensional garment seams, *Textile Res. J.*, 1999, **69**(7), 467–472.

24. Fan J, Liu F, Objective evaluation of garment seams using 3-D laser scanning, *Textile Res. J.*, 2000, **70**(11), 1025–1030.

25. Stylios G, Sotomi J O, Investigation of seam pucker in lightweight synthetic fabrics as an aesthetic property. Part I: A cognitive model for the measurement of seam pucker, *J. Text. Inst.*, 1993, **84**, 601–610.

26. Richard C, Pucker as fabric-thread machine mechanical instability phenomenon, *J. Fed. Asian Prof. Text. Assoc.*, 1995, **3**(1), 83.

27. Chong P T F, Colorimetry for textile applications, in *Modern Textile Characterization Methods*, ed. M. Rahed, 1996, Marcel Dekker, New York, pp. 355–388.

28. Ganz E, Whiteness: photometric specification and colorimetric evaluation, *Applied Optics*, 1976, **15**(9), 2039–2058.

29. Hunter R S, *The Measurement of Appearance*, 2nd edn. John Wiley & Sons, New York.

7

Fabric permeability testing

X DING,
Donghua University, China

Abstract: Fabric permeability is a property not well defined in the textbooks and not well understood by anyone outside the scientific fraternity. This chapter begins by introducing the terms and definitions of fabric permeability testing as well as testing principles. It then introduces the different techniques adopted in the measurement of this property which are divided into three types of permeation based on the different penetrants. It summarizes various applications of the test methods for different purposes as well. Finally, this chapter presents a review on innovative test methods for fabric permeability.

Key words: fabric permeability, water vapour permeability, air permeability, chemical permeability, test method.

7.1 Introduction: terms and definitions

Fabric permeability is a property of fabric which is used to assess the ability of fabric to allow a penetrant, such as a gas, liquid or solid material, to pass through such a barrier and then desorb into a specified medium. Since the 1960s and particularly during the past two decades, fabric permeability has been the subject of a great number of investigations that revealed very interesting properties and offered new insights into mass transfer through fabric materials. Such research leads to fabrics that can display very different properties and have various classical and non-classical applications, such as protective clothing, sportswear, laminated and coated fabrics, textiles for filtration, medical textiles, textiles for transportation, and other technical textiles (Raheel 1996; Byrne 2000).

Permeability is a property of a material, but the permeability of a body that performs like a material may be used. Permeability is the arithmetic product of permeance and thickness, which is often defined as the time rate of penetrant transmission through unit area of flat material of unit thickness induced by unit penetrant pressure difference between two specific surfaces. Fabric permeability is certainly the first species to have been considered for sorption, diffusion and permeation studies. Most investigations were undertaken to understand the basic relationships between the fabric structure and sorption or permeability, in order to control the permeable character by a proper structure design.

7.2   Aspects of wear comfort

Wear comfort is one of the most important topics in the field of textiles and clothing. Human beings cannot function efficiently if they are not comfortable, and if a person is operating machinery or driving a car, comfort becomes a factor determining safety. However, comfort has many different aspects, as summarized by one writer: 'Comfort is a complex matter, with physical, physiological, and psychological factors interrelated in an unpredictable combination which constantly undergoes variation.' (Saville 1999).

Clothing has a large part to play in the maintenance of wear comfort as it modifies the heat loss from the skin surface and at the same time has the secondary effect of altering the moisture loss from the skin. Perspiration is an important mechanism which the body uses to lose heat when its temperature starts to rise. Heat is taken from the body in order to supply the latent heat needed to evaporate the moisture from the skin.

There are two forms of perspiration:

- Insensible: in this form the perspiration is transported as a vapour and it passes through the air gaps between yarns in a fabric.
- Liquid: this form occurs at higher sweating rates and it wets the clothing which is in contact with the skin.

When perspiration takes place to cool the body, the water exuded through the skin appears initially as liquid which evaporates at once (in comfortable situation) and forms moisture vapour. This vapour is removed from the vicinity of the body, either by convection or through the clothing worn on the person, carrying heat away with it (Slater 1971).

When the moisture vapour reaches the inner surface of the fabric, several events can take place. The vapour may pass through the fabric system to its outermost surface, there to be carried away by the air. At the other extreme, it may be prevented from escaping through the fabric system if a component of the latter is impermeable, and will condense at some position in the system. The transfer of additional moisture vapour through the system will then be impeded by the liquid water layer so formed, and the plane of condensation will gradually move inwards form the impermeable barrier until eventually condensation takes place at the inner layer of the system, at the surface of the body, as soon as moisture is exuded through the skin, marking the onset of sensible perspiration.

There are two forms in which this discomfort is manifested. In hot weather, perspiration is the only problem, and the sensation of wetness, though a nuisance, does not lead to danger unless the temperature is so great that heat stroke or dehydration is possible. In cold weather, though, a more urgent risk arises. If, for instance, heavy work has been in progress and then is discontinued, the production of heat (and moisture) will continue for some time after work has ceased. If this moisture is not evaporated (by the heat generated in working, for instance), condensation will occur and the wetness will be evident, as before. Another hazard now comes into operation. The liquid moisture which forms inside the fabric acts as a much better conductor of heat than the air which it has displaced, so that the heat loss from the body increases by a very large factor. The resulting initial cooling effect of this liquid water by conduction is then increased by the enhanced loss of heat from the body surface in an effort to supply latent heat of vaporization to transform the liquid water into vapour. Because of this demand, heat losses from the body become immense and a grave risk of frostbite or hypothermia develops, possibly leading to irreparable damage or even death.

7.3   Principle of different test methods for fabric permeability properties

Various test methods are available for assessment of the fabric permeability. The most conventional method is to calculate the permeability by Equation 7.1:

$$P = D \cdot S \qquad\qquad 7.1$$

where P is permeability coefficient, D is diffusion coefficient and S is solubility coefficient. This equation has been widely used in the literature. However, the problem of penetrant diffusion in and permeation through inhomogeneous fabrics is more complex. This equation is only available in ideal permeation conditions such as at the limit of low permeability and diffusion concentration (Molyneux 2001). At the same time, a number of standard methods have been developed by several national and international organizations such as the American Society for Testing and Materials (ASTM), the American Association of Textile Chemists and Colorists (AATCC), the British Standard (BS), the Japanese Industry Standard (JIS), the International Standards Organization (ISO), and others to assess the permeability of fabric to air, water, chemical, etc. These test methods are able to meet the requirements for quality control as well as research work to a certain extent.

7.4   Types of fabric permeability tests

In general, fabric permeability tests include air permeability, water permeability and chemical (gaseous, liquid or solid chemical) permeability, which is related to the penetration of a gas, liquid or solid material to pass through such fabric barrier and then desorb into a specified medium. Each type of fabric permeability has different testing methods.

7.4.1 Air permeability

Air permeability is defined as velocity of an air flow passing perpendicularly through a test specimen under specified conditions of test area, pressure drop and time (BS 3424-16-1995). The principle of the test is that the rate of flow of air passing perpendicularly through a given area of fabric is measured at a given pressure difference across the fabric test area over a given time period. In the field of textiles and clothing, air permeability is often used in evaluating and comparing the 'breathability' of various fabrics (coated and uncoated) for such end uses as raincoats, tents, and uniform shirtings. It helps evaluate the performance of parachutes, sail cloth, industrial filter fabrics, and the covering fabrics of pillows and duvet covers.

Usually, an air permeability test apparatus consists of:

- A clamping device for securing the test specimen in a flat tensionless state.
- A device to prevent air leaking from the edges of the test area, usually called a guard ring.
- A pressure gauge or manometer to measure the pressure drop from one side of the specimen to the other.
- An air pump to draw a steady flow of air through the clamped specimen.
- A means of adjusting the rate of airflow to achieve and hold the specified pressure drop from one side of the specimens to the other.
- A flow meter to measure the actual rate of air flow through the specimen.

The Kawabata KES-F8-API Air Permeability Tester (Fig. 7.1) is one of the apparatus which is able to measure air permeability of fabric. In this apparatus, air is pumped by a piston at a constant volume of 8π cm^3/s. The velocity of the air is dependent on the plate chosen on the tester; each plate has a different aperture size, and air velocities of 0.4, 4 or 40 cm/s are

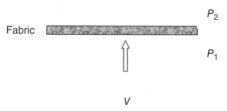

7.1 Schematic representation of Kawabata KES-F8-API air permeability tester.

possible. The pressure drop caused by the resistance of the specimen is measured by a differential pressure gauge. The output is the air resistance R, measured in kilopascals times seconds per metre (kPa·s/m), found from Equation 7.2:

$$R = \frac{P_1 - P_2}{V} = \frac{\Delta P}{V} \qquad 7.2$$

There are some defining equations of fluid flow that should be explained. If a specimen has small holes, the pressure drop is due to frictional loss and is defined as

$$\Delta P = KV \qquad 7.3$$

where ΔP is the pressure difference, V is the air velocity, and K is the constant for the specimen. This can also be expressed as

$$K = \frac{\Delta P}{V} \qquad 7.4$$

Here,

$$R = K \qquad 7.5$$

where R is resistance and is linear with respect to velocity for the specimen. A material with this response can be considered as a 'linear resistor'.

If the specimen exhibits large holes, then Bernoulli's law holds true, where

$$\Delta P = KV^2 \qquad 7.6$$

Rewritten,

$$KV = \frac{\Delta P}{V} \qquad 7.7$$

and

$$R = KV = \frac{\Delta P}{V} \qquad 7.8$$

Then, R is not constant because it is now a function of changing velocity. Such a material is considered as a 'non-linear resistor' (Dunn 2001).

Other test methods, such as ASTM D737-04 and BS EN ISO 9237:1995, are used extensively in the trade for acceptance testing. These two test methods apply to most fabrics including woven fabrics, non-woven fabrics, air bag fabrics, blankets, napped fabrics, knitted fabrics, layered fabrics, and pile fabrics. The fabrics may be untreated, heavily sized, coated, resin-treated, or otherwise treated (ASTM D737-04; BS EN ISO 9237:1995).

Construction factors and finishing techniques can have an appreciable effect upon air permeability by causing a change in the length of airflow paths through a fabric. Hot calendering can be used to flatten fabric components, thus reducing air permeability. Fabrics with different surface textures on either side can have a different air permeability depending upon the direction of air flow (ASTM D737-04).

For woven fabric, yarn twist also is important. As twist increases, the circularity and density of the yarn increase, thus reducing the yarn diameter and the cover factor and increasing the air permeability. Yarn crimp and weave influence the shape and area of the interstices between yarns and may permit yarns to extend easily. Such yarn extension would open up the fabric, increase the free area, and increase the air permeability (ASTM D737-04).

Increasing yarn twist also may allow the more circular, high-density yarns to be packed closely together in a tightly woven structure with reduced air permeability. For example, a worsted gabardine fabric may have lower air permeability than a woollen hopsacking fabric (ASTM D737-04).

7.4.2 Water permeability

Water permeability is used to assess the ability of a fabric to allow perspiration in its vapour or liquid form (which depends on the whole clothing system) to pass through it. Usually, several indexes can be applied to evaluate this ability, such as water vapour permeability, water repellency, water resistance and so on. These indexes have different applications respectively, which depend on different testing conditions and requirements.

On a normal condition, perspiration will pass through the clothing system in the vapour form. However, if the production of perspiration is greater than the amount the clothing system will allow to escape, moisture will accumulate at some point in the clothing system. If the outer layer is the most impermeable, moisture will accumulate in the inner layers. When excess moisture accumulates it causes a reduction in thermal insulation of the clothing and eventually condensation and wetting. The level of perspiration production is very dependent on the level of activity: clothing that may be comfortable at low levels of activity may be unable to pass sufficient moisture vapour during vigorous activity. However, when activity ceases, freezing can occur because the clothing is now damp and body heat production has been reduced, leading to after-exercise chill and, if the temperature is low enough, frostbite.

Therefore, it is important to be able to measure the rate at which a material can transmit moisture vapour if any assessment of the potential of that material in enhancing or reducing comfort needs to be made. A fabric of low moisture vapour permeability is unable to pass sufficient perspiration and this leads to sweat accumulation in the clothing and hence discomfort. As a result, the mechanism of permeability becomes of great theoretical interest, and direct measurement of permeability of fabric has been suggested. Some testing methods have been applied successfully in both research areas and industrial fields.

Water vapour permeability

For measurement of water vapour permeability of fabric, one of the main methods is the water cup method, which is especially suitable to evaluate fabrics with low sorption and permeability (Hsieh *et al.* 1990, 1991; Jeong *et al.* 2000a, 2000b). Such a comparatively simple method for testing the water vapour permeability of textiles will provide the manufacturer with a clearly recognized method for quality control within the plant.

ASTM E96-00 is based on this method to measure water vapour permeability of fabric. In this standard, two basic methods, the desiccant method and the water method, are provided for the measurement of permeance, and two variations include service conditions with one side wetted and service conditions with low humidity on one side and high humidity on the other.

In the desiccant method the test specimen is sealed to the open mouth of a test dish containing a desiccant, and the assembly is placed in a controlled atmosphere. Periodic weighings determine the rate of water vapour movement through the specimen into the desiccant. In the water method, the dish contains distilled water, and the weighings determine the rate of vapour movement through the specimen from the water to the controlled atmosphere. The vapour pressure difference is nominally the same in both methods except in the variation, with extremes of humidity on opposite sides. In this method, the water vapour transmission rate (WVT) is defined as the steady water vapour flow in unit area of a body, normal to specific parallel surfaces, under specific conditions of temperature and humidity at each surface, which can be calculated by Equation 7.9:

$$WVT = G/tA = (G/t)/A \qquad\qquad 7.9$$

where
   G = weight change (from the straight line), g
   t = time during which G occurred, h
   G/t = slope of the straight line, g/h
   A = test area (dish mouth area), m^2
   WVT = rate of water vapour transmission, g/h·m^2.

Water vapour permeance is defined as the time rate of water vapour transmission through unit area of flat material or construction induced by unit vapour pressure difference between two specific surfaces, under specified temperature and humidity conditions. It can be calculated using Equation 7.10 as follows:

$$\text{Permeance} = \text{WVT}/\Delta p = \text{WVT}/S(R_1 - R_2) \qquad 7.10$$

where

Δp = vapour pressure difference, mm Hg (1.333 × 102 Pa)
S = saturation vapour pressure at test temperature, mm Hg (1.333 × 102 Pa)
R_1 = relative humidity at the source expressed as a fraction
R_2 = relative humidity at the vapour sink expressed as a fraction.

Water vapour permeability (WVP) is defined as the time rate of water vapour transmission through unit area of flat material of unit thickness induced by unit area vapour pressure difference between two specific surfaces, under specified temperature and humidity. The average water vapour permeability can be calculated using Equation 7.11 as follows:

$$\text{WVP} = \text{permeance} \times l \qquad 7.11$$

where l = thickness of the membrane.

The purpose of these tests is to obtain, by means of simple apparatus, reliable values of water vapour transfer through permeable and semipermeable materials, expressed in suitable units. These values are for use in design, manufacture and marketing. A permeance value obtained under one set of test conditions may not indicate the value under a different set of conditions. For this reason, the test conditions should be selected that most closely approach the conditions of use (ASTM E96-00).

Water repellency

Water repellency in the field of textiles refers to the characteristic of a fibre, yarn or fabric to resist wetting. Water repellency of a fabric can be measured according to AATCC test method 70-2000 Water repellency: Tumble jar dynamic absorption test. In this test method, preweighed specimens are tumbled in water for a fixed period of time and are reweighed after the excess water has been removed from them. The percentage increase in mass is taken as a measure of the absorption or resistance to internal wetting. That is, the water absorbed for each specimen can be calculated using the following equation:

$$\text{WA} = \frac{W - C}{C} \times 100 \qquad 7.12$$

where:
WA = water absorbed, %

7.2 Dynamic absorption tester.

W = wet specimen weight, g
C = conditioned specimen weight, g.

This test method is applicable to any textile fabric, which may or may not have been given a water-resistant or water-repellent finish. It measures the resistance of fabrics to wetting by water. It is particularly suitable for measuring the water-repellence efficacy of finishes applied to fabrics, because it subjects the treated fabrics to dynamic conditions similar to those often encountered during actual use. It is not intended for use in predicting the probable rain penetration resistance of fabrics, since it measures absorption of water into, but not through, the fabric (AATCC TM 70-2000). A dynamic absorption tester is shown in Fig. 7.2.

Water resistance

Water resistance of fabric is the characteristic of this material to resist wetting and penetration by water. According to JIS L1092-1998, the tests can be classified as shown below.

Test for water penetration (hydrostatic pressure method)
This test applies mainly to textile fabrics with no air permeability.

* Method A (low hydraulic pressure method)
* Method B (high hydraulic pressure method) (This method usually applies to test specimens that can be tested by applying a hydraulic pressure exceeding 10 kPa.)

The apparatus shown in Figs 7.3 and 7.4 are applied to carry out tests for water penetration in according with method A (low hydraulic pressure

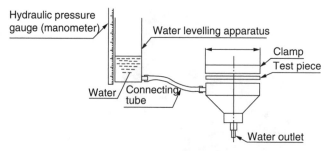

7.3 Water penetration test apparatus (for low hydraulic pressure).

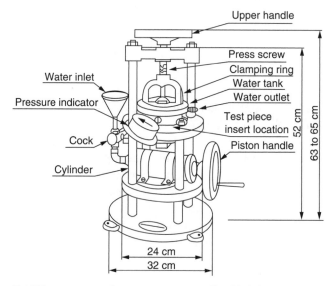

7.4 Water penetration test apparatus (for high hydraulic pressure).

method) and method B (high hydraulic pressure method) respectively. In these methods, the water level/water pressure is measured at the time when the water comes out from three places on the reverse surface of the test pieces (JIS L1092-1998).

Test for resistance to surface wetting (spray method)
This test applies to fabrics with air permeability. The apparatus in Fig. 7.5 is used to test for resistance to surface wetting (spray method). After dropping the excess water, the wet condition of the test specimen is assessed in comparison with the reference sample.

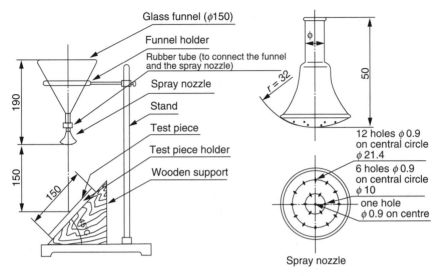

7.5 Apparatus for the test of resistance to surface wetting.

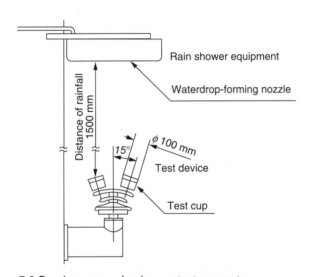

7.6 Bundesmann rain-shower test apparatus.

Rain test (shower test), method A

The apparatus in Fig. 7.6 is applied to the rain test (shower test). After exposure to a rain shower for 10 minutes (rain falling time may be 1 minute or 5 minutes), the wet condition of the test specimen is assessed in comparison with the reference sample. The amount of water absorption (g) and rate of water absorption (%) are calculated by means of Equations 7.13 and 7.14:

$$\text{Amount of water absorption(g)} = M - M_0 \qquad 7.13$$

$$\text{Rate of water absorption(\%)} = \frac{M - M_0}{M_0} \times 100 \qquad 7.14$$

where
  M_0 = mass of test pieces before the test, g
  M = mass of test pieces after the test, g.

7.4.3 Chemical permeability

Chemical permeability of fabric is used to assess the ability of fabric to allow molecular diffusion of a chemical (often referring to liquid and gaseous chemicals) through the fabric and its desorption into a specified medium. In general, two kinds of test methods are used to assess chemical permeability of fabric, that is:

- Determination of resistance of fabric to permeation by liquids and gases
- Measurement of repellency, retention, and penetration of liquids through fabric.

Resistance of clothing materials to permeation by liquids

BS 4724 specifies a laboratory test method that enables an assessment to be made of the resistance afforded by clothing materials to permeation by liquids. According to BS 4724, the test methods can be classified as shown below (BS 4724-1:1986; BS 4724: Part 2:1988).

Method for the assessment of breakthrough time
In this method, an indicator detects the presence in the vapour phase of a test liquid that has passed by permeation through the test specimen. If the volatility of the test liquid is not sufficient for its vapour to be detected directly by an appropriate indicator, a tracer, consisting of a volatile organic base or acid, is selected and mixed with the test liquid. The appropriate indicator solution is placed in a transparent glass cell and covered with a sample of the material under test. The test liquid, either alone or mixed with a tracer, is applied to the top surface of the material. The time for the test liquid itself or the tracer carried with the test liquid to cause the indicator solution to change colour is recorded. Therefore, breakthrough time in this method refers to the time interval between the application of a test liquid to the appropriate surface of the material and the detection, by any suitable method, of the test liquid on the other side of the material.

The test apparatus, as shown in Fig. 7.7, consists mainly of the following three parts:

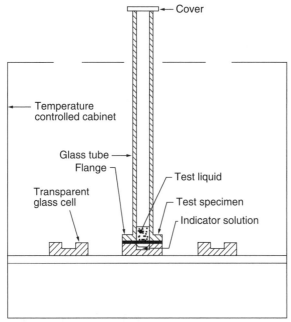

Note. The glass tube is integral with the flange.

7.7 Apparatus for assessment of breakthrough time.

- Glass tub, flanged at one end. The flange diameter is 30 mm. The tube has an internal diameter of 10 mm and a length of stem of not less than 100 mm and is provided with a loose glass cover.
- Transparent glass cell, with an overall height of 10 mm and with a central cavity 5 mm deep and 10 mm in diameter.
- Glass-fronted temperature-controlled cabinet to enable tests to be carried out at different temperatures if required.

Method for the determination of liquid permeating after breakthrough
In this method, the test specimen acts as a barrier between one compartment of a permeation cell, which contains the test liquid, and another compartment through which a stream of gas or liquid is passed for the collection of diffused molecules of the test liquid or its component chemicals for analysis. The mass of the test liquid or its component chemicals in the collecting medium is determined as a function of time after application to the test specimen, the breakthrough time and the masses permeating after breakthrough being derived graphically. Therefore, breakthrough time in this method refers to the elapsed time between the initial application of a test liquid to the appropriate surface of the material and its subsequent presence on the other side of the material.

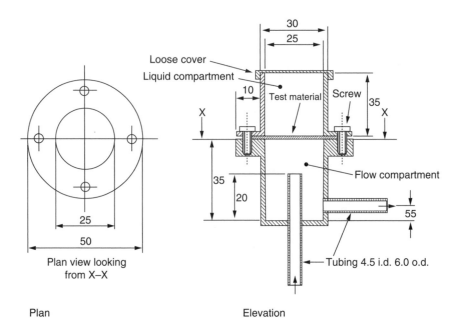

Note. Large arrows in the elevation denote direction of flow of gaseous or liquid collecting medium.

7.8 Permeation cell.

The permeation cell, as shown in Fig. 7.8, comprises two flanged compartments with dimensions forming a hollow cylinder when bolted together through the flanges. The upper compartment (or liquid compartment) for containment of the test liquid is fitted with a loose cover to avoid build-up of pressure and prevent excessive contamination of the immediate environment when volatile chemicals are under test. The lower compartment (or flow compartment) is fitted with pipework with dimensions to allow gas or liquid to circulate freely at the appropriate rates without build-up of pressure.

Measurement of repellency, retention, and penetration of liquids through fabric

BS ISO 22608:2004 specifies a test method to measure repellency, retention and penetration of a known volume of liquid pesticide when applied to protective clothing material. No external hydrostatic or mechanical pressure is applied to the test specimen during or after the application of the liquid pesticide. The degree of contamination depends on numerous factors such as type of exposure, application technique, and pesticide formulation.

As the level of exposure can vary considerably, this method is designed to rate relative performance of personal protective equipment (PPE) materials at two levels of contamination. Low level of contamination is achieved by applying 0.1 ml liquid formulation and high level by applying 0.2 ml (BS ISO 22608:2004).

In BS ISO 22608, a test liquid is applied using a pipette to the surface of the test assembly, which consists of single or multiple layer protective clothing material (test specimen) and an absorbent paper backed by polyethylene film (collector layer). After a specified time, another absorbent paper backed by polyethylene film (top layer) is placed on the surface of the test specimen to remove the remaining liquid. The top layer, the contaminated test specimen and the collector layer are separated. The amount of test liquid in each layer is measured either by gravimetric analysis (weighing) or by other appropriate analytical techniques. Method A is a gravimetric method that measures the mass of the test liquid in each layer, whereas method B is an analytical method that requires extraction of the test liquid and measures the mass of the active ingredient. Data is obtained to calculate percent repellency, pesticide retention, and penetration (BS ISO 22608:2004).

For method A, the percentage repellency (PR), the percentage retention (PLR) and the percentage penetration (PP) of the test liquid are calculated using Equations 7.15 to 7.17 respectively:

$$PR = (m_{ap}/m_t) \times 100 \tag{7.15}$$

$$PLR = (m_{pc}/m_t) \times 100 \tag{7.16}$$

$$PP = (m_{cl}/m_t) \times 100 \tag{7.17}$$

where

m_{ap} = mass of test liquid in 80 mm × 80 mm absorbent paper used to remove excess liquid pesticide after 10 min, mg

m_{pc} = mass of test liquid in the protective clothing material test specimen, mg

m_{cl} = mass of test liquid in the collector layer, mg

m_t = total amount of test liquid, mg.

The evaporation loss (EL) for each test specimen is calculated using Equation 7.18:

$$EL = 100 - (PT + PLR + PP) \tag{7.18}$$

For method B, the percentage repellency (PR), the percentage retention (PLR) and the percentage penetration (PP) of the test liquid are calculated using Equations 7.19 to 7.21 respectively:

$$PR = (m_{ap}/m_t) \times 100 \tag{7.19}$$

$$\text{PLR} = (m_{\text{pc}}/m_{\text{t}}) \times 100 \qquad\qquad 7.20$$

$$\text{PP} = (m_{\text{cl}}/m_{\text{t}}) \times 100 \qquad\qquad 7.21$$

where

m_{ap} = the mass of active ingredient in 80 mm × 80 mm absorbent paper used to remove excess liquid pesticide after 10 min, mg

m_{pc} = the mass of active ingredient in the protective clothing material test specimen, mg

m_{cl} = the mass of active ingredient in the collector layer, mg

m_{t} = total amount of active ingredient applied, mg.

The percentage extraction efficiency (EE) is determined using Equation 7.22:

$$\text{EE} = [(m_{\text{ap}} + m_{\text{pc}} + m_{\text{cl}})/m_{\text{t}}] \times 100 \qquad\qquad 7.22$$

where m_{t} is the total amount of active ingredient applied, mg.

7.5   Fabric permeability testing methods: applications

Testing methods for fabric permeability can have two types of applications:

- Quality assurance, i.e. routine quality control and marketing purposes
- Research and development, for innovative fabrics which can be applied to sportswear, protective clothing, smart textiles and other functional fabrics.

7.5.1 Quality assurance

Product testing is carried out for a number of reasons, the main one being to ensure complete customer satisfaction, thus making it very likely that repeat orders will follow. This is understandable, but monitoring of production during the manufacturing process is also important to ascertain if the product is suitable for the next stage in the production sequence (Fung 2002).

Quality assurance (QA) includes all factors which are relevant to quality and customer satisfaction, and has grown out of simple quality control. It goes from the earliest stages of product design, product development, purchase and monitoring of raw materials through to manufacturing, testing and inspection of the finished product. Quality assurance also involves contact with the customer, from the early stages of product design to meetings after delivery, to ensure customer satisfaction has been achieved. The QA department ensures that every member of the workforce and each member of staff is trained to regard quality as their duty and not just that of the quality department (Fung 2002).

The increasing popularity of sportswear, leisurewear, protective clothing and smart textiles and other functional clothing is reflected by the increasing trend in global sales. The variety of speciality fabrics used in these clothing sectors has expanded with the advance of new technology and consumer interest. An example is the recent success claimed by the breathable-waterproof fabric sector. The 'breathability' of a waterproof fabric has proved to be consumer desirable and can command a price premium. A variety of test methods have been developed to measure the fabric permeability and thus to communicate the fabric's potential to the would-be purchaser. The main commercial test methods in relation to fabric permeability are listed in Tables 7.1–7.3). These standards describe a comparatively simple method for testing the permeability of fabric that will provide the manufacturer with a clearly recognized method for quality control within the plant.

The test parameters generated from the above test methods are variable. Consequently, the results from different methods are not only not directly

Table 7.1 Test methods associated with air permeability (selected)

Standard code	Standard title
ASTM D6476-05	Standard Test Method for Determining Dynamic Air Permeability of Inflatable Restraint Fabrics
ASTM D737-2004	Test Method for Air Permeability of Textile Fabrics
ASTM D5886-1995 (Reapproved 2006)	Standard Guide for Selection of Test Methods to Determine Rate of Fluid Permeation Through Geomembranes for Specific Applications
ASTM D2752-1988 (Reapproved 2002)	Standard Test Methods for Air Permeability of Asbestos Fibers
BS ISO 7229-1997	Rubber- or Plastics-Coated Fabrics – Measurements of Gas Permeability
BS 3424-16-1995	Testing Coated Fabrics Part 16: Method 18: Determination of Air Permeability
BS EN ISO 9237-1995	Textiles – Determination of the Permeability of Fabrics to Air
BS EN ISO 4638-1995	Polymeric Materials, Cellular Flexible – Determination of Air Flow Permeability
BS 5636-1990	Determination of Permeability of Fabrics to Air
BS 3424-18:1986	Testing coated fabrics – Part 18: Methods 21A and 21B: Methods for determination of resistance to wicking and lateral leakage to air
NF G07-111-1995	Textiles – Determination of permeability of fabrics to air
NF G37-114-1983	Fabrics coated with rubber or plastics. Gas permeability test
ISO 9237-1995	Textiles – Determination of the permeability of fabrics to air

Table 7.2 Test methods associated with water permeability, water repellency and water resistance (selected)

Standard code	Standard title
AATCC 22-2005	Water Repellency: Spray Test
AATCC 193-2004	Aqueous Liquid Repellency: Water/Alcohol Solution Resistance Test
AATCC 70-2000	Water Repellency: Tumble Jar Dynamic Absorption Test
AATCC 35-1980	Water Resistance: Rain Test
ASTM E96/E96-M 05	Standard Test Method for Water Vapour Transmission of Materials
ASTM D6701-01	Standard Test Method for Determining Water Vapour Transmission Rates Through Nonwoven and Plastic Barriers
ASTM D5886-95 (Reapproved 2006)	Standard Guide for Selection of Test Methods to Determine Rate of Fluid Permeation Through Geomembranes for Specific Applications
ASTM D583-1963 (Withdrawn 1971)	Methods of Test for Water Resistance of Textile Fabrics
JIS L1092-1998	Testing methods for water resistance of textiles
JIS L1099-1993	Testing methods for water vapour permeability of textiles
BS ISO 8096:2005	Rubber- or plastics-coated fabrics for water-resistant clothing – Specification
BS EN ISO 15496-2004	Textiles – Measurement of water vapour permeability of textiles for the purpose of quality control
BS EN 13515-2002	Footwear – Test Methods for Uppers and Lining – Water Vapour Permeability and Absorption Chaussures
BS EN 13518-2002	Footwear – Test methods for uppers – Water resistance
BS 3546-2001	Coated fabrics for use in the manufacture of water penetration resistant clothing
BS EN 13073-2001	Test Methods for Whole Shoe – Water Resistance
BS ISO 7229-1997	Rubber- or Plastics-Coated Fabrics – Measurements of Gas Permeability
BS EN 29865:1993	Textiles – Determination of Water Repellency of Fabrics by the Bundesmann Rainshower Test
BS 3424-34-1992 (R1999)	Testing Coated Fabrics Part 34: Method 37: Method for Determination of Water Vapour Permeability Index
BS EN 20811-1992	Resistance of Fabric to Penetration by Water (Hydrostatic Head Test)
BS 3546-4-1991	Coated fabrics for use in the manufacture of water penetration resistant clothing – Part 4: Specification for water vapour permeable coated fabrics
BS 7209-1990	Specification for Water Vapour Permeable Apparel Fabrics. Appendix B. Determination of Water Vapour Permeability Index

Table 7.2 Continued

Standard code	Standard title
BS 3424-26:1990	Testing coated fabrics – Part 26: Methods 29A, 29B, 29C and 29D. Methods for the determination of resistance to water penetration and surface wetting
BS 5066:1974	Method of Test for the Resistance of Fabric to an Artificial Shower
NF G62-107-2002	Footwear – Test methods for uppers and lining – Water vapour permeability and absorption
NF G38-140-1999	Geotextiles and geotextile-related products. Determination of water permeability characteristics normal to the plane, without load
NF G07-058-1994	Textiles. Determination of water repellency of fabrics by the Bundesmann rain-shower test
NF G38-016-1989	Textiles: articles for industrial use; tests for geotextiles, measurement of water permeability ratio
NF G07-135-1978	Textiles. Tests of woven fabrics. Determination of impermeability of linen cloth for covers, tents and equipment. 'Pocket' method
EN 31092-1993E	Textiles. Determination of Physiological Properties. Measurement of Thermal and Water-vapour Resistance under Steady-state Conditions (Sweating Guarded-hot plate Test)
ISO 11092-1993	Textiles. Determination of Physiological Properties. Measurement of Thermal and Water-vapour Resistance under Steady-state Conditions (Sweating Guarded-hot plate Test)
ISO 9865:1991	Textiles. Determination of Water Repellency of Fabrics by the Bundesmann Rainshower Test

comparable but they may not even show a clear correlation. The differences in approach and in test conditions have an appreciable effect on the final result. To truly compare one fabric with another, it is essential to ensure that they have been assessed by the same method under the same conditions. To do otherwise could be misleading and have litiginous consequences.

7.5.2 Research and development of innovative materials and fabrics

New developments in fibre science and technology have resulted in fibres with tailored properties, thus expanding their uses beyond the domain of conventional textiles. A great deal of new technology was applied to research and develop innovative materials and fabrics for application in the field of

Table 7.3 Test methods associated with chemical permeability (liquid and gas) (selected)

Standard code	Standard title
ASTM F1461-07	Standard Practice for Chemical Protective Clothing Program
ASTM F903-03 (Reapproved 2004)	Test Method for Resistance of Materials Used in Protective Clothing to Penetration by Liquids
ASTM F1296-03	Standard Guide for Evaluating Chemical Protective Clothing
ASTM F1194-99 (Reapproved 2005)	Guide for Documenting the Results of Chemical Permeation Testing of Materials Used in Protective Clothing
ASTM F1154-99a (Reapproved 2004)	Practices for Qualitatively Evaluating the Comfort, Fit, Function, and Integrity of Chemical-Protective Suit Ensembles
ASTM F1001-99a (Reapproved 2006)	Guide for Selection of Chemicals to Evaluate Protective Clothing Materials
ASTM F1407-99a (Reapproved 2006)	Standard Test Method for Resistance of Chemical Protective Clothing Materials to Liquid Permeation – Permeation Cup Method
ASTM F1383-99a	Standard Test Method for Resistance of Protective Clothing Materials to Permeation by Liquids or Gases Under Conditions of Intermittent Contact
ASTM F739-99a	Test Method for Resistance of Protective Clothing Materials to Permeation by Liquids or Gases Under Conditions of Continuous Contact
ASTM D5886-95 (Reapproved 2006)	Standard Guide for Selection of Test Methods to Determine Rate of Fluid Permeation through Geomembranes for Specific Applications
ASTM-F739-91	Test Method for Resistance of Protective Clothing Materials to Permeation by Liquids and Gases
BS EN 14786:2006	Protective clothing – Determination of resistance to penetration by sprayed liquid chemicals, emulsions and dispersions – Atomizer test
BS EN 14605:2005	Protective clothing against liquid chemicals – Performance requirements for clothing with liquid-tight (Type 3) or spray-tight (Type 4) connections, including items providing protection to parts of the body only (Types PB [3] and PB [4])
BS EN 13034:2005	Protective clothing against liquid chemicals – Performance requirements for chemical protective clothing offering limited protective performance against liquid chemicals (Type 6 and Type PB [6] equipment)
BS EN ISO 6530:2005	Protective clothing – Protection against liquid chemicals – Test method for resistance of materials to penetration by liquids
BS ISO 22608:2004	Protective clothing – Protection against liquid chemicals – Measurement of repellency, retention, and penetration of liquid pesticide formulations through protective clothing materials

Table 7.3 Continued

Standard code	Standard title
BS EN 14325:2004	Protective clothing against chemicals – Test methods and performance classification of chemical protective clothing materials, seams, joins and assemblages
BS EN 374-3:2003	Protective gloves against chemicals and micro-organisms – Part 3: Determination of resistance to permeation by chemicals
BS EN 374-2:2003	Protective gloves against chemicals and micro-organisms – Part 2: Determination of resistance to penetration
BS EN 943-2:2002	Protective clothing against liquid and gaseous chemicals, including liquid aerosols and solid particles – Part 2: Performance requirements for 'gas-tight' (Type 1) chemical protective suits for emergency teams (ET)
BS EN 943-1:2002	Protective clothing against liquid and gaseous chemicals, including liquid aerosols and solid particles – Part 1: Performance requirements for ventilated and non-ventilated 'gas-tight' (Type 1) and 'non-gas-tight' (Type 2) chemical protective suits
BS EN ISO 6529:2001	Protective clothing – Protection against chemicals – Determination of resistance of protective clothing materials to permeation by liquids and gases
BS 2F 142:1999	Hydrolysis resistant, thermoplastic polyether polyurethane elastomer-coated nylon fabric for aerospace purposes
BS ISO 13994:1998	Clothing for protection against liquid chemicals – Determination of the resistance of protective clothing materials to penetration by liquids under pressure
BS EN 468:1995	Protective clothing – Protection against liquid chemicals – Test method: Determination of resistance to penetration by spray (Spray test)
BS EN 467:1995	Protective clothing – Protection against liquid chemicals – Performance requirements for garments providing protection to parts of the body
BS EN 463:1995	Protective clothing – Protection against liquid chemicals – Test method: Determination of resistance to penetration by a jet of liquid (Jet test)
BS EN 465:1995	Protective clothing – Protection against liquid chemicals – Performance requirements for chemical protective clothing with spray-tight connections between different parts of the clothing (type 4 equipment)

Table 7.3 Continued

Standard code	Standard title
BS EN 466:1995	Protective clothing – Protection against liquid chemicals – Performance requirements for chemical protective clothing with liquid-tight connections between different parts of the clothing (type 3 equipment)
BS EN 466-1:1995	Protective clothing – Protection against liquid chemicals – Part 1: Performance requirements for chemical protective clothing with liquid-tight connections between different parts of the clothing (type 3 equipment)
BS F 142:1995	Specification for hydrolysis resistant, thermoplastic polyether polyurethane elastomer-coated nylon fabric for aerospace purposes
BS EN 464:1994	Protective clothing – Protection against liquid and gaseous chemicals, including liquid aerosols and solid particles – Test method: Determination of leak-tightness of gas-tight suits (Internal Pressure Test)
BS EN 374-1:1994	Protective gloves against chemicals and micro-organisms – Part 1: Terminology and performance requirements
BS EN 368:1993	Protective clothing – Protection against liquid chemicals – Test method: Resistance of materials to penetration by liquids
BS EN 369:1993	Protective clothing – Protection against liquid chemicals – Test method: Resistance of materials to permeation by liquids
BS 7182:1989	Specification for air-impermeable chemical protective clothing
BS 7184:1989	Recommendations for selection, use and maintenance of chemical protective clothing
BS 4724: Part 2:1988	Resistance of clothing materials to permeation by liquids – Part 2: Method for the determination of liquid permeating after breakthrough
BS 4724-1:1986	Resistance of clothing materials to permeation by liquids – Part 1: Method for the assessment of breakthrough time
ISO 7229:1997(E)	Rubber- or plastics-coated fabrics – Measurement of gas permeability

sportswear, protective clothing, smart waterproof and breathable fabric and so on. Therefore, fabric permeability, which is related to the ability of fabric to allow a penetrant, such as a gas, liquid or solid material, to pass through, is the most important property because it has a very close relationship with clothing comfort and safety.

Sportswear

There has been a strong growth in the development and use of highly functional materials in sportswear and outdoor leisure clothing. The performance requirements of many such products demand balance of widely different properties of drape, thermal insulation, barrier to liquids, antistatic, stretch, physiological comfort, etc. The research in this field over the past decade has led to the commercial development of a variety of new products for highly functional end-uses (Buirski 2005).

Many smart double-knitted or double-woven fabrics have been developed for sportswear in such a way that their inner face, close to human skin, has optimal moisture wicking and sensory properties whereas the outer face of the fabric has optimal moisture dissipation behaviour. An example is Nike Sphere Dry, of which Nike supplied kits for the teams of the USA, Brazil, The Netherlands, Portugal, Korea, Mexico, Croatia and Australia in the World Cup 2006. It was reported that Nike Sphere Dry wicks sweat away from the body and through the shirt to keep the skin drier. The fabric's technology helps air move quickly through the garment and over the skin to assist the body's own natural cooling system and encourage the evaporation of sweat. At the same time, raised nodes on the underside lift the jersey away from the player's body and reduce clinging (McCurry and Butler 2006).

By designing new processes for fabric preparation and finishing, and as a result of advances in technologies for the production and application of suitable polymeric membranes and surface finishes, it is now possible to combine the consumer requirements of aesthetics, design and function in sportswear for different end-use applications. Hence, water permeability is one of the most important properties, which will affect the thermal insulation, quick liquid absorption and ability to evaporate water while staying dry to the touch, and be capable of transporting perspiration from the skin to the outer surface and then quickly dispersing it. Evaluation of water permeability becomes one of the most important requirements for research and development of sportswear (Buirski 2005).

Protective clothing

Scientific advancements made in various fields have undoubtedly increased the quality and value of human life. However, it should be recognized that the technological developments have also exposed us to greater risks and danger of being affected by unknown physical, chemical and biological attacks. One such currently relevant danger is from bioterrorism and weapons of mass destruction. In addition, we continue to be exposed to hazards from fire, chemicals, radiation and biological organisms such as bacteria and viruses. Fortunately, simple and effective means of protection from most of these hazards are available. Textiles are an integral part of most protective equipment. Protective clothing is manufactured using traditional textile manufacturing technologies such as weaving, knitting and non-wovens and also by specialized techniques such as 3-D weaving and braiding using natural and man-made fibres (Zhou *et al.* 2005).

Protective clothing is now a major part of textiles that are classified as technical or industrial textiles. Protective clothing refers to garments and other fabric-related items designed to protect the wearer from harsh environmental effects that may result in injuries or death. Today, the hazards that workers are exposed to are often so specialized that no single type of clothing will be adequate for protection. Providing protection for the general population has also been taken seriously in view of the potential disaster due to terrorism or biochemical attacks. Extensive research is being done to develop protective clothing for various regular and specialized civilian and military occupations (Zhou *et al.* 2005).

General requirements applicable to all types of personal protective equipment (PPE) concern design principles, innocuousness of the PPE, comfort and efficiency, and the information supplied by the manufacturer. Of these requirements, comfort and efficiency are among the most important and are related to water permeability and chemical permeability of fabric. Various test methods have been devised to measure the resistance of chemical protective clothing materials to penetration by liquid and gas, or qualitatively to evaluate the comfort, fit, function and integrity of chemical protective suit ensembles, etc. Moreover, new test methods will be established in parallel with progress in research and development of innovative protective clothing.

Waterproof and breathable fabric

Over the past few decades, there have been many advances in apparel textiles and clothing design that take account of the extremes of human thermoregulation and the environment. The main objective is to maintain the wearer in a state of thermo-physiological comfort under the widest possible range of workloads and ambient conditions. One approach has led to the proliferation of waterproof and breathable fabrics (WBF) for foul-weather clothing and other active sports and leisurewear. These materials have been scientifically engineered to balance the conflicting properties of high water vapour permeability (in order to expel perspiration) and waterproofness (to repel atmospheric precipitation). Therefore, as one type of interactive fabrics, waterproof and breathable fabrics could prevent the penetration of liquid water from outside to inside the clothing yet permit the penetration of water vapours from inside the clothing to the outside atmosphere. They are designed for use in garments that provide protection from the weather, that is, from wind, rain and loss of body heat (Lomax 1991; Holmes 2000; Fung 2002).

In general, polyurethane membranes used in breathable fabrics can be classified into three main groups (Lomax 1990):

• Microporous membranes and coatings
• Hydrophilic membranes and coatings
• Combined microporous and hydrophilic layers.

Gore-Tex™ and Sympatex™, as the representatives of commercial WBFs, have received wide interest. These two types of WBFs have been used to develop various applications in the textile industry. Gore-Tex is one type of WBF laminated with porous PTFE film, while Sympatex is another type of WBF laminated with dense hydrophilic polyester film and hydrophilic polyurethane film. Due to the obvious differences in water vapour penetrating mechanism, production technology and product properties, research and development in these two types of WBFs are still being undertaken throughout the world.

Porous membranes
For porous membranes applied to WBFs, the maximum size of micropore is between the diameter of a water vapour molecule and that of a water droplet. So the fabric laminated/coated with the porous membrane is able to separate water molecules from liquid water, and therefore can provide a good overall balance between breathability and waterproofness.

The surface and cross-section of Gore-Tex membrane were detected using SEM as shown as Figs 7.9 and 7.10. It is apparent that Gore-Tex

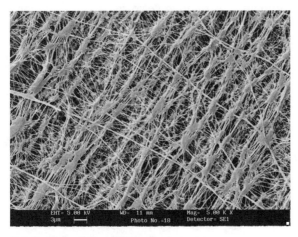

7.9 Surface of Gore-Tex membrane.

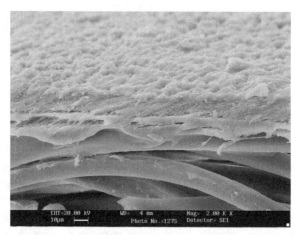

7.10 Cross-section of Gore-Tex membrane.

membrane is microporous and can provide the path for water vapour penetration (Ding *et al.* 2001). For the microporous membrane, if a pressure difference across the sample is present, convective gas flow through the sample carries water vapour along with the flow, which may add to or subtract from the diffusive flux, depending on the direction of the convective gas flow (Gibson 2000).

Dense membranes
Generally speaking, dense polymer membranes have no pores but there exists the thermally agitated motion of chain segments to generate penetrant-scale transient gaps in the matrix, that is, free volume of the membrane, allowing penetrants to diffuse from one side of the membrane to the other. Accordingly, it is reasonable to regard a dense polymer as a 'porous medium', where the 'pores' are gaps among the polymer matrix (Chen *et al.* 2001).

The surface and cross-section of Sympatex membrane were detected using SEM as shown in Figs 7.11 and 7.12. It is apparent from the SEM pictures that Sympatex membrane is dense and offers no path for water vapour penetration (Ding *et al.* 2001).

Smart/intelligent/adaptive materials are composed of three basic elements: sensors, actuators, and control processors (Liang *et al.* 1997). Such materials can undergo a response adaptively triggered by small changes in their environment, such as a change in temperature and loading. A famous example among textile applications is smart waterproof and breathable fabrics (SWBFs) using temperature-sensitive polyurethane (TS-PU). TS-PU is one type of smart polymer that is able to sense and respond to external tem-

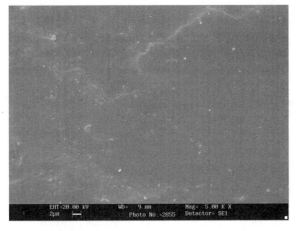

7.11 Surface of Sympatex membrane.

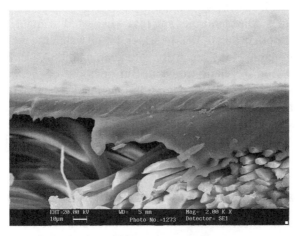

7.12 Cross-section of Sympatex membrane.

perature in a predetermined way. That is, this material can sense the change of external temperature, which will lead to a significant increase in water vapour permeability (WVP) of the polymeric membrane. Particularly, by means of appropriate molecule design, the abrupt change in water vapour permeability of TS-PU membrane can be controlled within desired temperature ranges such as room temperature range (Ding *et al.* 2003, 2004a, 2004b, 2005, 2006, 2008).

Such a property enables TS-PU materials to have broad application to the textile industry, medicine, environmental projects and so on, as shown in Fig. 7.13. The figure shows the structure of an amphibious diving suit that

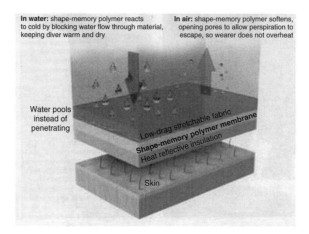

In **water:** shape-memory polymer reacts to cold by blocking water flow through material, keeping diver warm and dry

In **air:** shape-memory polymer softens, opening pores to allow perspiration to escape, so wearer does not overheat

Water pools instead of penetrating

Low-drag stretchable fabric
Shape-memory polymer membrane
Heat reflective insulation

Skin

7.13 Fabric structure of dual-purpose clothing.

was designed by the US Army's Soldier and Biological Chemical Command Laboratory in Natick, Massachusetts, and that is expected to enable wearers to be comfortable both in and out of the water. In the water, the amphibious diving suit performed like any other dry suit, keeping the wearer warm by preventing water from reaching the skin. But once out of the water, the structure of its novel three-layer membrane changed to let perspiration escape, preventing the wearer from overheating. Therefore, the amphibious suit was considered suitable for divers from the US Navy's Sea-Air-Land (SEAL) division who can get out of the water ready for action in light-weight garb (Graham-Rowe 2001).

7.6   Innovative test methods for fabric permeability

The movement of moisture, air and chemical liquids and gases is a complex series of processes. Over many years, a number of innovative ideas and novel test methods for fabric permeability have been created by constant exploration and modification.

7.6.1 Dynamic testing methods

Laboratory testing is usually a necessary first step in evaluating the comparative water vapour transport properties of candidate materials for new clothing system designs. However, comparison of material properties often becomes complex due to changes in tested properties under different test conditions. One material may be rated better than another material at one particular set of test parameters, yet the ranking may reverse under a different set of conditions. The two effects that are usually responsible for changes in ranking of materials are concentration-dependent permeability and temperature-dependent permeability (Gibson 2000).

Concentration-dependent permeability

Membranes that contain a continuous hydrophilic component, such as Gore-Tex and Sympatex, change their transport properties based on the amount of water contained in the hydrophilic polymer layer. The magnitude of the relative changes in water vapour transfer rate as a function of membrane water content are quite large for several common clothing materials and systems. The water content of these materials is a function of the water vapour content (humidity) of the environment on either side of the clothing layer. Test methods that evaluate concentration-dependent permeability need to be capable of independently varying the relative humidity of the environment on the two sides of the material.

Temperature-dependent permeability

Some polymer membranes may exhibit lower intrinsic water vapour transfer properties at low temperatures. This effect is of practical importance for the ability of cold-weather clothing to dissipate water vapour during active wear, or for boots, gloves and sleeping bags to dry out under cold conditions. Knowledge of temperature-dependent permeability is also important when comparing test results between test methods or laboratories that may conduct standard testing at different temperatures. Analysis of temperature-dependent permeability must distinguish between changes in the intrinsic transport properties of the material, and the apparent decrease in water vapour transport rates due simply to the lower vapour pressure of water at lower temperatures (Gibson 2000).

It is reported that one test method, the dynamic moisture permeation cell (DMPC), can control the humidity and flow rate on the two sides of the test sample, and hence control the temperature of the test system (Fig. 7.14). This allows temperature-dependent effects to be separated from concentration-dependent effects on mass transfer phenomena. The DMPC permits the experimenter to explore the temperature dependence of the diffusion behaviour at different points on the vapour sorption isotherm of the hydrophilic polymer component of a polymer film or membrane laminate (Gibson 1993, 1999, 2000; Gibson *et al.* 1995).

7.6.2 Whole assessment methods

Moisture transfer properties of textile fabrics and garments are important to the thermal comfort of clothed persons. A number of test methods have

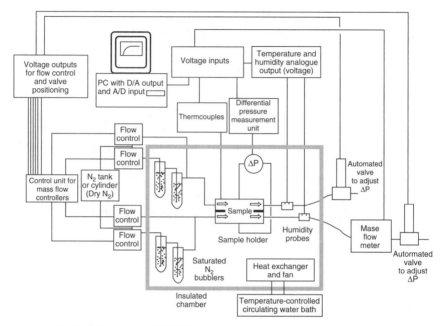

7.14 Schematic of DMPC test arrangement.

been developed to evaluate the moisture transfer properties of textile fabrics and garments. In these test methods, different techniques and testing conditions were used in order to try to provide complete assessment of moisture transfer properties of textile products. Although every test method was close to real use conditions at some point, no one test method has been able to simulate the complicated process. Therefore, many researchers are trying to investigate the differences and interrelationships between the results from these different test methods.

Dolhan (1987) compared two Canadian standards (CAN2-4.2-M77 and CAN/CGSB-4.2 No. 49-M91) and the ASTM E96 test methods for measuring the water vapour transmission properties and found that the results of these tests were not directly comparable because of the differences in the water vapour pressure gradients driving the moisture transmission in the different test methods.

Gibson (1993) conducted an extensive investigation on the relationship of the test results from the sweating guarded hot plate (ISO 11092) and those from the ASTM E96 Cup Method. In his work, permeable materials, hydrophobic and hydrophilic membrane laminates were tested and the results were standardized in the units of air resistance and water vapour transmission rate. It was found, except for the hydrophilic samples, that there is a clear correlation between the results from the two tests. As the test condition in the guarded sweating hot plate tests resulted in much higher equilibrium water content in the hydrophilic polymer layer, which influences the polymer's permeability, the water vapour transmission rate through the hydrophilic membrane is greater when tested using the sweating guarded hot plate.

Gretton *et al.* (1996) classified the fabric samples into four categories, including air permeable fabrics, microporous membrane laminated fabrics, hydrophilic membrane laminated/coated fabrics and hybrid coated/laminated fabrics, in investigating the correlation between the test results of the sweating guarded hotplate (ISO 11092) and the evaporative dish method (BS 7209). They showed that there is a good correlation between the two test methods for all fabrics except for the hydrophilic coated and laminated fabrics that transmit water vapour without following the Fickian law of diffusion.

McCullough *et al.* (2003) measured the water vapour permeability and evaporative resistance of 26 different waterproof, windproof and breathable shell fabrics using five standard test methods. The water vapour transmission rate (WVTR) was measured using the ASTM E96 upright and inverted cup tests with water, the JIS L1099 desiccant inverted cup test and the new ASTM F2298 standard using the dynamic moisture permeation cell (DMPC). The evaporative resistance was measured using the ISO 11092 sweating hot plate test. The WVTRs were consistently highest when measured with the desiccant inverted cup, followed by the inverted cup, DMPC and upright cup. The upright cup was significantly correlated with the DMPC (0.97), and the desiccant inverted cup was correlated to the sweating hot plate (−0.91).

7.6.3 Modern characterization methods

Measurement of water vapour transport property of
fabrics by differential scanning calorimeter

Measurement of the rate of water vapour evaporation can be carried out using a TA Instruments model 2920 Modulated Differential Scanning Calorimeter (DSC). Experiments are carried out using dry nitrogen as the carrier gas at a flow rate of 40 mL/min to remove the evaporated water from the cell environment. The standard sample pans are replaced with brass containers (one reference assembly and the other containing the test sample). The special container assembly (Fig. 7.15) was designed based on Day's technique (Day and Sturgeon 1986) and it essentially consists of a small water-holding brass cup and a brass retainer. The fabric sample to be tested is placed over the 'O' ring which in turn is fitted to the inside of the groove. A brass retainer of diameter slightly less than the inner diameter

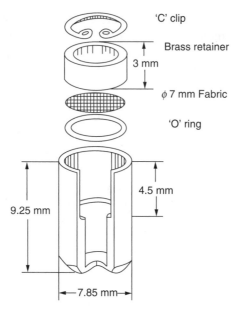

7.15 Container assembly used in modulated DSC.

of the brass cup is placed over the test specimen and properly secured in place by a 'C' clip. A similar arrangement for the empty reference cup is used for comparative purposes. The rate of evaporation of samples is measured with the help of a DSC curve choosing time versus heat flow. The quantity of water taken in the container for all the samples is constant (5 mg) and the exact quantity of water is placed in the containers with the help of a micro-burette. The experiments are conducted with temperatures programmed from room temperature to 40°C and a heating rate of 20°C/min is used. The isothermal temperature of 40°C is chosen so as to simulate the normal skin temperature of the human being (Indushekar *et al.* 2005).

The water vapour flux through the fabric occurs in four different ways (Chen *et al.*, 2001). They are:

- Diffusion of water vapour through the air spaces between the fibres
- Absorption, transmission and desorption of the water vapour by the fibres
- Adsorption and diffusion of the water vapour along the fibre surface (wicking)
- Diffusion of the water vapour between yarn spaces.

Therefore, heat flow behaviour can be exhibited by DSC curves as shown in Fig. 7.16. The first part of the curve (A) indicates the quantity of heat flow required to heat the specimen cups from room temperature up to the isothermal temperature of 40°C. The curve then follows a steady flow path

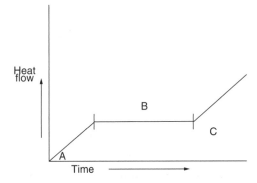

7.16 Heat flow behaviour exhibited by DSC curves.

(B) indicating that the equilibrium has been established and gives an indication of the heat flow and time required for evaporating the water. The curve then again follows an upward path (C) indicating that there is a sudden change of the heat flow as the water in the cup is completely evaporated and the measured heat flow reverts to that of the two empty cups. From the initial weight of the water (5 mg) and the curve path (A–B), i.e., the observed time required for complete evaporation, the evaporation rates for the specimen are computed. The evaporation rates are generally expressed as a function of the specimen area and hence the results are also expressed in $g/m^2/h$ (Indushekar *et al.* 2005).

The DSC employing special specimen cups offers a simple experimental procedure for measuring the water vapour transport property of fabric. The results can be applied to evaluating the relative evaporative cooling efficiencies of fabrics as well as providing indications of the potential metabolic heat problem associated with impermeable vapour barriers. However, Indushekar *et al.* compared the water vapour transmission rates measured by a modulated differential scanning calorimeter (MDSC) and those by the conventional dish technique as specified in BS7209 for a wide range of woven-based fabrics used in cold-weather protective clothing. The study showed that results from these two test methods differ widely due to the differences in the water vapour gradients that occurred in the two methods (Indushekar *et al.* 2005).

Measurement of liquid moisture management property using the moisture management tester

Moisture management properties influence the human perception of moisture and comfort. The moisture management properties of fabrics depend on their water resistance, water repellency, water absorption, wicking of the

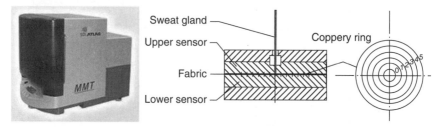

7.17 Schematic of Moisture Management Tester (MMT).

fibres and yarns, as well as the geometric and internal structures of constituting materials such as fibres, yarns and fabrics. The moisture management tester (MMT) provides a procedure for the evaluation of the dynamic movement of liquid moisture in porous fabrics.

A schematic of the moisture management tester (MMT) is shown in Fig. 7.17 (Hu *et al.* 2005). Liquid moisture management properties of fabrics are tested by placing a sample of the fabric between upper and lower concentric moisture sensors. A predefined amount of test solution (synthetic sweat) is introduced onto the upper side of the fabric, and then the test solution will transfer onto the fabric in three directions:

• Spreading outward on the upper surface of the fabric
• Transferring through the fabric from the upper surface to the bottom surface
• Spreading outward on the lower surface of the fabric and then evaporating.

The liquid moisture content measured can be applied to assess the dynamic liquid moisture transport behaviours in these multiple directions inside the material.

Measurement of moisture vapour resistance using thermal manikin

There are relatively few sweating manikins available for measuring the evaporative resistance or vapour permeability of clothing. Some manikins are covered with a cotton knit suit and wetted out with distilled water to create a saturated sweating skin. However, the skin will dry out over time unless tiny tubes are attached to the skin so that water can be supplied at a rate necessary to sustain saturation. Other manikins have sweat glands on different parts of the body. Water is supplied to each sweat gland from inside the manikin, and its supply rate can be varied. A new type of sweating manikin (as shown in Fig. 7.18) uses a waterproof but moisture-permeable fabric skin, through which water vapour is transmitted from the inside of the body to the skin surface (Fan and Chen 2002; Fan and Qian 2004). Some

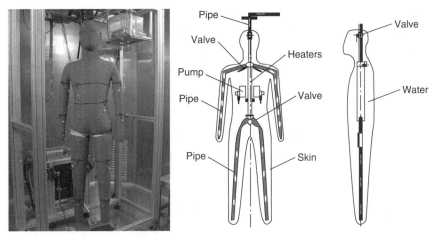

7.18 The perspiring fabric thermal manikin and the inner construction.

manikins keep the clothing from getting wet by using a microporous membrane between the sweating surface and the clothing, but this configuration may increase the insulation value of the nude manikin (McCullough 2005).

ASTM F2370 specifies procedures for measuring the evaporative resistance of clothing systems under isothermal conditions, i.e. where the manikin's skin temperature is the same as the air temperature so that there is no temperature gradient for dry heat loss. An alternative protocol in the standard allows the clothing ensemble to be tested under environmental conditions that simulate actual conditions of use; this is called the non-isothermal test. The same environmental conditions are used for the insulation test and the non-isothermal sweating manikin test. The air temperature is lower than the manikin's skin temperature, so dry heat loss is occurring simultaneously with evaporative heat loss, and condensation may develop in the clothing layers. The evaporative resistance determined under non-isothermal conditions is called the apparent evaporative resistance value. The apparent evaporative resistance values for ensembles can only be compared to those of other ensembles measured under the same environmental conditions (McCullough 2005).

Kar *et al.* (2007) investigated the correlations between the moisture vapour resistances/transmission rates measured using the newly developed sweating fabric manikin (Walter) (Fig. 7.18), the moisture transmission test (Model CS-141), the ASTM E96 testing method and the sweating guarded hot plate method. For the range of air-permeable knitted fabrics tested, it was found that good interrelationships exist between the results from the four types of test methods, although some discrepancies exist between different tests due to differences in testing conditions. Test results from different moisture transfer test methods can therefore be convertible with due consideration.

7.7   Conclusions

Fabric permeability is an important factor in the performance of most textile materials. It is related to wear comfort and wear safety, as well as having various applications in the industrial field. At the same time, progress in textile science and technology results in the continued appearance of novel testing instruments and technologies for fabric permeability. Therefore, innovative test methods for fabric permeability will continue to be developed.

7.8   References

Books

Buirski, D. (2005). Current sportswear market, in *Textiles in Sport*, ed. R. Shishoo, Woodhead Publishing, Cambridge, UK, and CRC Press, Boca Raton, FL, pp. 15–84.

Byrne, C. (2000). Technical textiles market – an overview, in *Handbook of Technical Textiles*, ed. A. R. Horrocks and S.C. Anand, Woodhead Publishing, Cambridge, UK, in association with the Textile Institute, pp. 1–23.

Chen, C. X., Han, B. B., Li, J. D., Shang, T. G., Zou, J. and Jiang, W. J. (2001). A new model on the diffusion of small molecule penetrants in dense polymer membranes, *Journal of Membrane Science* **187**(1–2): 109–118.

Day, M. and Sturgeon, P. Z. (1986). Water vapor transmission rates through textile materials as measured by differential scanning calorimetry, *Textile Research Journal* **56**(3): 157–160.

Ding, X. M., Hu, J. L., Tao, X. M. and Hayashi, S. (2001). Effect of crystal structure of shape memory polyurethane film on water vapour permeability, *6th Asia Textile Conference*, Hong Kong.

Ding, X. M., Hu, J. L. and Tao, X. M. (2003). Temperature-sensitive polyurethane properties, *Textile Asia* **34**(4): 42–45.

Ding, X. M., Hu, J. L. and Tao, X. M. (2004a). Effect of crystal melting on water vapor permeability of shape memory polyurethane film, *Textile Research Journal* **74**(1): 39–43.

Ding, X. M., Hu, J. L., Tao, X. M. and Hu, C. P. (2004b). Microstructure and water vapor permeability of temperature-sensitive polyurethane, *83 World Textile Conference*, Shanghai, China, pp. 1279–1285.

Ding, X. M., Hu, J. L., Tao, X. M., Wang, Z. F. and Wang, B. (2005). Free volume and water vapor transport properties of temperature-sensitive polyurethanes, *Journal of Polymer Science: Part B: Polymer Physics* **43**: 1865–1872.

Ding, X. M., Hu, J. L., Tao, X. M. and Hu, C. P. (2006). Preparation of temperature-sensitive polyurethanes for smart textiles, *Textile Research Journal* **76**(5): 406–413.

Ding, X. M., Hu, J. L., Tao, X. M., Hu, C. P. and Wang, G. Y. (2008). Morphology and water vapor permeability of temperature-sensitive polyurethanes, *Journal of Applied Polymer Science* **107**(6): 4061–4069.

Dolhan, P. A. (1987). A comparison of apparatus used to measure water vapour resistance, *Journal of Coated Fabrics* **17**: 96–109.

Dunn, M. (2001). Design and permeability analysis of porous textile composites formed by surface encapsulation, in *Surface Characteristics of Fibers and Textiles*, ed. C. M. Pastore and P. Kiekens, Marcel Dekker, New York.

Fan, J. and Chen, Y. S. (2002). Measurement of clothing thermal insulation and moisture vapour resistance using a novel perspiring fabric thermal manikin, *Measurement Science and Technology* **13**: 1115–1123.

Fan, J. and Qian, X. (2004). New functions and applications of Walter, the sweating fabric manikin, *European Journal of Applied Physiology* **92**: 641–644.

Fung, W. (2002). Testing, product evaluation and quality, in *Coated and Laminated Textiles*. Woodhead Publishing, Cambridge, UK, and CRC Press, Boca Raton, FL, pp. 250–306.

Gibson, P., Kendrick, C., Rivin, D., Sicuranza, L. and Charmchi, M. (1995). An automated water vapor diffusion test method for fabrics, laminates, and films, *Journal of Coated Fabrics* **24**: 322–345.

Gibson, P. W. (1993). Factors influencing steady-state heat and water vapor transfer measurements for clothing materials, *Textile Research Journal* **63**(12): 749–764.

Gibson, P. W. (1999). Water vapor transport and gas flow properties of textiles, polymer membranes, and fabric laminates, *Journal of Coated Fabrics* **28**(4): 300–327.

Gibson, P. W. (2000). Effect of temperature on water vapor transport through polymer membrane laminates, *Polymer Testing* **19**(6): 673–691.

Graham-Rowe, D. (2001). Going commando, *New Scientist* March (31): 22.

Gretton, J. C., Brook, D. B., Dyson, H. M. and Harlock, S. C. (1996). A correlation between test methods used to measure moisture vapour transmission through fabrics, *Journal of Coated Fabrics* **25**: 301–310.

Holmes, D. A. (2000). Waterproof breathable fabrics, in *Handbook of Technical Textiles*, ed. A. R. Horrocks and S. C. Anand, Woodhead Publishing, Cambridge, UK, in association with the Textile Institute, pp. 282–315.

Hsieh, K. H., Tsai, C. C. and Tseng, S. M. (1990). Vapor and gas permeability of polyurethane membranes. Part 1. Structure–property relationship, *Journal of Membrane Science* **49**: 341–350.

Hsieh, K. H., Tsai, C. C. and Chang, D. M. (1991). Vapor and gas permeability of polyurethane membranes. Part 2. Effect of functional group, *Journal of Membrane Science* **56**(3): 279–287.

Hu, J. Y., Li, Y., Yeung, K. W., Wong, A. and Xu, W. L. (2005). Moisture management tester: a method to characterize fabric liquid moisture management properties, *Textile Research Journal* **75**(1): 57–62.

Indushekar, R., Awasthi, P. and Gupta, R. K. (2005). Studies on test methods used to measure water vapour transmission of fabrics by DSC and conventional dish techniques, *Journal of Industrial Textiles* **34**: 223–242.

Jeong, H. M., Ahn, B. K. and Kim, B. K. (2000a). Temperature sensitive water vapour permeability and shape memory effect of polyurethane with crystalline reversible phase and hydrophilic segments, *Polymer International* **49**(12): 1714–1721.

Jeong, H. M., Ahn, B. K., Cho, S. M. and Kim, B. K. (2000b). Water vapor permeability of shape memory polyurethane with amorphous reversible phase, *Journal of Polymer Science: Part B: Polymer Physics* **38**(23): 3009–3017.

Kar, F., Fan, J. and Yu, W. (2007). Comparison of different test methods for the measurement of fabric or garment moisture transfer properties, *Measurement Science and Technology* **18**: 2033–2038.

Liang, C., Rogers, C.A. and Malafeew, E. (1997). Investigation of shape memory polymers and their hybrid composites, *Journal of Intelligent Material System and Structure* **8**(4): 380–386.

Lomax, G. R. (1990). Hydrophilic polyurethane coating, *Journal of Coated Fabrics* **20**(10): 88–107.

Lomax, G. R. (1991). Breathable, waterproof fabrics explained, *Textile* **20**(4): 12–16.

McCullough, E. A. (2005). Evaluation of protective clothing systems using manikins, in *Textiles for Protection*, ed. R. A. Scott, Woodhead Publishing, Cambridge, UK, and CRC Press, Boca Raton, FL, pp. 217–232.

McCullough, E. A., Kwon, M. and Shim, H. (2003). A comparison of standard methods for measuring water vapour permeability of fabrics, *Measurement Science and Technology* **14**: 1402–1408.

McCurry, J. and Butler, N. (2006). World Cup 2006 – a global showcase for textile technologies, *Technical Textiles International* **15**(5): 11–14.

Molyneux, P. (2001). Transition-site model for the permeation of gases and vapors through compact films of polymers, *Journal of Applied Polymer Science* **79**(6): 981–1024.

Raheel, M., ed. (1996). *Modern Textile Characterization Methods*, International Fiber Science and Technology, Marcel Dekker, New York.

Saville, B. P. (1999). *Physical Testing of Textiles*, Woodhead Publishing, Cambridge, UK, and CRC Press, Boca Raton, FL.

Slater, K. (1971). Thermal comfort properties of fabrics, in *Progress in Textiles: Science and Technology*, ed. V. K. Kothari, IAFL Publications, **1**.

Zhou, W., Reddy, N. and Yang, Y. (2005). Overview of protective clothing, in *Textiles for Protection*, ed. R. A. Scott, Woodhead Publishing, Cambridge, UK, and CRC Press, Boca Raton, FL, p. 1.

Standards

ASTM D737-04 Standard test method for air permeability of textile fabrics.

BS EN ISO 9237:1995 Textiles – Determination of the permeability of fabrics to air.

ASTM E96-00 Standard test method for water vapour transmission of materials.

AATCC TM 70-2000 Water repellency: Tumble jar dynamic absorption test.

JIS L1092-1998 Testing methods for water resistance of textiles.

BS 4724-1:1986 Resistance of clothing materials to permeation by liquids – Part 1: Method for the assessment of breakthrough time.

BS 4724: Part 2:1988 Resistance of clothing materials to permeation by liquids – Part 2: Method for the determination of liquid permeating after breakthrough.

BS ISO 22608:2004 Protective clothing – Protection against liquid chemicals – Measurement of repellency, retention, and penetration of liquid pesticide formulations through protective clothing materials.

8
Testing for fabric comfort

P BISHOP,
University of Alabama, USA

Abstract: Comfort, a key quality of clothing and other fabric applications, has been shown to impact user performance. For the purposes of this review, comfort is categorized as thermal and sensory. Thermal comfort is primarily an issue of environmental conditions, metabolic rate, fabric characteristics, and clothing construction. Sensory comfort is primarily an issue of pressure, tactile, and psychological perceptions. Methods for evaluating fabric comfort are described. It is important to note that fabric characteristics and comfort can change over time while in use. Although objective fabric comfort testing and characterization are valuable, the most accurate means for assessing comfort are human-use tests under ecologically valid conditions.

Key words: comfort evaluation, fabric comfort characteristics, thermal comfort, sensory comfort, clothing comfort wear tests.

8.1   Introduction: defining comfort

Comfort may be among the most complex characteristics of clothing. Fourt and Hollies (1970) describe comfort in terms of (1) thermal, (2) non-thermal, and (3) wear conditions. Slater's description of comfort (Slater, 1985) takes a slightly different approach, emphasizing three components: (1) physiological, (2) psychological, and (3) physical, and considering also their impact on, and interaction with, the 'harmony' between humans and the environment. According to Slater's description, the physiological and physical aspects are interrelated and refer to homeostasis (physiological aspect) in the face of a varying environment (physical aspect). The psychological aspect of comfort refers to satisfactory mental function within a given environment. However, Li (2001) has recently suggested that comfort is the subjective integration of visual, thermal, and tactile sensations, psychological status, body–clothing interactions, and the external (macro-) environment.

Although comfort can be defined in various ways, for the purposes of this review, clothing comfort is operationally defined as general human approval of all aspects of a clothing system while being worn (as opposed to a fabric sample). Assessing discomfort may be a more practical way of evaluating clothing systems, in that comfort exists when perception of discomfort is minimal or non-existent. It follows, then, that there are relative degrees of comfort which are practically evaluated by relative degrees of the absence of discomfort. We will consider comfort as being comprised of two major components, thermal comfort and sensory comfort.

Comfort is affected by the interaction of the human body with the clothing, as well as the environment which stimulates thermal, mechanical, and even visual sensations (Li and Wong, 2006). Such sensations are a product of the simultaneous integration of multiple clothing system characteristics and the related physiological and psychological responses. Among the many sensations are thermal comfort/discomfort, or how warm/cool the clothing system feels. Thermal comfort is determined in part by neural thermosensory monitoring of the micro-environment, that area between the wearer's skin and the outermost layer of clothing (Bishop *et al.*, 2000, 2003a; Muir *et al.*, 2001).

In turn, the micro-environment is a product of both the metabolic heat and moisture generation of the subject and the macro-environment. All readers will recognize that clothing thermal comfort changes if the metabolic heat production or the environment immediately next to the skin changes. In active humans, this can be affected by moving in and out of environmentally controlled and uncontrolled areas, or by changing the intensity of physical activity, both of which often change rapidly. To further complicate matters, the micro- and macro-environmental humidity and air motion also influence thermal comfort.

Beyond thermal comfort, there is also tactile and pressure comfort, as well as psychological comfort. All of these combine to determine sensory comfort, which can also interact with thermal comfort (i.e., clothing that is tactilely comfortable in a cool environment may be uncomfortable in a warm environment). Greater complexity arises when clothing becomes damp or wet, which further impacts comfort in most cases. Moreover, duration of wear is a factor, in that clothing may be evaluated differently after some period of wear. For example, our ability to feel pressure is known to fade in only a few minutes in order to avoid overloading the sensory system.

On top of these complex issues are cultural and individual aesthetics and the human ability to relate to prior experiences. Humans are known to evaluate current comfort in relation to significant prior experiences. Prior experiences can thus impact present perceptions. To borrow from another sensory experience, most of us have had unfortunate results with particular foods. We may have become ill soon after we ate a particular food, and although the food was not the proximate cause of our illness, the association of that particular food with that negative experience impacts our attitude towards that food. The same thing can happen with clothing system comfort; but the experiences are typically more subtle. That is, clothing can be associ-

Table 8.1 Some key comfort variables

Thermal	Sensory
Clothing insulation	Pressure
Air permeability	Perceived and actual weight
Vapor permeability	Absorbency
Metabolic rate	Roughness/abrasiveness
Macro-environment	Rigidity
– Humidity	Human mood
– Radiant heat gain/loss	Other non-clothing comfort factors
– Convective heat gain/loss	Aesthetics/social expectations
– Conductive heat gain/loss	Stretch
– External convection	Cling
Micro-environment	Prior experiences
– Clothing fit	
– Internal convection	
– Sweat rates	
Internal blood circulation (convection)	
Environmental stability	

ated with a bad experience that had nothing to do with clothing, but does impact the wearer's 'comfort' in a clothing style or fabric. Additionally, preconceived notions manifested by input from others' experience could bias (positively or negatively) our expectations of a clothing system before we ever don a garment.

The purpose of this review is to consider the role of comfort in clothing and describe some basic principles for evaluating comfort. We will also consider applications of comfort principles in clothing and the anticipated future of clothing comfort evaluation. Because of the complex nature of comfort, first we will deal with each of the major classes of comfort separately and then we will combine them. Some key thermal and sensory factors that impact comfort are summarized in Table 8.1.

8.1.1 The importance of comfort

Comfort is rated by users as one of the most important attributes of clothing (Wong and Li, 2002; Li and Wong, 2006). Almost everything consumers do in terms of selecting, sizing, and modifying clothing could be considered efforts to maximize comfort (Li and Wong, 2006). Most of us suspect that comfort is important not only in the marketplace but also in the utility of the garment. A recent study (Bell *et al.*, 2005) demonstrated that performance on a cognitive exam was associated with the comfort rating of the clothing. Although not well-controlled, Bell *et al.*'s findings suggested that comfort plays an important role in other aspects of human performance. For example, the comfort of firefighter protective clothing may impact firefighter performance. Clearly thermal comfort impacts physical performance, sometimes profoundly, as will be discussed further in the next section.

8.2   Evaluating thermal comfort

Evaluating thermal comfort is complex, but there are more concrete methods for evaluating thermal comfort compared to sensory comfort. Thermal comfort is more concrete because the constructs which determine thermal comfort have been mostly elucidated. The macro-environment can be readily measured to within about 0.1°C and within a few millimeters of vapor pressure (typically about 2% relative humidity up to near saturation where the error increases to about 5%). Once conduction, convection, radiation, and evaporation are fully described, then the macro-environment is reproducible. Likewise, the metabolic rate, and therefore metabolic heat production, can be accurately measured to within about 5%. Because the micro-environment and the metabolic status can be fully described, the impact of the macro-environment on an inert micro-environment can be modeled. The accuracy of the model depends on how completely all variables are identified and quantified. As might be anticipated, this is no easy task.

For example, permeability of fabrics and clothing is just one aspect of thermal comfort. Yet, a whole sub-specialty of fabric characterization has developed which can describe the permeability to air and water vapor (as well as chemical permeability) of single and composite fabrics. These evaluations are typically conducted on dry and wet heat sources, including static and moveable manikins. These mechanical characteristics of clothing's water and air permeability are discussed in another chapter. For the purposes of comfort evaluation, it is critical to note that vapor permeability is a key fabric characteristic of clothing comfort in most situations.

Fabric characteristics are a major contributor to clothing comfort, but not the sole contributor. For example, we recently analyzed another laboratory's data on firefighter clothing (unpublished). We noted that a large variability in total heat transfer (the combination of water vapor and dry heat transfer) of the thermal barrier (the thermal barrier, which is the key element on turnout clothing, is composed of a multi-layer fabric) made a very small difference in physiological heat strain. Large variations in total heat transfer on average accounted for less than 0.1°C of rectal temperature and a few beats of heart rate. Perhaps this was due to the complex nature of clothing systems, wherein seams, fabric overlap, clothing venting, addition of different fabrics at wear points and other characteristics make a large contribution to the clothing thermal function, and thereby to clothing comfort. So, clothing testing can yield different results from fabric testing. On the other hand, not all fabric is worn as clothing. Bedding and towels, for example, would be ideal candidates for precise comfort characterization.

We would argue that fabric testing is foundational to, but not sufficient for, tests of intact clothing systems. Likewise, modeling and testing on artificial systems are foundational to, but insufficient for, determining human comfort responses to clothing. This is primarily due to the lack of a precise integrating center (i.e. brain) when testing on artificial systems. Where testing on artificial systems may involve more than one factor, it is likely that no experimental set-up incorporating artificial systems would have the capacity of the human brain to consider virtually all factors influencing comfort and do this simultaneously and maintain its effectiveness as the environment undergoes acute changes (which, as noted, it often does). Also, without a human brain to formulate a broad range of subjective responses, a great deal of ecological validity is lost.

8.2.1 Thermal comfort test methods

Thermal comfort can be predicted within reasonable accuracy using computer modeling. Prek (2004) provides a recent example of this using a thermodynamic approach to modeling, while Newsham (1997) offers an approach to modeling incorporating human behavior with respect to comfort. Laboratory simulations of humans using guarded hot plates, manikins of various types and other objective tests are useful in establishing basic fabric and clothing system characteristics. However, the comfort range ultimately must be established by human users.

NASA uses a definition of thermal comfort such as to maintain a ±69 kJ (19 Wh or 65 BTU) band of heat storage (Cline *et al.*, 2004). Humans typically require a micro-environment temperature of about 28–30°C to maintain thermal balance at rest (Astrand and Rodahl, 1986). For our purposes, the key issue is the narrow band of comfort. Fanger's text on thermal comfort (Fanger, 1970) is one of the foundations of the field. For a thorough discussion of thermal comfort see Charles (2003).

There is some question as to the human sensory inputs that determine thermal comfort. Frank *et al.* (1999) reported that the tympanic temperature and the mean skin temperature of their experimental subjects had about an equal impact on thermal comfort. However, from our view, the statistical error (discussed later) should be corrected to show closer to a three times greater contribution for tympanic compared to mean skin temperature. This useful experiment did not include the factor of time. Clearly, the skin is able to detect temperature changes much more rapidly than the core of the body because it is closer to the environment and has a smaller mass. From the Frank *et al.* (1999) experiment, it would appear that the core temperature determines long-term comfort, overruling the immediate comfort response of the skin. However, acute comfort responses might be more strongly related to the skin (before the core has had adequate time to change in response to a heat load).

There are several factors to consider in human evaluations of thermal comfort:

- Metabolic rate
- Macro-environmental conditions
- Macro-environmental air movement
- Air movement caused by human movement
- Clothing fabrics and construction.

The first consideration is to recognize that, for a given clothing system, as metabolic rate changes, thermal comfort will change. Thus clothing which is comfortable in cool temperatures at a low metabolic rate (i.e., rest) may be too warm for moderate or high metabolic rates. So, thermal comfort must first be described in relation to a specified metabolic rate. For example, in cold temperatures, clothing that is uncomfortable for a low metabolic rate will likely prove comfortable when the metabolic rate increases sufficiently. Metabolic rate is most accurately measured via direct calorimetry or from spirometry. It can also be estimated from tables (Ramsey and Bishop, 2003) or from a user's heart rate (Astrand and Rodahl, 1986).

The second consideration in thermal comfort is the macro-environment. This includes temperature, relative humidity and air movement. Temperature can be expressed more specifically in terms of radiant and dry energy gain. These characteristics along with relative humidity and air movement can be measured in a number of ways, perhaps the most popular being the wet-bulb-globe-temperature (WBGT) (Ramsey and Bishop, 2003).

Air movement has further application when considering the contribution of the human body. In motionless air, the body releases heat and humidity to the micro- and macro-environments. This impacts the macro-environment and causes local convection depending on the macro conditions. When there is macro air motion, the impact of human production of heat and humidity is less in proportion to the air speed. When clothing is permeable to air movement and there are openings in the garment (which permit exchange with the macro-environment), the air motion also impacts the micro-environment. When natural air movement is simulated by fans, the distance from the fan and the cross-sectional area of the air flow is very important because air flow rate (wind speed) varies accordingly.

Another factor in thermal comfort is the movement of humans wearing the clothing system, because human movement generates heat and causes air flow between the skin and the fabric. With encapsulating clothing, the micro-environment is the only environment the body senses, except at a few places. The macro-environment can be very cold and dry, and yet with good clothing and sufficient metabolic activity, the micro-environment near the skin can be very comfortable or even uncomfortably hot and humid. Furthermore, loose clothing with much ventilation will induce air both through the fabric and through openings. Very snug permeable clothing will reduce, but not eliminate, air movement through both sources. This air movement is not due to the macro air motion but to the movements of the human wearing the clothing. It could be described as 'relative' wind. So clothing thermal comfort as well as sensory comfort will only be absolutely accurate for a given set of movements. For example, a human walking or riding a bicycle even in still macro air will experience air motion that can impact the micro-environment.

In our research we have tried to simulate the motions involved in different tasks appropriate to the clothing. Havenith *et al.* (2002) emphasize that the contributions of body and air motions to the micro-environmental humidity and thermal comfort are so large that they must be incorporated into comfort predictions.

The last aspect of thermal comfort is the fabric characteristics and construction of the clothing. This includes the fabric itself, other construction materials, seams, overlaps (e.g. pockets, collars, and cuffs), the location and type of openings, clothing fit at various points on the body, and the closures. Clothing, unlike fabric, is highly variable in all of these characteristics. Each of these can impact thermal comfort.

Once the parameters of metabolic rate, environment, movement, and fabric and construction/fit are specified, then two types of thermal comfort measurements can be made. First, we can objectively measure the nature of the micro-environment in terms of temperature and relative humidity. Because human thermal comfort limits are relatively narrow in terms of local (micro-) environment, comfort can be predicted. However, human variability, as mentioned earlier, should never be ignored (Bishop, 1997). Second, because of this, to accurately describe comfort we also need assessment by a human test panel actually wearing the clothing system.

Thermal measurements have been detailed in a series of useful standards. Some of these include ISO 8320, ASTM E 1530, ASTM D 1518, and ASTM C 518. All of these describe aspects of using guarded hot plates and other methods to measure thermal conduction in fabric samples. ASTM F 1154 provides standard methods for qualitatively evaluating comfort in chemical protective clothing (in humans). Standards for thermal comfort measured with manikins can be found in ASTM F 1291, ASTM F 1720, ISO 7920, and ASHRAE HI-02-17-4. ASTM D 5725-99 and 4772-07 give the procedures for measuring fabric absorbency of sheeted material and terry fabrics, respectively. ASTM D 737-96 gives techniques to evaluate air permeability through fabrics, and ASTM F 2298-03 gives techniques to evaluate water vapor diffusion and air flow resistance of fabrics. ASTM D 1388 and ASTM

D 4032 give techniques to measure fabric stiffness, and ASTM D 6571 measures compression resistance and recovery of high-loft non-woven materials. It is important to note that ASTM D 1776 covers the conditioning and testing of textiles, which is important in most fabric comfort tests.

8.2.2 Thermal and sensory rating scales

The human thermal comfort rating can be assessed by asking subjects to rate their perceptions of comfort on visual analog scales. In devising thermal and sensory rating scales, there are a number of considerations. One approach is the 'Likert' scale in which a single descriptor is given and respondents are asked to choose one of five options between 'strongly agree' and 'strongly disagree'. This is useful for characteristics without clear polar opposites. The major disadvantage of this type of rating is that the scores are ordinal data and technically should not be mathematically manipulated, but rather should be analyzed using appropriate statistics for ordinal numbers. For these reasons, we do not use Likert-type scales in our laboratory.

The Likert approach can be modified to allow the respondents to choose more than five options, which would provide greater precision; however, if discrete choices are utilized, the data are still ordinal. Winakor et al. (1980) have used a 99-point certainty scale. We have taken a similar approach in measuring thermal comfort in a locally developed visual scale based on a 100-mm line (see Fig. 8.1). This approach offers several advantages. First, it is easy to create and modify, it is easy to score, and it is very easy to explain to test participants. Second, it is a continuous measurement allowing test subjects to choose whatever degree of opinion suits them. Third, it is inter-val data, which can be correctly analyzed using higher-order statistics. Lastly, the increased precision of this scale reveals differences too small to be detected using a Likert approach. Whatever number of choices is provided to the respondents, having an equal number of 'favorable' and 'unfavorable' choices is less biased.

Figure 8.1 shows an example of a scale we have used in several clothing comfort studies. After explaining the process, we administer this scale to our research test panel. Each panel member marks the scale as desired. We then apply a ruler, measuring from left to right to obtain the score to the nearest millimeter. A clear ruler works well. We record the score and often administer the comfort scale several times under similar and variable condi-tions. In devising scales we recommend always making the more desirable characteristic on the right, so that a higher numerical score represents more of the positive quality. When we have mixed the directions, it hasn't been any harder for the study participants, though it is much harder for us as investigators, and it would be likely to generate more errors by respondents

Questionnaire for all work bouts

Time POINT: ____ initial ____ 25 min ____ 45 min (at end of work bout)

Please rate the work experience on the following characteristics:

Mark the line anywhere with the left end representing the least amount of that quality and the right end representing the most amount of that quality.

Overall comfort, stickiness, wetness, coolness, hotness

Overall comfort
    Very uncomfortable _____ Very comfortable

                                  ←————————100 mm————————→

Sticking to skin
    Not sticky _____ Very sticky

Wetness
    Not wet _____ Very wet

Coolness
    Not cool _____ Very cool

Hotness
    Not hot _____ Very hot

8.1 Comfort scale utilizing a 100-mm line.

and by investigators. We have found it very confusing to use comfort measures with mixed directions. These scales can also be used for sensory comfort measurements, as will be discussed shortly.

8.2.3 Thermal comfort and performance

Much of the research in our laboratory has examined the thermal effects of clothing on performance, particularly in hot environments. In fact, most of our efforts have been devoted to reducing the thermal discomfort associated with protective clothing. As noted earlier, even in a thermally comfortable environment, performance can be affected by comfort. The study of Bell *et al.* (2005), cited earlier, illustrates this observation. In a very hot environment, comfort is compromised by clothing which reduces heat loss, and promotes sweat accumulation and body heat storage. Heat storage can precipitate heat stroke and death. Short of that, it has been demonstrated that heat storage can also reduce attention and concentration, and other cognitive performance variables (Ramsey and Kwon, 1992), as well as lead to early fatigue. This, in turn, likely reduces safety. For example, a police officer who is over-heated, dehydrated and uncomfortable may be less safe for himself and others. It has also been demonstrated that heat storage can substantially reduce fine motor performance (Robinson *et al.*, 1988, Ray and Bishop, 1991) and physical work capacity (Bishop *et al.*, 1991, 1993, 1995; Smith *et al.*, 1994; Solomon *et al.*, 1994).

Thermal homeostasis requires that the heat loss through the clothed and unclothed parts of the body (ΔH) equals the heat produced metabolically (MH). A person may also gain heat from an environment hotter than the body, requiring that heat losses be very high. Because the human in motion is typically less than 20% efficient, over 80% of the metabolic energy goes into heat production. The total heat loss or gain is the sum of the dry heat loss/gain (through conduction, convection and radiation) through fabric and clothing openings, plus the heat loss though evaporation from the skin and subsequent transmission of water vapor through fabric and out of openings. Therefore the heat loss or gain through clothing can be represented by the following equation:

$$\Delta H = MH + \text{dry heat gain} - (\text{dry heat loss} + \text{evaporative heat loss}) \quad 8.1$$

Likewise in cold macro-environments, to maintain a comfortable micro-environment, the heat loss must be reduced to match the heat production. Barker (2002) has attempted to model the comfort limits by employing Woodcock's energy loss formulas (Woodcock, 1962a, 1962b). Although a generalized model would be very useful, it is important to bear in mind that exact values for metabolic rate, body surface area, sweat production, sweat distribution, and sweat evaporation can vary greatly among humans even for standardized conditions (Bishop *et al.*, 1991) and usually vary across time.

8.3   Moisture and comfort

Moisture has a big impact on thermal comfort, but also on sensory comfort. This sensory comfort may change with different activity rates and environmental conditions, along with different garment designs. Most obviously, moisture from sweat or from external sources (e.g. rain) will impact both the micro- and macro-environments, which in turn impacts the comfort perception of both hot and cold conditions. Less obviously, the permeability of the fabric to gas movement is impacted a great deal in some fabrics such as wool and cotton, but very little in polyester, and intermediately in rayon (Wehner *et al.*, 1987). This occurs in part because moisture causes individual fibers to swell. Also, as in the case of rain, liquids can actually fill the voids between fibers and thus impact permeability directly.

In many situations, the impact of liquids on the fabric, and therefore comfort, depends upon the volume of free liquid, absorbed liquid in the fibers, and water vapor. Some fabrics can handle small amounts of liquid or water vapor well, but above a certain level the fabric is overwhelmed. For example, it is a common observation that some multi-layer fabrics (e.g. Gore-Tex®) do well at transferring water vapor outwards under mild conditions, but become 'overwhelmed' under conditions with very high micro-environmental vapor pressures.

Liquid volume and liquid location both impact comfort. Liquid conducts heat much more readily than air, so if the volume of liquid is sufficient, and the fabric characteristics are such as to form a continuous medium of thermal conduction from the skin to a cooler substance (macro-air for example), then the liquid will facilitate heat loss. The volume of liquid required to saturate a fabric will depend upon fabric moisture characteristics, fabric thickness, and rate of loss due to evaporation. This is true even for 'wicking' fabrics such as polyester. Liquids may also reduce insulation if material loft is reduced due to wetting (e.g. wet goose-down insulation is greatly reduced if it becomes wet). The impact of the fabric on comfort can be different when the volume of liquid is less. For example, wicking fabrics such as polypropylene are able to move moisture off the skin and towards the micro-environment. Moisture wicking away from the skin is most useful in cold environments where sweat evaporating from the skin can cause high rates of undesirable heat loss. If reducing the loss of body heat is important to comfort, as it is in cold macro-environments, then evaporating sweat some distance from the skin aids thermal and sensory comfort. If the fabric becomes totally wet, however, conduction occurs and the advantage may be lost.

The moisture transmission capabilities of fabrics can be different from that of clothing systems. A semi-permeable multi-membrane fabric might have a low vapor transmission rate, but if the garment is a poncho, or other design which ventilates well due to clothing pumping, the comfort and protection afforded by the garment may be quite acceptable (Barker, 2002). Although articulated moving manikins may serve as good means to characterize garments, the ultimate test of comfort is to obtain measures of comfort from human subjects performing realistic activities in realistic environments including wet and dry conditions, as appropriate. Barker (2002) gives an example of a practical test of surgical gowns that used alternating rest and varying physical activities. We have done the same in our tests of protective clothing (Bishop et al., 1991, 1993, 1995, 2003b; Smith et al., 1994; Solomon et al., 1994).

In very warm environments, when sweat rates are much higher than evaporation rates, the fabric may become saturated, negating the value of the wicking but facilitating conductive heat loss. In this case, fabric wetting occurs to an extent that varies from body region to body region. At any rate, fabrics which are effective in wicking moisture away from the skin are generally perceived as more comfortable (Wickwire *et al.*, 2007). However, these fabrics have not been found to improve heat loss (Ha *et al.*, 1995; Gavin *et al.*, 2001; Wickwire *et al.*, 2007), though others have shown some fabrics to have potential heat transfer advantages (Yasuda *et al.*, 1994; Kwon *et al.*, 1998). Even though a wicking fabric may not offer any advantage regarding heat transfer, the comfort may nonetheless be enhanced under certain circumstances.

Thoroughly wetting fabrics may also change their mechanical characteristics and generally reduce sensory comfort. For example, as moisture content increases, the friction with the skin increases and reduces comfort (Li and Wong, 2006, p. 131). As water vapor moves though fabrics, fabric fibers absorb and desorb moisture, which may impact the fiber characteristics. The degree of impact depends on several factors, the most important of which is the physical properties of the fiber composition (Li and Wong, 2006, pp. 78–79).

Given the issues that arise when fabric absorbs liquid, several studies have evaluated the effect of clothing made with hydrophobic fabrics. Kwon *et al.* (1998) found no difference between three different clothing ensembles for rectal, skin, or mean body temperature in warm environments (30°C at 50% relative humidity), while there was no air motion. However, with air motion (1.5 m/s), rectal temperature was significantly lower while wearing a high moisture-regain fabric (wool–cotton blend) than for polyester. In that study, fabrics with high moisture regain could reduce heat stress relative to the low moisture-regain material but only when air motion impacted the micro- and macro-environments. This underscores the importance of the micro- and macro-environmental conditions on fabric testing.

However, Yasuda *et al.* (1994) found that the surface temperature of the fabric rose more quickly with wool than with polyester, perhaps due to better sweat evaporation from the surface of the polyester. Similarly, Ha *et al.* (1995) showed that clothing surface temperatures were higher in cotton than in polyester. But, four out of five subjects felt wetter in polyester during the latter half of the experiment. The authors speculated that the fabric wetness insulation due to the absorption of moisture in cotton fabrics reduced the thermal insulation and accelerated the dry heat loss. This slowed the rate of increases in both core temperature and pulse rate. Gavin *et al.* (2001) found no advantages for rectal, skin, or mean body temperature while wearing synthetic clothing purported to enhance sweat evaporation. However, in this study only three out of eight skin temperature measurements were made under the clothing.

Hu *et al.* (2006) report some comparative data on moisture handling among eight different knitted fabrics. A plain knitted fabric of 92% nylon and 8% Spandex had the highest liquid moisture-management capacity, one-way transfer capacity, a large spreading rate, and a medium wetted radius. This suggests that this fabric can readily transfer sweat from the skin to the surface and will also dry relatively quickly. In contrast, two knitted fabrics of 92% cotton and 5% Spandex had the worst moisture handling ability, low wetted radii and spread rates coupled with negative one-way transport capacities, suggesting that these fabrics would likely be uncomfortable in high-sweat conditions. Another rib-knitted fabric of 94% cotton and 6% Spandex had limited ability to move sweat or other liquids away from the skin and to the surface for evaporation and were slow in drying. This same study demonstrated that a plain-knitted fabric of 88% polyester and 12% Spandex had intermediate ability to transfer liquid away from the skin and good drying capacity. Another polyester fabric (88% polyester and 2% Spandex) could move liquid away from the skin well, but was not as quick to dry.

The drying of wet fabrics during wear also plays a role in comfort. Fabric wicking impacts the rate of drying and the comfort during drying. As water evaporates, it impacts the relative immediate macro-humidity. Likewise, the water absorbed by the fabric is evaporated last (Li and Wong, 2006, p. 79), so the moisture continues to impact comfort until the fabric is totally dry. For all comfort characteristics, including moisture, the degree of fabric–skin contact plays a major role. Fabric that does not contact the skin will have only an indirect impact on comfort through its influence on the micro-environment.

One other issue with moisture is condensation in the clothing. When a temperature gradient exists in single or multiple layers of clothing, it is possible for moisture to evaporate from the skin or an inner layer due to the temperature of the micro-environment. The moisture will then condense in a cooler outer layer once the dew point temperature is reached. This condensed moisture will act in accordance with the principles outlined earlier.

Using double jersey fabrics of similar yarn count and weight and somewhat similar thickness, Li and Luo (2000) reported that wool had greater water vapor sorption and led in initial rate of sorption, followed by cotton, porous acrylic, and polypropylene. Wool also showed the highest initial temperature increase due to the sorption of water vapor, followed by cotton and porous acrylic. Polypropylene again had the lowest temperature increase. Wool and cotton are considered to be highly hygroscopic and the artificial fibers are classed as weakly hygroscopic.

Our laboratory has considerable experience in measuring the thermal comfort conditions of humidity and temperature of the micro-environment. Formerly, we used a small capacitive humidity meter and thermocouple attached underneath the clothing for the duration of the observation to measure the micro-environment with the output shown on a remote monitor connected by a small cable. That system is no longer available, so we have switched to a portable humidity device so that we can temporarily locate the pickup for the meter under the clothing, and then take a humidity reading. We now use a thermocouple placed under the clothing for the duration of the observation to assess micro-temperature.

Skin temperature is readily measured through the use of thermocouples (bimetallic devices which develop an electrical voltage in relation to temperature) or thermistors (devices whose electrical resistance responds to temperature) applied to the skin under the clothing. The thermometers must be attached in a manner that does not impact the thermal characteristics of the clothing or skin. This is best done by fabricating small elastic holders that cover minimal skin and minimally overlay the actual measurement sensor, with a light large mesh or other material that does not insulate. For areas of skin where this is difficult (e.g. the chest or back) we recommend a single layer of sticky heat-conductive tape. We typically use a piece of surgical dressing obtained from medical supply stores.

To conclude, the influence of moisture on comfort is complex. Liquids and vapors can influence both fabric characteristics and comfort. The volume of liquids and vapors, locations, distributions, fabric composition and construction, along with the macro-environment and physical activity, all determine the impact of moisture on comfort of clothing and fabrics. There remains much work for investigators in this area.

8.4   Ease of movement

The ease of movement of the clothed human is not always considered a major factor in clothing comfort. Certainly, when clothing binds or constricts the user, discomfort is perceived. This is why most users find loose clothing to be more comfortable than tight clothing. However, the psychological discomfort arising from style/fashion requirements can override the desire for comfort. It is often overlooked that clothing also contributes to total energy costs for activity. This indirectly influences comfort because it adds to metabolic rate and increases heat storage under some clothing and environmental conditions (Bishop and Krock, 1991). The manufacturers of firefighter protective garments understand this and work to develop styles which require less energy for normal firefighter movements.

8.5   Evaluating sensory comfort

Simply put, the feel of fabric is usually described as 'handle' or 'hand' and can be evaluated in several ways. The most commonly used objective evaluation is by the Kawabata Evaluation System (KES) (Kawabata, 1980; Barker, 2002). The intention of evaluating hand is to predict the mean human sensory comfort perception. The Kawabata Evaluation System attempts to measure a combination of softness, stiffness, and smoothness. The KES is also specific to the end-use of the clothing (Barker, 2002). This means that the KES rating applies to the clothing system and not to the fabric particularly. Barker (2002) illustrates that the KES changes for woven versus knit fabric and provides equations for calculating the hand for men's suits, single and double knit fabrics, and sheeting.

This latter observation raises the earlier issue regarding thermal comfort. The comfort of a fabric may be quite different from the comfort of a garment made from that fabric. A fabric's hand would be a major variable in clothing comfort, but fabric hand can never fully predict the comfort of a clothing system with multiple pieces, differences in fit, differences in type and location of seams, total weight of the clothing, movement in the clothing and aesthetic comfort. For example, Schutz et al. (2005) found fit to be the most important 'customer satisfaction' factor in military daily-wear uniforms. They also found that tactile rating and appearance were as important as functional factors (e.g. 'protection') (Schutz et al., 2005).

Fabric weight is a common sensory characteristic of interest. The actual physical weight of the garment also has a major impact on comfort, particularly for prolonged wear. Firefighters are conscious of the overall garment weight and prefer lighter garments for ease of movement and because they must carry the weight of the garments in addition to the other equipment they carry. The total weight of the worn garment can also affect pressure comfort.

Measuring sensory comfort is even more difficult than measuring thermal comfort. Whereas thermal comfort has a fairly narrow set of concrete conditions that are comfortable, sensory comfort depends on the interactions of numerous abstract characteristics. The same hierarchy of comfort testing mentioned for thermal comfort measurement also applies to sensory comfort. Bench tests of fabric characteristics including hand, stickiness, and other qualities shown in Table 8.2, are the most basic sensory comfort measurement techniques. Barker (2002) supports the notion that subjective evaluation by human wearers represents the only way to accurately measure the combined effects of all relevant variables of clothing systems on comfort.

One dimension of sensory comfort that is overlooked on occasion is psychological or aesthetic comfort. Though subject to both cultural and time-period influences, some people find some clothing to 'feel comfortable', not from a tactile or physical pressure sense but from a societal pressure sense. For example, baggy short trousers worn very low on the hips

Table 8.2 Sensory attributes identified by numerous authors and
approximately grouped by type

Stiffness/crispness/pliability/flexibility/limpness
Softness/harshness/hardness
Thickness/bulkiness/sheerness/thinness
Weight/heaviness/lightness
Warmth/coolness/coldness (thermal characteristics)
Anti-drape/spread/fullness
Tensile deformation/bending/surface friction/sheer
Compressibility
Snugness/looseness
Clinginess/flowing
Dampness/dryness/wetness/clamminess
Quietness/noisiness
Prickliness/scratchiness/roughness/coarseness/itchiness/tickliness/stickiness/
   smoothness/fineness/silkiness
Looseness/tightness

---

would be extremely uncomfortable for me and for many people around the
world; however, young men in the USA seem very comfortable in such
attire.

Our understanding of sensory comfort has evolved over time. Sensory
data can be described in several ways, starting most basically with Weber's
Law. More recently investigators have found that Stevens' Power Law
(Stevens, 1946) better describes some sensory comfort variables. Weber's
Law states:

$$\Delta Sp/Sp = K \qquad\qquad 8.2$$

where K is a constant describing the ability of a human to detect a sensation, and Sp is the magnitude of the sensation. ΔSp is the smallest detectable
difference. Thus, we see that the precision of human detection of a given
sensation is highly dependent upon the magnitude of the sensation. For
example, the magnitude of thermal sensations is relatively large, since
humans can detect a range of temperatures from very cold to very hot. In
contrast, in rating clothing such as how clingy or how flowing, the ability to
differentiate these is limited.

This relationship between precision of detection and magnitude is
expressed in more stark terms in Fechner's Law, which is:

$$Rs = k \log Sp \qquad\qquad 8.3$$

where Rs represents the perceived sensation, k is the lowest detectable
stimulus (stimulus threshold), and Sp is magnitude of the sensation.
Therefore, the sensation must increase logarithmically for the perception
to increase arithmetically.

Finally, Stevens proposed that the relationship between stimulus sensation and human detection is not always logarithmic but may be a power function. Thus the 'Stevens' Power Law' is expressed as:

$$Rs = c\,Sp^x \qquad\qquad\qquad 8.4$$

where Rs represents the perceived sensation, c is a constant scaling factor, and Sp is the magnitude of the sensation which is raised to some power x, which is dependent on the nature of the sensation being measured.

These laws are shown here in the order in which they were developed. As experience with comfort measurement has grown, the relationship has been found to be more complex and has been described accordingly by the various researchers. For example, Li (2005) conducted a series of trials with a test panel comparing objective measurements with subjective ratings to elucidate the mechanisms involved in temperature and moisture sensations in clothing systems during environmental changes. The mean subjective perceptions of warmth seemed to follow Fechner's law and Stevens' power law, whereas dampness seemed to follow Fechner's law most closely. Comfort was positively related to the skin temperature and negatively related to the relative humidity of the microclimate.

8.5.1 Objective tests of sensory comfort

Just as there have been objective laboratory tests developed to establish the thermal characteristics of fabrics, there have also been objective bench tests developed to describe the sensory characteristics of clothing. Barker (2002) summarizes concisely the objective measurement of fabrics. Barker notes that Kawabata's system (Kawabata, 1980) is the most detailed. Kawabata (1980) reports that The Hand Evaluation and Standardization (HES) Committee determined seven characterizations of fabric for men's suits. These were smoothness, crispness, stiffness, anti-drape (spread), fullness and softness, appearance of the surface, and 'others'. The HES Committee determined that different fabric applications required different descriptors, which complicates sensory measurement. For example, for men's summer and winter suits these descriptors were given different percentage weightings which summed to 100%. For detail and descriptions see Kawabata (1980). Likewise in the same publication, the key mechanical properties are delineated and described including measurement methods. The HES Committee identified six mechanical-property groups: tensile deformation, bending, surface friction and roughness, shear, compressibility, and weight/thickness. Kawabata describes his own development of objective test equipment for measuring fabric characteristics (Kawabata, 1980). For example, the Kawabata Thermolab measurement of warm/cool feeling of fabrics is well known.

Methods of objectively assessing specific sensory comfort fiber characteristics are as follows.

- *Prickliness* can be measured with a KES-FB compression tester modified to measure initial bend in protruding fibers, plus some method to quantify the number of protruding fibers per area of fabric, plus quantification of skin–fabric contact. Obviously many factors impact the fabric prickliness.
- *Stiffness* can be measured with a Shirley cantilever, a Shirley cyclic bending tester, or a Cusick Drapemeter. For details see Kenkare and May-Plumlee (2005).
- *Softness* has been measured with a compression load cell in an Instron tensile tester. Softness has also been measured as friction on an Instron load cell (Li and Wong, 2006, p. 134). One of the more innovative approaches to measuring softness was performed by taking stereo photographs and using computer software to generate topographical maps of the fabric samples (Li and Wong, 2006, p. 134). Friction, as a key characteristic of softness, can be measured with a Kawabata surface tester. Softness is associated with objective physical properties such as roughness, friction, prickle, shear and bending stiffness, thickness and areal density.
- *Garment pressure* can be most readily assessed in clothing by direct measurement of pressure by way of pressure transducers of various types. For example, Chan and Fan (2002) reported the use of a digital pressure meter to objectively assess clothing pressure. One crucial issue that must be taken into account is that the thickness of the device can contribute to the pressure measured. Small fiber optic pressure measurement devices offer an option for clothing pressure measurements. Key areas seem to be the knees, elbows, seat, and back. One way to measure garment pressure is to draw a series of spaced perpendicular lines on the skin. The percentage of displacement of the line associated with various movements (e.g. standing to sitting) can be recorded as a measure of skin strain as reflective of garment pressure on the skin. Additionally, clothing fit would impact pressure sensory measurement.

Just as there have been attempts to model thermal comfort, there have been numerous attempts to model sensory comfort often using the above fabric characteristics. Wong *et al.* (2004) used different combinations of inferential statistics, fuzzy logic, and neural networks to predict the sensory responses derived from data from 28 young females on a test panel. Using a relatively small test panel ($n = 10$), the derived models yielded correlations of $r = 0.61$ to 0.82.

8.5.2 Subjective measures of sensory comfort

Hollies (1997) gives a general outline for developing subjective sensory comfort measurements of clothing by means of the following steps.

1. Delineation of clothing attributes
2. Descriptors of these attributes
3. Development of a sufficiently precise scale to assess perceptions of key attributes
4. An appropriate human sample to assess the clothing
5. Accurate data analysis techniques.

An alternative procedure is to allow a test panel to develop a list of descriptors in their own terms. Then these descriptors can either be qualitatively grouped, or alternatively the panelists can come together as a group and provide their own consensus as to the best descriptors, and these descriptors for a given fabric can be arranged as polar-pairs and scored using a scale chosen by the investigator. Alternatively, as shown in Fig. 8.1, research subjects can rate a given clothing system individually using sensory polar-pairs and a 100-mm line. Each subject rates a clothing system under a given set of environmental conditions and work rate, and these scores are treated as interval data.

These descriptor approaches are most useful when a few fabrics are being considered for a particular application. For example, if several different fabrics are being considered to comprise the inner liner of a jacket, the test panel would use the candidate fabrics in the situations and with the movements in which it would be employed. They would then rank the fabrics on criteria such as hotness/coolness, or abrasiveness/smoothness, etc. The same scale would be used for each rating, with the highest summed rating being the most favorable. If differing scales are used, then the data would have to be normalized before being combined for comparative purposes.

For both thermal and sensory measurements, there are numerous subjective comfort rating scales. Winakor et al. (1980) gave a classic example of subjective assessment of fabrics. They used a 99-point certainty scale with a large panel of 59 judges. Large F values were found for discrimination between fabrics evidencing different levels of stiffness, roughness, and thickness, suggesting good sensitivity in the rating scale.

It is important to recognize and examine the wide range of applications for a given fabric. The environmental interaction may be substantial. The duration of the trial is also important, in that a fabric that is initially comfortable may become increasingly uncomfortable over a few hours. Likewise, a clothing system may have initial pressure discomfort that decreases to acceptable pressure over time. Laundering also impacts comfort. For example, it is well known that some cotton denims change their characteristics after repeated launderings.

So, practically speaking, comfort measurements only apply to a set of conditions which should always be clearly and completely specified. Issues such as test duration, pre-laundering, and panelists' experience are among the many issues which must be specified.

8.6   Statistical considerations in comfort measurement

The ultimate goal of textile comfort research would be to fully account for the many variables along with their interactions which describe clothing comfort across all reasonable activity levels and environments. Much of the prior modeling has advanced us towards this goal. However, as befits good science, we must continue to be appropriately critical of modeling.

Establishing the relationship between fabric characteristics and human reports of comfort seems a logical approach to modeling. Our recent experience is that error analysis would sometimes be more useful than correlation analyses in evaluating objective surrogates for human comfort factors (Wickwire et al., 2007). However, it is common in human modeling to see a good R-square for the derivation model, only to see that explained variation to decline precipitously in the calibration model. In addition, we typically derive our initial model on a very diverse sample of fabrics and conditions, which are not normally distributed. As pointed out by Altman and Bland (1983), such a situation leads us to overestimate the true relationship. For example, Figure 5 in Barker (2002) compares the ranking data for the Kawabata Thermolab measurement of warm/cool feeling of fabrics with human subjective assessment of the same quality.

We reanalyzed the Kawabata Thermolab data in terms of error (see Fig. 8.2). The Pearson correlation coefficient ($r = 0.93$, $R^2 = 0.85$) is very similar to the Spearman correlation coefficient for ranked data ($r = 0.92$). This correlation looks very good and there is little bias in the Thermolab data, but the error is pretty large. If you look at the 95% boundaries of agreement, you can see that the confidence interval is about 8 units out of a total ranking of 16 (all scores are between 2 and 18) units or about 50% of the usable scale. This is to say, a Thermolab ranking of 10 would yield a human ranking between 6 and 14 about 95% of the time. There are some contexts where such a large variation would be acceptable and some where it would not be.

Wolins and Dickinson (1973) suggest that weighting sensory scores based on deviations from the mean improves the validity of the scores. Extreme scores get higher weights because those raters appeared more certain about

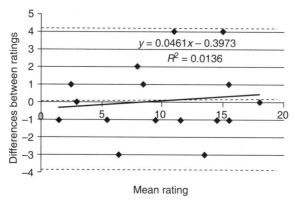

8.2 Errors between cool touch and subjective ratings (adapted from Barker, 2002).

their choices (Wolins and Dickinson, 1973). This approach seems a matter of opinion. It could be argued that this procedure merely gives more weight to more expressive or passionate panel members rather than to the actual perception. We would argue that all scores be rated equally, because an appropriately drawn sample should be representative of the actual user population.

In their well-designed study, Frank et al. (1999) made a statistical error that all of us should consider. Considering Fig. 2 of that paper, we see two temperature measures, tympanic temperature and mean skin temperature. The problem is that these two data sets have very different distributions. That is, rectal temperature is constrained between about 35 and 37.5°C, whereas mean skin temperature varies between 28 and 37°C, as is apparent in their figure. When data from two different distributions are used in a multiple regression, the beta weights are not comparable. If beta weights are to be compared among variables with different distributions, all the data must first be normalized.

Although it is not possible to do this accurately without the raw data from that study, if we approximate the distribution, we would expect that each 3°C of mean skin temperature corresponds to approximately 1°C of tympanic temperature. Therefore, when the ratios appear to be 1:1, as reported, a better estimate would be about 3:1, with tympanic temperature making three times the contribution that mean skin temperature makes. When computing derivations such as total heat storage and others, it is important to consider the distributions of the data and use appropriate transformations when necessary. Likewise, when we choose to mathematically manipulate different sensory or thermal comfort variables, we must consider the data distributions.

8.7   Comfort measurement summary

In this review we have sought to address 'objective' measures of comfort. Whereas we always seek that all of our measurements be as objective as possible, it may be mis-communication to speak of objective measures of comfort and more correct to say that we are objectively measuring fabric or clothing characteristics, rather than comfort. That is, we are quite capable of objectively characterizing important comfort factors of fabrics, but a bit less capable of objectively characterizing comfort factors for clothing systems. Subjective mean comfort ratings can sometimes be predicted with acceptable accuracy using these objective fabric characteristics. Assuming that we can often accurately and objectively characterize or predict human responses to clothing systems seems ill advised in view of the observed variable nature of humans with respect to comfort (Bishop *et al.*, 1991).

Objective measurement of fabric/clothing characteristics forms a necessary but insufficient foundation for measuring human comfort. As has been pointed out, humans are the ultimate determination of comfort and human variability is large. As we have also discussed, it is important to describe comfort in terms of all relevant factors including varying fits, activities, environments, durations, and aesthetic comfort, among others.

Reaching some consensus on comfort scales would be beneficial to all investigators because it would allow us to more readily compare findings among studies. Scientific history suggests that this is unlikely, at least in the near future. Fortunately, there is a wealth of published research that has done much to advance our knowledge of clothing comfort. Also fortunately for those of us employed in this field, there is still much work to be done.

8.8   Applications of comfort assessments

Most practical applications of comfort measurement involve intact clothing systems. Our laboratory has collected comfort data on non-woven encapsulating coveralls, military chemical protective clothing, and fire protective clothing. Obviously, comfort data can be gathered for both clothing and non-clothing situations. When clothing is used, it is important to evaluate the entire ensemble because there will be interactions among clothing parts. In each comfort assessment, it is important to include all clothing parts that are likely to be used together, using activities that are likely to be performed, for durations long enough to allow for some plateau in comfort, and in environments users are likely to encounter. For example, we are about to begin an evaluation of firefighter protective clothing. In this particular study, we are trying to assess the role of each component in the overall thermal comfort, energy costs, and heat storage.

8.9   Future trends

The future of comfort measurement is exciting. The move towards active fabrics that respond to activity and environment will have a great impact on comfort. Perhaps one day we will see fabrics and clothing systems that sense the environment and respond appropriately to maximize comfort. Clothing systems will automatically adjust gas transmission capacity both through the fabric and at openings. Clothing systems will also actively supply and remove heat as the activity and environments demand. As these clothing systems progress, a range of adjustments will be possible to allow each user to individualize their preferences to personalize their clothing comfort. Fashion and style, although less technical in nature, will also change periodically.

To facilitate these advancements in clothing, textile scientists must continue to explore the art and science of clothing comfort. As I have tried to portray in this chapter, this is a great challenge. To progress in clothing comfort, continued efforts must be made to quantify and qualify human comfort. Because of the complexity of both clothing systems and humans, I strongly doubt that comfort will ever be fully predictable from purely objective characteristics of fabric or clothing. Perhaps the development of rather sophisticated models will permit specification of some parameters and identification of the range of adjustment needed to accommodate 95% of the user population. The attempt by Wong *et al.* (2004) to use a combination of approaches to modeling comfort probably offers the best hope. Fashion and style may never be predictable.

Clearly those scientists who work in textiles and clothing systems will have to constantly reassess their approaches and adapt to advancing knowledge. It is a great challenge, but that is one of the things that keeps textile research appealing.

8.10   Sources of further information and advice

For further information and advice on comfort, I recommend foremost the text, *Clothing Biosensory Engineering*, edited by Y. Li and A.S.W. Wong, Woodhead Publishing, Cambridge, 2006. For further information on thermal comfort, the foundation work would be Fanger (1970), *Thermal Comfort*, Danish Technical Press, Copenhagen. For a summary that would be easy to obtain, see Charles (2003), Fanger's Thermal Comfort and Draught Models, *IRC Research Report RR-162*, Institute for Research in Construction, NRC Canada (available at http://www.nascoinc.com/standards/breathable/PO%20Fanger%20Thermal%20Comfort.pdf).

Havenith, Holmer and Parsons (2002) give good insight into the roles of body and air motions on micro-environmental humidity and thermal comfort. For those unfamiliar with characterizing the thermodynamics of humans in clothing, they give some of the key equations.

If you are interested in Kawabata's objective hand evaluation, the obvious source is *The Standardization and Analysis of Hand Evaluation*, 2nd edition, The Textile Machinery Society of Japan, Osaka, 1980. A concise summary of comfort is available in Roger Barker's (2002) 'From fabric hand to thermal comfort: the evolving role of objective measurements in explaining human comfort response to textiles', *International Journal of Clothing Science and Technology*, 14: 181–200.

8.11   References

Altman DG, and Bland JM (1983) 'Measurements in medicine: the analysis of method comparison studies', *The Statistician* **32** 307–317.

Astrand P-O, and Rodahl K (1986) *Textbook of Work Physiology*, McGraw-Hill, New York.

Barker RL (2002) 'From fabric hand to thermal comfort: the evolving role of objective measurements in explaining human comfort response to textiles', *International Journal of Clothing Science and Technology* **14** 181–200.

Bell R, Cardello AV, and Schutz HG (2005) 'Relationship between perceived clothing comfort and exam performance', *Family Consumer Science Research Journal* **33**(4) 308–320.

Bishop P (1997) 'Thermal environments: Applied physiology of thermoregulation and exposure control', in *The Occupational Environment – Its Evaluation and Control*, ed. SR DiNardi, American Industrial Hygiene Association, Fairfax, VA, Chapter 24.

Bishop P, and Krock L (1991) 'Energy costs of moderate work activities in protective clothing', *Advances in Industrial Ergonomics and Safety III*, 623–628.

Bishop P, Pieroni R, Smith J, and Constable S (1991) 'Limitation to heavy work at 21°C of personnel wearing the U.S. military chemical defense ensemble', *Aviation Space Environmental Medicine* **62**(3) 216–220.

Bishop P, Ray P, Smith J, Constable S, and Bomalaski S (1993) 'Ergonomics of work in encapsulating protective clothing – An analysis of nine studies with similar subjects and clothing', *Advances in Industrial Ergonomics and Safety V*, 443–450.

Bishop P, Ray P, and Reneau P (1995) 'A review of the ergonomics of work in the US military chemical protective clothing', *International Journal of Industrial Ergonomics* **15**(4) 278–283.

Bishop P, Gu D, and Clapp A (2000) 'Climate under impermeable protective clothing', *International Journal of Industrial Engineering* **25**(3) 233–238.

Bishop P, Church B, and Jung A (2003a) 'An alternative approach to clothing adjustment factors for protective clothing', *Proceedings of the 2nd European Conference on Protective Clothing*, Montreux, Switzerland, May.

Bishop P, Jung A, and Church B (2003b) 'Micro-environmental responses to five protective suits in two environments', *Proceedings of the 2nd European Conference on Protective Clothing*, Montreux, Switzerland, May, pp. 292–296.

Chan AP, and Fan J (2002) 'Effect of clothing pressure on the tightness sensation of girdles', *International Journal of Clothing Science and Technology* **14**(2) 100–110.

Charles KE (2003) 'Fanger's Thermal Comfort and Draught Models', *IRC Research Report RR-162*, Institute for Research in Construction, NRC Canada.

Cline CHO, Thornton SB, and Nair SS (2004) 'Control of human thermal comfort using digit feedback set point reset', *Proceedings of the 2004 American Control Conference*, Boston, MA, pp. 2302–2307.

Fanger PO (1970) *Thermal Comfort*, Danish Technical Press, Copenhagen.

Fourt L, and Hollies NRS (1970) *Clothing: Comfort and Function*, Martin Dekker, New York.

Frank SM, Srinivasa NR, Bulcao CF, and Goldstein DS (1999) 'Relative contribution of core and cutaneous temperatures on thermal comfort and autonomic responses in humans', *Journal of Applied Physiology* **86**(5) 1588–1593.

Gavin TP, Babington JP, Harms CA, Ardelt ME, Tanner DA, and Stager JM (2001) 'Clothing fabric does not affect thermoregulation during exercise in moderate heat', *Medicine Science in Sports and Exercise* **33**(12) 2124–2130.

Ha M, Yamashita Y, and Tokura H (1995) 'Effect of moisture absorption by clothing on thermal responses during intermittent exercise at 24°C', *European Journal of Applied Physiology* **71** 266–271.

Havenith G, Holmer I, and Parsons K (2002) 'Personal factors and thermal comfort assessment: clothing properties and metabolic heat production', *Energy and Buildings* **34**(6) 581–591.

Hollies NRS (1997) 'Psychological scaling in comfort assessment', in *Clothing Comfort*, ed. NRS Hollies and RF Goldman, Ann Arbor Science Publishers, Ann Arbor, MI, pp. 107–120.

Hu JY, Li YI, and Yeung KW (2006) 'Liquid moisture transfer', in *Clothing Biosensory Engineering*, ed. Y Li and ASW Wong, Woodhead Publishing, Cambridge, UK, pp. 229–231.

Kawabata S (1980) *The Standardization and Analysis of Hand Evaluation*, 2nd edition, The Textile Machinery Society of Japan, Osaka.

Kenkare N, and May-Plumlee T (2005) 'Evaluation of drape characteristics in fabrics', *International Journal of Clothing Science and Technology* **17**(2) 109–123.

Kwon A, Masako K, Kawamura H, Yanai Y, and Tokura H (1998) 'Physiological significance of hydrophilic and hydrophobic textile materials during intermittent exercise in humans under the influence of warm ambient temperature with and without wind', *European Journal of Applied Physiology* **78** 487–493.

Li Y (2001) 'The science of clothing comfort', in JM Layton (ed.), *Textile Progress*, **31**(1/2), The Textile Institute, Manchester.

Li Y (2005) 'Perceptions of temperature, moisture and comfort in clothing during environmental transients', *Ergonomics* **48** 234–248.

Li Y, and Luo ZX (2000) 'Physical mechanisms of moisture diffusion into hygroscopic fabrics during humidity transients', *Journal of the Textile Institute* **91**(2) 302–316.

Li Y, and Wong ASW, eds (2006) *Clothing Biosensory Engineering*, Woodhead Publishing, Cambridge, UK.

Muir I, Bishop P, and Kozusko J (2001) 'Microenvironment changes inside impermeable protective clothing during continuous work exposure', *Ergonomics* **44**(11) 953–961.

Newsham GR (1997) 'Clothing as a thermal comfort moderator and the effect on energy consumption', *Energy and Buildings* **26** 283–291.

Prek M (2004) 'Thermodynamical analysis of human thermal comfort', *Proceedings of the Second ASME–ZSIS International Thermal Science Seminar (ITSS II)*, Bled, Slovenia, June.

Ramsey JD, and Bishop PA (2003) 'Hot and cold environments', in *The Occupational Environment: Its Evaluation, Control and Management*, 2nd edn, ed. SR DiNardi, American Industrial Hygiene Association, Fairfax, VA, pp. 612–645.

Ramsey JD, and Kwon YG (1992) 'Recommended alert limits for perceptual motor loss in hot environments', *International Journal of Industrial Ergonomics* **9** 245–257.

Ray P, and Bishop P (1991) 'Effects of fatigue and heat stress on vigilance of workers in protective clothing', *Proceedings of the Human Factors Society 35th Annual Meeting*, San Francisco, September, pp. 885–889.

Robinson MC, Bishop P, and Constable S (1988) 'Influence of thermal stress and cooling on fine motor and decoding skills' (Abstract), *International Journal of Sports Medicine* **9** 148.

Schutz HG, Cardello AV, and Winterhalter C (2005) 'Perceptions of fiber and fabric uses and the factors contributing to military clothing comfort and satisfaction', *Textile Research Journal* **75** 223–232.

Slater K (1985) *Human Comfort Vol III*, Charles C Thomas, Springfield, IL.

Smith G, Bishop P, Beaird J, Ray P, and Smith J (1994) 'Physiological factors limiting work performance in chemical protective clothing', *International Journal of Industrial Ergonomics* **13** 147–155.

Solomon J, Bishop P, Bomalashi S, Beaird J, and Kime J (1994) 'Responses to repeated days of light work at moderate temperatures in protective clothing', *American Industrial Hygiene Association Journal* **55**(1) 16–19.

Stevens SS (1946) 'On the theory of scales of measurement', *Science* **103** 677–680.

Wehner JA, Miler B, and Rebenfeld L (1987) 'Moisture induced changes in fabric structure as evidenced by air permeability measurements', *Textile Research Journal* **57** 247–256.

Wickwire PJ, Bishop PA, Green JM, Richardson RT, Lomax RG, Casaru C, and Curtner-Smith M (2007) 'Physiological and comfort effects of commercial "wicking" clothing under a bullet proof vest', *International Journal of Industrial Ergonomics* **37**(7) 643–651.

Winakor G, Kim CJ, and Wolins L (1980) 'Fabric hand: tactile sensory assessment', *Textile Research Journal* **50**(10) 601–610.

Wolins L, and Dickinson TL (1973) 'Transformations to improve reliability and/or validity for affective scales', *Educational Psychology Measurement* **33** 711–713.

Wong ASW, and Li Y (2002) 'Clothing sensory comfort and brand preference', *Proceedings of the 4th IFFTI International Conference*, Hong Kong, pp 1131–1135.

Wong ASW, Li Y, and Yeung PKW (2004) 'Predicting clothing sensory comfort with artificial intelligence hybrid models', *Textile Research Journal* **74**(1) 13–19.

Woodcock AH (1962a) 'Moisture transfer in textile systems, Part I', *Textile Research Journal* **32**(8) 628–633.

Woodcock AH (1962b) 'Moisture transfer in textile systems, Part II', *Textile Research Journal* **32**(9) 719–723.

Yasuda T, Miyama M, Muramoto A, and Yasuda H (1994) 'Dynamic water vapor and heat transport through layered fabrics. III Surface temperature change', *Textile Research Journal* **64** 457–461.

9
Dyeing and colouring tests for fabrics

C HURREN,
Deakin University, Australia

Abstract: This chapter discusses the types of tests undertaken to evaluate fabric colour. The chapter first reviews the assessment of colour and colour change with detailed information on the visual and automated assessment. The chapter then examines in detail the factors involving colour fastness testing. Tests examined include light fastness, wash fastness, fastness to environmental factors, fastness to manufacturing processes, and tests specific to printed materials.

Key words: colour assessment, light fastness, wash fastness, colour fastness, staining, shade change.

9.1 Introduction: key issues in the testing of dyes and colours

A myriad of factors can affect the performance of a colour in a textile fabric. Colour performance may be assessed in many ways. These include levels of fading, change of hue, change of saturation and staining of other items. Knowing the correct test to perform and the most accurate measurement system to adopt can put the colour performance of a fabric ahead of the competitors. This chapter looks at the measurement of colour and colour change. It is designed to show how to adopt the correct test method to evaluate coloured fabric. It examines each type of test that can be performed and provides a detailed overview of variations between test methods.

9.2 Assessing colour and colour change

The assessment of colour and colour change is the most important part of testing dyes and colours. Incorrect colour measurement wastes time and money and can result in a substantial claim for compensation from a customer. Colour is measured by measuring the reflected light from a sample over a variety of wavelengths. Each colour has its own reflectance fingerprint defined by the percentage of light reflected at a given wavelength. A reflectance curve is measured in the visible region for a colour and is plotted as percentage reflectance (%R) versus wavelength. To simplify the description of colour, The International Committee on Illumination

(Commission Internationale de l'Eclairage or CIE) set a formulated system for the definition of colour in terms of 'tristimulus values' X, Y, Z. The tristimulus values of a sample represent the amounts of red (X), green (Y) and blue (Z) primary colours which are necessary to produce the 'colour' of the sample. They are determined from the reflectance value (R_λ), spectral radiant flux per unit area for the source light (E_λ) and the tristimulus eye sensitivity functions of the CIE standard observer ($\bar{x}_\lambda$, $\bar{y}_\lambda$ and $\bar{z}_\lambda$). The integration is usually performed over the wavelength (λ) range of 380–740 nm. A constant (k) is used to normalise the results. Equations 9.1, 9.2, 9.3 and 9.4 show the formulas for calculating the tristimulus values.

$$X = k \int_{\min \lambda}^{\max \lambda} E_\lambda \bar{x}_\lambda R_\lambda \, d\lambda \qquad \qquad 9.1$$

$$Y = k \int_{\min \lambda}^{\max \lambda} E_\lambda \bar{y}_\lambda R_\lambda \, d\lambda \qquad \qquad 9.2$$

$$Z = k \int_{\min \lambda}^{\max \lambda} E_\lambda \bar{z}_\lambda R_\lambda \, d\lambda \qquad \qquad 9.3$$

$$k = \frac{100}{\int_{\min \lambda}^{\max \lambda} E_\lambda \bar{y}_\lambda \, d\lambda} \qquad \qquad 9.4$$

In 1976 the CIE introduced the L^*, a^* and b^* and the cylindrical L^*, C^* and h chromaticity coordinates and these parameters are now widely used. These values are derived from the original X, Y, Z tristimulus values using equations 9.5, 9.6, 9.7, 9.8 and 9.9. L^* is defined as the lightness of the colour, a^* is the axis that extends from red (positive) to green (negative), b^* is the axis that extends from yellow (positive) to blue (negative), C^* is the intensity of chroma, h is the angle of hue, and X_n, Y_n and Z_n are the tristimulus values for the relevant standard illuminant and observer. Each of these parameters and the relationship between them is shown using the CIE colourspace described in Fig. 9.1.

$$L^* = 116(Y/Y_n)^{1/3} - 16 \qquad \qquad 9.5$$

$$a^* = 500\left[(X/X_n)^{1/3} - (Y/Y_n)^{1/3}\right] \qquad \qquad 9.6$$

$$b^* = 200\left[(Y/Y_n)^{1/3} - (Z/Z_n)^{1/3}\right] \qquad \qquad 9.7$$

$$C^* = \sqrt{\left((a^*)^2 + (b^*)^2\right)} \qquad \qquad 9.8$$

$$h = \arctan\left(\frac{b^*}{a^*}\right) \qquad \qquad 9.9$$

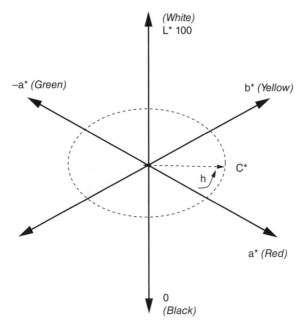

9.1 CIELAB colourspace.

Decomposition of dye molecules or physical removal of dye from the fibre are two mechanisms that cause colour change in a fabric. Where a single dye is removed or degraded, the result is generally a reduction of colour depth and/or a reduction of colour purity. The colour change in a commercial fabric can be a little more complex, as most dyes and dye recipes are made of a combination of dyes with different hues and concentrations. Each dye will behave differently under different environmental influences, therefore a change in colour could be caused by a change in hue as one dye is affected more than another.

There are two systems used for the assessment of colour change. These are visual assessment and computer aided assessment. Visual assessment has been in use since the performance of colour was first considered and is still used widely today. Visual assessment is a subjective measurement system and is significantly influenced by the person undertaking the assessment. The development of computer assessment systems and customers' requirement for repeatability and accountability are driving the shift to computerised measurement.

9.2.1 Visual assessment

The most well-known and used system of visual assessment is the grey scale. Grey scales are used as the rating system for most standard test methods

9.2 ISO grey scales.

as they are widely available, low cost and easily used. There are two types of grey scales. One set measures the change in shade of a coloured textile and the other measures the degree of staining in an adjacent fabric. The two grey scales are shown in Fig. 9.2. Grey scales have a rating of 1 to 5 with 1 being the worst colour performance and 5 being the best. Each rating can be split so that there are nine available ratings within the grey scale system.

The type of light used to illuminate the sample is important when visually assessing colour. Each light source produces a different emission spectrum and this influences the colour seen by the observer. Visual assessment of colour is normally carried out in a light box under a specified illuminant. The most common illuminants are artificial daylight, incandescent light, fluorescent light, horizon light, point of sale light and ultraviolet light. Each illuminant can then be broken down into individual types. An example is artificial daylight that has multiple source types including D50, D65 and D75. Most test methods specify the light source under which the samples should be rated. If the test method does not state the light source, a light source is agreed on and fixed by all stakeholders in the colour measurement.

The angle of the observer and the incident light are important in visual assessment. The illuminant light should hit the sample at 45°. This is achieved by resting the samples on a table set at a 45° angle to the light. This presents a perfect angle for the viewer to observe the samples perpendicular to the fabric surface. It is also important to exclude any light from external sources. Lights from external sources include room lighting or light from a window. Placing the light box in a dark room or placing a curtain around the light box will combat stray light.

The colour of the viewing surface is also important. Most standard test methods recommend a matt grey finish. The matt grey colour does not distract the viewer from the colour being assessed. It is important to keep the light box viewing surface clean and free of defects or imperfections. Damaged light box surfaces should be repaired with a paint that has the correct colour and gloss level.

9.2.2 Automated assessment

The use of spectrophotometers for fabric colour measurement has been adopted widely in the last 10 years. This technique has not changed much over the last 30 years; however, data collection and management of the spectrophotometer has. The development of low-cost high-powered desktop computers has allowed the quick acquisition, manipulation and quantification of colour information. Information obtained is both qualitative and quantitative, and can be efficiently stored for future reference or use. Information obtained using one spectrophotometer can be compared with results from another without any error. A spectrophotometer can provide a huge amount of measured and calculated information including ΔE^* values, multiple light source colour information, comparisons with measured or inputted standards, colour histograms of multiple batches, reflectance versus wavelength graphs and recipe advice.

CIELAB colour difference (ΔE^*) is the most common system used in automated colour assessment for defining a difference in colour. ΔE^* is the difference in colour between two samples (1 and 2), with the coordinates L^*_1, a^*_1, b^*_1 and L^*_2, a^*_2, b^*_2, and is calculated using equation 9.10. A CIELAB ΔE^* value of one unit represents the smallest colour difference that can be visually detected. Subsequent experience has shown that the visual detection limit is more like 0.8 of a unit. ΔE^* is most commonly used in fabric testing to electronically specify the change in shade or degree staining of a sample or adjacent fabric after fastness testing. Grey scales have been assigned fixed CIELAB ΔE^* values by the standard boards. The CIELAB-assigned ΔE^* values for each of the grey scales are reproduced in Table 9.1.

$$\Delta E^* = \sqrt{\left((L^*_1 - L^*_2)^2 + (a^*_1 - a^*_2)^2 + (b^*_1 - b^*_2)^2\right)} \qquad 9.10$$

Colour data management has improved markedly since the development of automated assessment. Databases attached to the computer colour measurement systems allow for the efficient storage of colour data without the need to store the physical test sample. Specialised computer screens allow on-screen reproduction of the test standard and are accurate enough for a person to perform a visual assessment. Results can be sent worldwide using the Internet and appraised visually by the customer immediately after the test is completed.

Table 9.1 CIELAB Δ*E** assigned values for grey scales

CIELAB colour difference for fading		Colour fastness grade	CIELAB colour difference for staining	
Value	Tolerance		Value	Tolerance
0.0	<0.40	5	0.0	<1.10
0.8	0.40–1.25	4–5	2.2	1.10–3.25
1.7	1.25–2.10	4	4.3	3.25–5.15
2.5	2.10–2.95	3–4	6.0	5.15–7.25
3.4	2.95–4.10	3	8.5	7.25–10.25
4.8	4.10–5.80	2–3	12.0	10.25–14.45
6.8	5.80–8.20	2	16.9	14.45–20.45
9.6	8.20–11.60	1–2	24.0	20.45–29.05
13.6	>11.60	1	34.1	>29.05

9.3   Change in shade and staining tests

Change in the shade of a coloured fabric, or the staining of a fabric in the proximity of the coloured fabric, are performance problems associated with coloured fabrics. An example of change of shade is seen during light fastness testing of fabric. During exposure to light, most dyes degrade and change or lose their colour, causing a change of shade in the fabric. Staining of a fabric is seen during laundering when a white garment turns pink. This is due to migration of red dye from a garment that is also in the wash bath.

9.3.1 Reversible colour change

Some colours undergo reversible colour change. The light-initiated version of reversible colour change is called photochromism. Photochromism occurs because the photons of light striking the coloured surface induce a structure change in the dye instead of degradation of the dye. A change in dye structure results in a change in colour. After duration of no exposure to light the structure reverts back to its original form and colour. This type of colour change can also occur due to exposure to heat or chemicals.

9.3.2 Metamerism

Metamerism is the colour change seen in a coloured item because of different spectral emissions from different light sources. Each light source has its own emission spectrum (colour) so when a light source is projected onto a surface, the surface colour is influenced by the colour of the light. This produces a different colour to the observer for the same item when the light source type is changed. An example of this is the bluer tinge of a coloured sample when observed under the fluorescent light TL84 compared to daylight. A standard light source is important when viewing, rating and specifying colour change to minimise the effects of metamerism.

9.3.3 Optical brightening agents

Optical brightening agents (OBAs) can cause serious error in the evaluation of colour and hence in the evaluation of colour fastness. OBAs are normally used to enhance the whiteness of a textile. They convert ultraviolet light into a wavelength in the visible spectrum. Most OBAs used for improving the white effect of a textile emit light in the blue spectrum, as most off-white colours reflect higher in the red/yellow end of the spectrum. The addition of blue to a white with a red/yellow base causes a flattening of the reflectance curve, resulting in the colour looking whiter. The addition of an OBA to a dyed colour will change the observed colour.

Most commercial laundry detergents contain an OBA to make whites washed in them look whiter/cleaner. All colours washed in these detergents adsorb OBAs, altering the colour even though the fabric's dye may not have been affected by the washing process. Most wash fastness test methods stipulate if OBAs are to be present in the wash liquor to limit the associated colour measurement problems. The presence of an OBA is easily detected in a fabric or detergent by placing it under an ultraviolet light source. The blue ultraviolet light will be reflected from the surface in the colour of the OBA.

Light fastness testing can be influenced by the presence of OBAs. OBAs have poor fastness to light and generally fade at a higher rate than most dyes. The chemical bond structure of the OBA provides the mechanism that allows ultraviolet light to be converted to visible light. This bond structure is easily changed by the ultraviolet light rendering the OBA colourless. OBA-induced light fastness problems are normally seen in whites and pastels.

9.4   Test standards

There are many different standard test methods for colour fastness testing of fabrics. The key standard setters for textile colour fastness are the Society of Dyers and Colourists (SDC) and the American Association of Textile Chemists and Colorists (AATCC). These two associations have spent many years developing standards and provide excellent information and advice on their websites. Other standard test setters in this area are the International Organisation for Standardisation (ISO) and the International Wool Textile Organisation (IWTO). These organisations provide standards to cover all facets of textile processing. Their colour standards are quite often based on the SDC or AATCC equivalent. There are some country-based standard organisations, such as the British Standard (BS), American Standard Test Method (ASTM) and Australian Standard (AS). These are generally based on the ISO, SDC or AATCC test method with slight changes made to account for cultural or environmental differences. For example, it is not suitable to light fastness test fabrics for an Australian market under European natural daylight as the incident angle and intensity are different.

Consumers are increasingly looking for textile products that are environmentally friendly and have neutral health effects. Standard setters are creating new test standards to measure this. The Oeko-Tex Association has followed this theme and developed the Oeko-Tex 100 standard that is based around a human and environmentally friendly product. This standard looks at reducing harmful processing methods such as formaldehyde-based finishes. It also looks at minimising the environmental impact of textile processing by reducing environmentally harmful chemicals, waste and processes.

9.5   Light fastness

There are a large number of different light fastness tests available on the market. Each has its advantages and disadvantages. The most commonly used are the xenon arc and MBTF lamp; however, carbon arc and natural sunlight are also used. A fabric exposed to light is influenced by its depth of shade, the intensity of the light, the wavelength profile of the light, the temperature of the fabric, contaminants within the fabric, and the moisture content of the fabric. Other factors such as exposure cycling, exposure time and substrate colour change are also influential.

The light fastness rating system is based on the rate of fading of eight blue-dyed wool samples (blue chips) which are rated from 1 (poor) to 8 (excellent), with each successive standard dyeing taking twice as long to fade as the previous one in the series. The blue chips are placed into the light box with the samples to be tested and faded in parallel with the test samples. This is done as the light output of the light source can vary from test to test. Most test methods assess the light fastness when the fabric being faded exhibits a change in shade equal to 4 on the grey scale for colour loss.

9.5.1 Depth of shade

The depth of shade of a colour has a significant effect on the light fastness of a product. Light fastness is the degradation of a fixed number of dye molecules per exposure to a fixed intensity of light. A deeper shade is

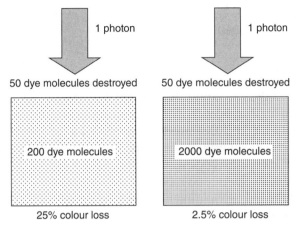

9.3 Depth of shade effect on light fastness.

affected less than a pale shade, as a smaller percentage of the overall dye molecules are degraded per light exposure. Figure 9.3 represents this.

9.5.2 Intensity

Most accelerated tests for exposure to light fastness use high-intensity light to minimise the test time. High-intensity light can reduce the amount of time to undertake the test, though it can also influence the result. Higher-intensity light generally results in higher sample temperatures, causing the reaction rate of dye degradation to increase. The rate of dye degradation with respect to temperature is not linear. At elevated temperatures one dye may fade at a quicker rate than another, giving a hue change not seen at lower temperature fading.

Most high-energy light sources have varying intensity depending on the age of the light source. Most have a run-in time before they can be reliably used and most have a set number of hours of use before they should be replaced. Most standard test methods have details of the run-in times and maximum number of running hours of a light source.

9.5.3 Wavelength

Different light sources have a different wavelength profile. Some are close to natural daylight; however, the majority have wavelength peaks of high intensity that are different from natural daylight. Daylight itself varies depending on the latitude of the viewer and the time of the year. The wavelength of the incident light on the dye bonds significantly influences the rate of degradation of those bonds. A wavelength profile different from natural daylight could show an increased or decreased fastness rating, depending on the dye tested.

9.5.4 Temperature

The temperature of the test sample will influence the rate of light-induced degradation of a dye. Most light fastness tests have black body temperature measurement and a method of sample cooling so that temperatures do not become too high. High-intensity light sources produce increased sample temperatures, so increased cooling capacity is required to lower test temperatures. Tests for automotive fabrics are conducted at higher test temperatures, as the fading environment within a car interior can involve elevated temperatures not normally seen in the fading environment for clothing.

9.5.5 Moisture

The level of moisture in the fabric can influence the rate of colour degradation. The presence of moisture during a light fastness test can lead to the generation of peroxide radicals that significantly influence the results of the test. Moisture content is hard to control as the heat from the light source tends to decrease the moisture content of the fabric. Moisture levels can be monitored during a test by using azoic-dyed cotton with specified fading properties which vary with the amount of relative humidity. Some tests involve starting the test samples wet to simulate line drying of fabrics. Some tests involve intermittent jets of water to simulate rain on an exterior fabric. Tests that involve the use of water sprays are generally referred to as weathering tests.

9.5.6 Contaminates

Some fabrics are exposed to chemical contaminates when they are in use. These contaminates can include salt and chlorine. Tests have been developed to intermittently spray the fabric with chemical-contaminated water during the test. Like water, chemical contaminates can become involved in the chemical degradation of the dye molecules.

9.5.7 Test cycling

Sometimes exposing the fabric to a light source for a fixed duration is not enough to see the behaviour of the colour. Cycling of the light source on and off is sometimes used to simulate night and day. Some dyes can degrade to a certain point and degrade no further with continuous exposure to light.

Switching the light off for a period of time allows the energies within the dyes to return to their ground state. This can initiate a second round of fading when the light is restarted.

9.5.8 Substrate colour change

Substrate colour change is quite common in pastel wool colours. Wool initially photo-bleaches when it is first exposed to light; however, it yellows quite rapidly with continued exposure. The colour of the dye might not be affected by the incident light, though the change of the substrate colour will result in a change in the fabric shade.

9.5.9 Photochromism

Photochromism is commonly caused by exposure to light. The level of photochromism can be determined by placing a colour in an intense light source for a short period of time and then immediately rating it for colour change. If subsequent return to normal colour occurs after conditioning in the dark then the colour is photochromic. Some colours require a period of time to condition in the dark after light fastness testing before colour assessment can be conducted to avoid any photochromic effect.

9.6   Wash fastness

Consumers launder their fabric at some time in the lifespan of the textile. Change of colour or staining of another garment during laundering is generally immediately evident to the consumer and has a high impact on consumer satisfaction. Like light fastness testing there are a huge number of different iterations of the wash fastness test. The wide variety of test methods have mostly arisen due to the variety of washing methods available, cultural practices, the material being washed, and the end use of the product. The development of detergents and bleaches has influenced the development of new wash fastness test methods.

It is important when selecting a wash fastness test method to choose one that best simulates the washing environment of the fabric's end use. If no end use has yet been selected then it is important to clearly label the level of wash fastness testing that has been conducted on the fabric. The most common washing methods in use today are dry cleaning, hand washing, gentle machine washing, machine washing, permanent press and industrial laundering. Each of these methods has one or more wash fastness tests to determine fabric colour suitability. Table 9.2 shows the variety of conditions in the first five ISO wash fastness tests for domestic and commercial washing.

Table 9.2 The first five ISO wash fastness test conditions

Test	Temperature (°C)	Time (min)	Number of steel balls	Chemicals
ISO 1	40	30	0	Soap
ISO 2	50	45	0	Soap
ISO 3	60	30	0	Soap + Na_2CO_3
ISO 4	95	30	10	Soap + Na_2CO_3
ISO 5	95	240	10	Soap + Na_2CO_3

Each of these tests would be carried out at a 50:1 liquor ratio in a 2.0 litre wash wheel vessel with 5.0 g/l standard soap solution.

9.6.1   Equipment

Most wash fastness tests are carried out in enclosed 2000 ml vessels that are rotated at a constant speed and at a constant temperature in a wash wheel. An Atlas laundrometer is the most common make of equipment used for undertaking this test. The wash wheel is commonly referred to as a laundrometer. The fabric, adjacent material and wash liquor are placed into the test vessel before it is sealed. Some tests require the addition of stainless steel balls or discs to the wash liquor to increase the severity of the test.

9.6.2   Soaps and detergents

In a wash fastness test the soap or detergent is used to remove unfixed dye from the fabric. The use of a soap or detergent can also cause a breakage of bonds that hold the dye on the fibre. The pH of the soap or detergent has a major influence on the movement of dye from the fabric into the wash liquor and from the wash liquor to adjacent fabrics. Since the invention of detergents, the blend of chemicals used for domestic and commercial laundering has become quite complex. Most detergents contain mild oxidising agents, softeners, optical brightening agents, salts and other fillers. These detergent auxiliaries can increase the amount of dye removed from the fabric and in the case of oxidising agents cause degradation of the dye molecules. The dispersing nature of different detergents can also reduce the level of cross-staining to an adjacent fabric as the dye is held in the wash liquor by the detergent and is not allowed to redeposit. Some test detergents are specifically designed to contain bleaches and bleach activators like sodium perborate tetrahydrate and tetraacetylethylenediamine (TAED).

Table 9.3 SDC multifibre strips

SDC multifibre DW	SDC multifibre TV
Secondary cellulose acetate	Triacetate
Bleached unmercerised cotton	Bleached unmercerised cotton
Nylon 6,6	Nylon 6,6
Polyester (Terylene)	Polyester (Terylene)
Acrylic	Acrylic
Unbleached wool	Viscose rayon

9.6.3 Test fabrics

Cross-staining of a third-party fabric is assessed using an adjacent fabric fixed to the fabric test sample. Adjacent fabrics come in many types and many forms, including single-component woven fabrics and multifibre woven fabrics. The single-component adjacent fabrics commonly used are cotton, wool, polyamide, acrylic, viscose rayon, polypropylene and polyester. Tests involving single-component adjacent fabrics have the adjacent fabric fixed to one or both sides of the fabric test sample. Sometimes a different adjacent fabric composition is used for each side to show staining on two fabric types. Nylon 6,6 is commonly used in staining tests as it tends to scavenge any free dyestuff from the wash liquor better than any other fabric composition.

Multifibre adjacent fabrics allow staining exposure to a range of different fabric types during one test. The most common supplier of multifibre fabric is the Society of Dyers and Colourists (SDC). The SDC produces two multifibre fabrics; one contains wool and one does not (Table 9.3). Multifibre fabric is affixed to one side of the test specimen and the other side is normally a polypropylene fabric in a staining test.

9.6.4 Agitation time

The agitation time of a laundering test can significantly affect the test results. Short test times limit the dissolution of unfixed or poorly fixed dye into the liquid or onto the adjacent fabric. The dye has more time to escape from the fibre and to migrate onto the adjacent fabric in longer tests. Longer test times can allow the dye to deposit on and migrate into the fibres of adjacent fabrics following a mechanism similar to dyeing, leading to an increase in the level of staining.

9.6.5 Temperature

The results of a wash fastness test are significantly influenced by the test temperature. Lower test temperatures are used where the end-use fabric requires low-temperature washing. Low-temperature washing is normally seen for delicate fabrics like wool, silk and viscose. Low test temperatures limit the migration of dye from the fabric surface and generally cause low levels of staining on adjacent fabrics. There is not as much energy available at low temperatures, and energy is needed for the dye to attach and penetrate the adjacent fibre. High test temperatures should be used for fabrics that could be warm or hot washed during their lifetime. Higher test temperatures provide the energy required to swell fibres and migrate dyes. At higher temperatures adjacent fabrics are more likely to be stained by any free dyes in the wash liquor. Fabrics that are going to be used in a product destined for industrial laundering are generally tested at higher temperatures than normal, as industrial laundering temperatures are generally higher.

9.7   Fastness in relation to environmental factors

Wash fastness testing looks at the loss of colour or staining of adjacent fabrics due to laundering. However, there are numerous other environmental influences that may cause colour performance problems, most commonly perspiration. Perspiration from the human body is a complex chemical containing large quantities of salts; depending on the human metabolism, it can be either acidic or alkaline. Most of the tests for perspiration fastness are based on a solution containing the chemical histidine.

The fabric and adjacent fabric are generally soaked in the test solution before being placed under pressure in a perspirometer and incubated at body temperature for a period of time. Figure 9.4 shows an illustration of the general layout of a perspirometer. The test fabrics with adjacent material attached are sandwiched between Perspex plates under a fixed mass.

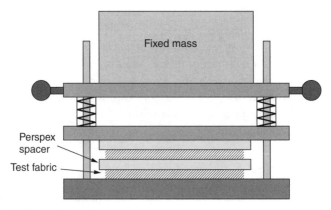

9.4 Perspirometer.

The perspirometer is also used for other tests including testing for fastness to water and seawater. Fastness to water testing simulates the effect of leaving washed fabric sitting in a wet pile after laundering. The staining assessment is the most important part of this test, as the colour can easily transfer from article to article under wet pressurised conditions. Fastness to seawater looks into the same effect as fastness to water but includes sodium chloride in the test solution, as salt can cause increased migration of dyes.

Fastness to chlorinated water is used to evaluate the colour fastness of swimwear, towels, deck furniture webbing or other articles that may be exposed to large amounts of chlorinated water. The test is normally carried out using a wash wheel under similar conditions to a wash fastness test; however, the temperatures used are selected to reflect pool water. It is important to check the active chlorine levels of the test solution before the test, as chlorine can reduce in strength over time.

Spot testing is used to determine the effects of spotting chemicals onto a fabric to remove a point stain or to simulate spot staining. An adjacent fabric can be fixed behind the spot to assess staining; however, the test is normally used to assess spot migration of the dyes within the fabric. Spotting tests include fastness to water, acid, alkali, dry cleaning fluid, and white spirits.

Fastness to rubbing is used to ensure that fabrics do not transfer their colour when rubbed against another layer of fabric. This test is also known as crocking and is carried out using a crockmeter. A crockmeter is a piece of equipment that applies a constant force on the test fabric against the tested specimen as it is rubbed back and forth. Rub fastness is carried out with either a dry or wet cotton fabric that is rubbed against the surface of the dyed fabric to remove unfixed dyestuff. Rub fastness using a wet test fabric tends to show higher colour transfer than when using a dry test fabric.

9.8   Fastness in relation to manufacturing processes

There are a range of tests that are based around the mechanical processing of textile fabrics during and after the manufacturing process. The tests are described only briefly here as books on processing, product circulars and test standards describe these test methods more specifically. The tests for processing are based around three main concepts: the application of heat, gases and chemical processes.

9.8.1 Colourfastness to heat

Heat-based tests look at the change of colour due to heat. Some dyes sublime under extreme heat and can be evaporated from or heat-transferred from the fabric. Some dyes are degraded by extreme heat and can change or lose their colour. The presence of moisture during heating can increase swelling of the fibre and thus increase the transfer of colour into an adjacent fabric. Heat can be applied in various ways, so the test selected must reflect the way in which the heat is applied. The four ways in which a heat can be applied are dry heat without pressure, steaming without pressure, dry heat with pressure, and steaming with pressure. The test for dry heat without pressure is used to simulate drying of fabrics. The test for steaming without pressure looks at the colour change due to steam relaxation or steam setting. The test for dry heat under pressure is a test to simulate hot pressing, ironing and calendaring. The test for steaming with pressure is used to replicate steam pleating, steam pressing and decatising.

9.8.2 Colourfastness to chemical processing

There are a large number of chemical processes that expose coloured fabrics to chemicals. Most chemical processes in manufacturing involve chemical finishes applied to the fabric after dyeing. A large number of these finishes have specific tests defined by the company that look at the level of colour change occurring during the finishing process. Chemical finishes can also affect light fastness and rub fastness results. Therefore for some fabrics these parameters should be measured after the finish has been applied.

During manufacture a dyed fabric may be exposed to a number of chemicals. Standard test methods have been developed to assess the changes that these factors can cause. Testing includes bleaching with different bleaching agents including chlorite, hypochlorite and hydrogen peroxide. Wet processing of fabrics such as milling, carbonising and crabbing can also affect fabric colour. The test for crabbing is also referred to as potting, as it involves boiling the fabric under tension for a period of time.

9.8.3 Gas exposure

There are a whole range of tests that are based around drying fabrics. The tests for oxides of nitrogen and burnt gas fumes are used to evaluate the effects of inefficient and badly regulated direct-fired drying equipment. Ozone is also a chemical that is generated in the environment or during combustion. Ozone can rapidly break down dye, so there are a number of standards that relate to the effects of ozone in the presence of a textile fabric.

There are also tests that look at the influence of residual chemicals in the fabrics during drying. The most common of these tests look at the presence of residual hardness salts, acid and alkali. Specific tests can also be conducted to look at the effects of aftertreatments applied during or subsequent to the drying process.

9.9   Printing tests

The colour fastness of printing is slightly different from that of dyeing, as most printing techniques use a pigment to colour the surface of the fabric generally by forming a pattern on the fabric. The pigment is bound to the fabric surface using a polymer binder, most commonly acrylic. The main problems associated with prints are poor registration of the pattern, wicking of some or all of the colours and poor rub fastness. Printing is commonly done over the top of a previously dyed fabric and the printing process and chemicals can have an effect on this colour. The printing process can also affect fabric properties other than colour and these should also be measured. Most of the standard tests used for assessing colour in dyed goods are also used for printed goods.

9.9.1   Registration

Registration refers to the alignment of a single print colour on the fabric with reference to the fabric and other colours in the print. A malfunction or poor setup of the printing process can result in poor registration of a print. This will be seen in the final product as a misalignment of part or all of the individual colours of the print. Testing for registration can be conducted visually with deviation from registration measured with a ruler or callipers. The development of image processing software has resulted in several good automated print registration test apparatus being available in the market.

9.9.2   Wicking

Wicking is the transfer of some or the entire print colour along the fibres in a fabric due to a capillary action. Wicking can be seen as a reduction of sharpness of a printed edge and can be assessed visually or with the same software as is used for the testing of print registration. Sometimes the printing ink can cause bleeding of the fabric base colour. Bleeding of the base colour looks blotchy along the edge of the print or, when printing is done in a garment, a transfer of dye onto the fabric occurs adjacent to the print face.

9.9.3   Rub fastness

Rub fastness is of significant importance to prints, as the colour is provided by a pigment that is fixed to the exterior of the fibre. The pigment on the surface of the fabric is the first component to come under attack when a fabric is rubbed or abraded. Crocking is a simple method for determining fastness to rubbing. The standard 10 cycles employed in the crocking test may not be enough to break down a faulty pigment binder system, so some rub fastness test methods involve an increased number of rubbing cycles. There are test methods and testing equipment developed that have an increased abrasive effect on the fabric surface. These include oscillating drum, wire mesh and emery abrasion testers. Each of these testers abrades the surface more than the crock meter and they are used for textiles that require high resistance to abrasion such as military fabrics.

After rub fastness testing the rubbed surface is often appraised for colour change or frosting. Frosting is common in prints and is the pigment rubbing away from the surface of the fabric exposing the natural fibre colour below. This can also be called fibrillation, as single fibres poke through the surface of the print. Frosting is easily identified in a loss of depth of shade of the print.

9.9.4 Fastness to steaming

Most prints are steamed to improve penetration into the fabric or to assist in fixation of the dye or binder system. It is important that dyes or pigments used in the printing industry are fast to steaming. The test should not just be limited to the effect of the steaming on the print but should be expanded to include the fabric base colour. A change in fabric base colour can occur even though the print is unaffected. The tests undertaken for steaming are generally the same as the tests explained earlier for steam without pressure.

9.9.5 Light fastness testing

When testing light fastness it is important to assess all of the colours in the print. Light fastness test samples should be selected to maximise the area of each of the colours within the print, and more than one light fastness sample per print may be required to achieve this. Pigments generally have better light fastness than dyes, as the chromophore does not need to be selected to fit into the fibre matrix or to have specialised bonding and solubilising groups.

9.9.6 Plastisol prints

Plastisol prints pose their own unique fastness problems. Inaccurately cured pigment can result in cracked and peeling prints or bleeding of the colour into other fabrics. Prints can be tested for curing by a simple domestic washing followed by a tumble dry. Other curing tests can involve the use of solvents and pressure. The print is spotted with small amounts of solvent and then pressed against a fabric to assess the colour transfer. Dry heat is normally used to apply a plastisol print, so fabrics should be evaluated for dry heat in pressing to ensure that there is no colour change to the fabric base colour.

9.10   Applications

There are many different applications for the testing of coloured fabrics. The textile colourist utilises colour testing to confirm a colour matching formula before it is used in manufacture. Most textile mills will implement quality control testing of colours during manufacture to reduce the chance of faulty work being processed further than it needs to be. It is hard to improve the wash fastness of a fabric that has been cut into a garment.

It is important to measure the colour fastness properties of a dyed fabric after manufacture. Most customers require a quality control certificate for the fabric that they are purchasing to confirm that it meets their target limits. Testing will highlight possible problems and avoid despatch of under-specification fabric. A manufacturer should measure a fabric that has come from a supplier or commission dye house if the fabric does not already have a set of test statistics. When a fabric does come with test results, it is advisable to double-check test results for the first few deliveries from a new supplier and then randomly audit deliveries as purchases continue. Suppliers that have their own test laboratories can be production-biased when rating fastness results and can pass samples that have not met the testing requirements. Double-checking the results will give confidence in the skill and accuracy of the supplier's test house.

Quite often a manufacturer will be interested in replicating a competitor's product. Careful testing of the product can reveal the type of colouration method used and the level of fastness required to duplicate the product. Testing results of a product can be used to exploit the marketing potential of a product. The development of machine wash fast colours was originally a strong selling point for a textile product. Proof of meeting Oeko-Tex 100 environmental and health standards is an example of a new selling point that can be confirmed by accurate product testing.

9.11   Future trends

In the future we will see increased use of electronic measurement of colour and colour fastness test results. The development of data handling, transfer and storage has revolutionised the way in which test results are measured and conveyed within the mill and to the customer. The Internet transfer of colour and colour details is becoming adopted by more processors and customers, and speeds up the test path as a customer can approve a colour test just minutes after it has been conducted. The use of fuzzy logic mathematics helps to analyse all of the testing data and enables the manufacture to optimise processing and reduce reject rates.

Consumer requirements for environmentally friendly, neutral health effect and neutral environmental impact textiles will increase. Some companies are leading the way by meeting standards like Oeko-Tex 100; however, more stringent standards will be developed. Environmentally friendly coloured textiles will see significant development in test methods and accreditation over the next decade.

9.12 Sources of further information and advice

Society of Dyers and Colourists, www.sdc.org.uk
American Association of Textile Chemists and Colorists, www.aatcc.org
Pantone, www.pantone.com
Oeko-Tex Association, www.oeko-tex.com
International Organisation for Standardisation, www.iso.org
International Wool Textile Organisation, www.iwto.org
Ecological and Toxicological Association of Dyes and Organic Pigment Manufacturers, www.etad.com

9.13 Bibliography

Broadbent A D (2001), *Basic principles of Textile Colouration*, Cambridge, Woodhead.
Jones E B (1978), *Textile Progress; Chemical Testing and Analysis*, Manchester, The Textile Institute.
Kadolph S J (1998), *Quality Assurance for Textiles and Apparel*, New York, Fairchild.
Saville B P (1999), *Physical Testing of Textiles*, Cambridge, Woodhead.
SDC Enterprises (2005), *Colour Fastness and Testing Products Catalogue*, Bradford, SDC Enterprises.
Slater K (1993), *Textile Progress; Chemical Testing and Analysis*, Manchester, The Textile Institute.
Xin J H (2006), *Total Colour Management in Textiles*, Cambridge, Woodhead.

10
Testing intelligent textiles

J HU and K M BABU,
The Hong Kong Polytechnic University, China

Abstract: Intelligent textiles represent the next generation of fibers, fabrics and products made from them. In the last decade, research and development in smart/intelligent materials and structures have led to the birth of a wide range of novel smart products in aerospace, transportation, telecommunications, homes, buildings and infrastructures. New methods of testing and evaluation procedures for testing intelligent and smart fabrics are becoming extremely important in the industry as the future lies in these textiles. In this chapter the principles and some of the important test methods for testing intelligent fabrics, such as shape memory fabrics, phase change materials and self-cleaning fabrics, are explained.

Key words: Intelligent testing, testing shape memory fabrics, testing phase change materials, electronic response testing, self-cleaning tests.

10.1 Introduction: role of intelligent textile testing

Today's textile industry is in the transition zone between a traditional textile production and the realization of highly focused design and production of added value textiles. The innovative field of smart and intelligent textiles is becoming increasingly popular and commercially successful because it combines product use with new material properties. Intelligent textiles represent the next generation of fibers, fabrics and products made from them. They can be described as textile materials that think for themselves. This means that they may keep us cool in hot environments or warm in cold conditions. Apart from the clothing sector, they are gaining popularity in various other fields, such as biomedical materials, electronics, the automobile industry, protective clothing, etc. They provide ample evidence of the potential and enormous wealth of opportunities still to be realized in the textile industry, in the fashion and clothing sector, as well as in the technical textiles sector. Integrating intelligent textiles into clothing is an exciting new field, which opens up a vast arena of applications. With revolutionary advancements occurring at an unprecedented rate in science and electronics, the possibilities offered by wearable technologies are tremendous. As the technologies become more refined so that complex systems can be embedded unobtrusively in everyday clothing they will soon be more and more commonplace in commercial products. It is anticipated that the results will likely support the possibility of creating and producing 'intelligent clothing' and smart technologies in the clothing sector, and offer services in the integration of intelligence with clothing.

Working closely with the clothing industry will develop the base that is needed to offer developments in intelligent clothing with huge commercial potential at minimum risk. At a later stage of development, this is likely to create more solid product assortments. Those experienced in the industry expect that technologies in smart clothing will be launched in the market within the next five to ten years. Hence there exists an absolute necessity to understand these new fabrics and their technologies. Testing of fabrics hitherto limited to traditional fabrics, such as apparel, home furnishings and some varieties of technical textiles, may not help in the long run in understanding the properties of these new fabrics. New methods of testing and evaluation procedures for testing intelligent and smart fabrics are becoming extremely important in the industry as the future lies in these textiles.

10.2   Understanding existing materials and technologies

Textiles are used in a number of applications from apparel to technical textiles. The era of technical textiles has opened up a new challenge for design of new type of textiles for protective clothing, sports, pleasure, social promotion, etc. Before any such new development takes place it becomes absolutely important to understand the existing types of textile fabrics and their technologies. Currently fabrics are being used in protective clothing for protection against vagaries of weather, fire, severe thermal effects, etc. Technical textiles in the form of fabrics are widely used in a number of industries such as aerospace, military, sports, composites, agriculture, automobiles, etc. A large variety of fabrics are being used in geotechnical applications as geotextiles for reinforcement of soil, filtration and drainage purposes. The medical field has opened up new avenues for functional textiles for biomedical applications such as sutures, scaffolds, stents, etc. All the above existing textiles form the basis for the new developments. But the future trends lie in making a fabric more multifunctional so that it acts as a smart fabric, finding its own applications in new environments.

The textile and garment industry in the developing countries is much more challenging. The companies that concentrate on specialty products and technical development still believe in the future. In order to survive, the companies have to follow the developments in the market and must come up with new ideas with high added value.

10.3 Development of new products

Intelligent textiles and wearable technology are of great interest today. The new ideas and products developed in this area are expected to be the winners of tomorrow. Various types of fibers, textiles and garments that have interactive properties, i.e. that react with logic to information received from the environment, are so-called intelligent textiles. By integrating electronics and sensors in textiles, new types of cross-scientific products can be created.

The most important intelligent materials at present are phase change materials, shape memory materials, chromic materials and conductive materials. Phase change materials are compounds which melt and solidify at certain temperatures and in doing so are capable of storing or releasing large amounts of energy. Phase change materials can be incorporated into a thermal storage system in order to store daytime solar energy to provide space heating. Shape memory materials are those that can revert from the current shape to a previously held shape, usually due to the action of heat. When these shape memory materials are activated in garments, the air gaps between adjacent layers of clothing are increased in order to give better insulation, thus conferring greater versatility in the protection of the garment against extremes of heat and cold. Chromic textiles change their color reversibly according to external environmental conditions; for this reason they are called chameleon fibers. The color change in these materials can be stimulated by light, heat, electricity, pressure and energy, liquids or electronic beam. Conductive fabrics combine the latest high wicking finishes with high metallic content in textiles that still retain the comfort required for clothing. With the addition of nickel, copper and silver coatings of varying thickness, these fibers provide a versatile combination of physical and electrical properties for a variety of demanding applications. For example, the thousand-fold increase in thermal conductivity of metal over conventional polymers used in clothing offers sports apparel with the minimum of thermal insulation (Karthik, 2006; Murthy *et al.*, 2003).

In the last decade, research and development in smart/intelligent materials and structures have led to the birth of a wide range of novel smart products in aerospace, transportation, telecommunications, homes, buildings and infrastructures. Although the technology as a whole is relatively new, some areas have reached the stage where industrial application is both feasible and viable for textiles and clothing (Tao, 2001). Many exciting applications have been demonstrated worldwide. Extending from the space program, heat generating/storing fibers/fabrics have now been used in ski wear, shoes, sports helmets and insulation devices. Textile fabrics and composites integrated with optical fiber sensors have been used to monitor the soundness of major bridges and buildings. The first generation of wearable motherboards has been developed, which have sensors integrated inside garments and can detect information regarding injury to and health of the wearer, and transmit such information remotely to a hospital. Shape memory polymers have been applied to textiles in fiber, film and foam forms, resulting in a range of high-performance fabrics and garments, especially sea-going garments. Fiber sensors, which are capable of measuring temperature and strain/stress, and sensing gas, biological species and smell, are typical smart fibers that can be directly applied to textiles (Boczkowska and Leonowicz, 2006). Conductive polymer-based actuators have achieved very high levels of energy density. Clothing with its own senses and brain, such as shoes and snow coats which are integrated with Global Positioning System (GPS) and mobile phone technology, can tell the location of the wearer and give him or her directions. Biological tissues and organs such as ears and noses can be grown from textile scaffolds made from biodegradable fibers. When integrated with nanomaterials, textiles can be imparted with very high-energy absorption capacity and other functions such as stain proofing, abrasion resistance, light emission, etc. Incorporating electronic devices into textiles leads to a new branch of science called textronics (Gniotek and Krucińska, 2004). This is a new research area which so far has not produced very many applications, though expectations are high.

The diversity in the application of electronic textiles (e-textiles) is increasing and becoming interesting. The textile clothes, being lightweight, strong and bendable, can be stretched over any frame into a desired shape with a concept to change the size, shape and style of clothes by weaving 'muscle wires' into the fabric. The wires are made of shape-memory alloys that change length according to the small current passed through them. Electronic wires and sensors woven into the fabric can perform the function of listening for faint sounds. That means people resting in tents or camouflage nets may hear the distant sounds of vehicles or stepping/movement of people, animals, enemies, etc. The use of fabric to deploy electrical components results in wearable electrical/computing devices and makes it easier to move with computing devices with less consumption of human energy and effort. Moreover the flexibility of fabric provides the opportunity to modify the shape to conform to new requirements of applications. The relative position of components, including sensors, actuators and processing elements, can be altered (Uddin, 2006).

In a nutshell, there are some exciting developments happening in the textiles area, and the whole wellbeing sector – although still a niche – is becoming stronger. Current trends lie in developing 'functionalized' surfaces, using new materials and coating technologies as well as integrating active ingredients. Work is also progressing to develop extremely light, highly breathable, elastic, water-repellent as well as absorbent textiles, as barriers against infectious agents and solid particles. Apart from sensor technology, research is also being carried out on the use of radio frequency identification technology (RFID) integrated into textiles. This technology supports hospital logistics, for instance the location of equipment or even people.

10.4   Research and development in new products

The potential of intelligent textiles is huge. One can think of many applications for each of the examples given earlier. The other way around, starting from an application, the basic concepts have to be defined and evaluated for their use in or as a textile product. Selection of materials, structures and production technologies is the first step in the design. The actual research phase will be long and hard in many cases. Basic items that need to be addressed to come to a real breakthrough and to innovate are as follows:

- Transformation and conversion mechanisms to define the basic concept
- New materials
- New structures that can offer the requested functions.

Conductive materials, metals as well as conducting polymers, are already being used in many applications: antistatic working, EMI shielding, heating, transport of electrical signals, etc. Inherently conducting polymers (ICPs) are fascinating, dynamic, and molecular systems suitable for application in many domains of intelligent clothing: polymer batteries, solar energy conversion, biomechanical sensors, etc. Some materials are already available, be it at laboratory level. Some substantial disadvantages are the instability of the polymer in the air, the weak mechanical properties and the difficult processing. However, in the United States one laboratory has managed to spin the first polyaniline fiber (Santa Fe Science and Technology Inc., Santa Fe, NM).

Another class of materials that will play a major role without any doubt in many intelligent clothes is optical fibers. They are well known from applications in electronics, but the range of deformations to deal with in textile applications is of a different order and causes problems that restrict the number of applications at present. In all these cases, control of quality and certification becomes much more important. Although the development of intelligent fabrics and smart fabrics is in the infancy stage, spread all over the world in different organizations and industries, the level of quality products produced and their quality certification, etc., all become crucial for further research and development in these new products.

10.5   Types of testing: shape memory effect

10.5.1 Definition and significance of shape memory effect

The 'shape memory effect' (SME) is a special behavior of temperature-stimulating shape memory polymers, which are a special class of adaptive materials that can convert thermal energy directly into mechanical work. This phenomenon occurs when one of these special class of polymers is mechanically stretched at high temperature, typically 20°C above the glass transition or crystal melting temperature (T_g/T_{ms}), and cooled down to low temperatures, then heated above the critical transition temperature (T_g/T_{ms}), which results in the restoration of the original shorter 'memory' shape of the specimen (Lendlein and Kelch, 2002). The important properties which are considered for this shape memory effect are the thermomechanical properties of the thermoplastic polymer, such as shape fixity, shape recovery, and recovery stress and recovery speed (Tobushi *et al.*, 2001). These materials have two phase structures, namely the fixing phase, which remembers the initial shape, and the reversible phase, which shows a reversible soft and rigid transition with temperature.

10.5.2 Shape memory effect of fabrics

Materials with the unique property of reverting to their original, permanent shape from a fixed temporary shape only upon being triggered by an external stimulus are classified as demonstrating the shape memory effect. Fabrics treated with these materials can be called shape memory fabrics. The distinction between shape memory fabrics and wrinkle-free fabrics can be explained as follows. A wrinkle-free fabric has good elasticity because it can recover after release of the forces that cause deformations. Thus wrinkle-free finishing improves the elasticity of a fabric and can be called elastic finishing. Wrinkle-free fabric is not temperature sensitive and cannot recover its original shape when its temperature has been changed. On the other hand, shape memory fabrics, depending on the chemical used, can have lower, similar or higher (different levels of) elasticity than wrinkle-free fabrics at room temperature such as home laundering and body temperature environment, and recover their original shape, such as a flat appearance on which the fabric has no wrinkle. Shape memory fabrics can recover their original shape based on the use of shape memory polymers (SMPs) under certain conditions (temperature variation). This phenomenon can be called thermal elastic finishing or shape memory finishing, because the fabric can recover its original shape (elasticity) at higher temperature (thermal energy triggering), which leads to the shape memory effect. At the same time, shape memory fabrics probably have even better elasticity at higher temperature than below the switch temperature. Thus, when a fabric has some residual wrinkles after removing external forces at room temperature, it can further recover its original shape at higher temperatures in tumble drying and/or machine laundering. So shape memory fabrics can doubly ensure the recovery of wrinkles when they are properly finished.

As a type of smart materials, shape memory polymers (SMPs) are increasingly drawing attention. At present, a good variety of polymers have been reported to show shape memory properties. But the different characterization or evaluation methods of shape memory properties and the varying conditions employed by different researchers have resulted in the reported properties of SMPs being basically not comparable. Therefore the relationship between shape memory properties and structures is not completely known for some species of SMPs. This may hinder the development of high-performance SMPs eventually. Furthermore, in contrast to the rapid increase in the numbers of SMPs, their application lags far behind, perhaps mainly because the current characterization of their properties cannot provide a comprehensive understanding for researchers. Therefore the evaluation of SMPs is crucial for their development and application.

10.6   Evaluation methods for shape memory fabrics

10.6.1 Wrinkle recovery, smoothness appearance, and crease retention

Several methods are available for measuring the wrinkled appearance of fabrics. For example, a subjective grading of wrinkles can be referenced from the AATCC test method 124-2001, 'Appearance of Fabrics after Repeated Home Laundering'; crease retention can be also subjectively graded by the AATCC test method 88C-2001, 'Retention of Crease in Fabrics after Repeated Home Laundering', and an objective evaluation of wrinkles can be carried out by the AATCC test method 66-1998, 'Wrinkle Recovery of Woven Fabrics: Recovery Angle'.

The AATCC test method 66-1998 (Wrinkle Recovery of Woven Fabrics: Recovery Angle) for accessing wrinkle recovery is applicable to fabrics made from any fibers, or combination of fibers. In this experiment, a test specimen is folded and compressed under controlled conditions of period and applied force to create a folded wrinkle. The test specimen is then suspended on a test instrument for a controlled recovery period, after which the recovery angle is measured (Fig. 10.1). The use of the wrinkle recovery test for measuring the crease angle and its recovery for shape memory fabrics is helpful for accessing the shape memory effect, because the original angle of the crease, and the recovery crease angle after the fabric is deformed,

10.1 Accessories for the AATCC test method 66-1998.

10.2 Accessories for the AATCC test method 124-2001.

10.3 Accessories for the AATCC test method 88C-2001.

must be obtained in advance. In our case, a function taking into account the temperature effect is needed to describe the shape memory effect of the fabrics, due to the fact that they have the recovery effect triggered by the change in environmental temperature.

The AATCC test method 124-2001 (Appearance of Fabrics after Repeated Home Laundering) is designed to evaluate the smoothness appearance of flat fabric specimens after repeated home laundering by simply looking at them. The fabric is graded by comparing with some scored standard fabrics with different surface appearances (Fig. 10.2). Any washable fabric and fabrics having different structures (woven, knit and non-woven) can be evaluated for smoothness appearance using this method.

The AATCC test method 88C-2001 (Retention of Crease in Fabrics after Repeated Home Laundering) is designed to evaluate the retention of pressed-in creases in fabrics after repeated home laundering (Fig. 10.3). Any washable fabric and fabrics of any structure may be evaluated for crease retention using this method. The techniques for creasing are not outlined, since the purpose is to evaluate fabrics as they are supplied from manufacture or as ready for use. Furthermore, application of the creasing technique depends upon the fabric properties.

In recent years, Yang and Huang (2003) have developed a method for fabric 3-D surface reconstruction using a photometric stereo method. The 3-D surface of an AATCC standard wrinkle pattern is reconstructed and its wrinkle degree is measured using four index values indicating the variation of the surface height. The result suggests that there is a good linear correlation between the index value and the wrinkle degree of the pattern.

However, many evaluations of shape memory fabrics still rely on the traditional standard methods. Hashem *et al.* (2003) reported that the crease angle recovery and strength date were correlated with the amount of add-on polyelectrolyte. The authors developed methods of forming the ionic cross-links of cotton to provide crease angle recovery performance without the potential for releasing low molecular weight reactive materials like formaldehyde. The result was evaluated by crease angle according to the AATCC standard test method 66, and the breaking strength from Instron tensile tester according to the ASTM test method D1682 (Standard Methods of Test for Breaking Load and Elongation of Textile Fabrics, 1972). It was found that crease angle recovery could be imparted to cellulose fabric by the application of ionic cross-links. Carboxy-methylated woven fabric treated with cationized chitosan shows significant increases in wrinkle recovery angle without any strength losses. Also, fabric treated simultaneously or sequentially with 3-chloro-2-hydroxypropyl trimethyl ammonium chloride (CHTAC) and either chloracetic acid (CAA) or sodium chloromethyl sulfonate (CMSA) improved the wrinkle recovery angle and strength.

10.6.2 Primary parameters for testing of shape memory polymers

To characterize the shape memory properties of polymers, a set of parameters are desired. Firstly, the parameters should be able to reflect the nature of polymers. Secondly, distinguished from other properties of materials, shape memory properties are shown through a series of thermomechanical cyclic processes. Therefore the parameters should be able to define the whole shape memory processes as well. Lastly, the design of the parameters should pay attention to the potential applications. In view of these considerations, some parameters have been proposed and quantified (Tobushi *et al.*, 2001; Kim and Lee, 1996; Li *et al.*, 1998). In the following, these parameters are introduced.

Shape fixity

As has been described in the foregoing sections, when a shape memory polymer is heated up to a temperature above the transition temperature for triggering shape memory behavior (T_{trans}) (Lendlein and Kelch, 2002), it can develop large deformations which can be mostly fixed by cooling to a temperature below T_{trans}. This parameter was proposed to describe the extent of a temporary shape being fixed in one time of shape memorization (Tobushi *et al.*, 2001; Kim and Lee, 1996; Lendlein and Kelch, 2002). It should be noted that various confusing usages of notation or even expression have taken place in the characterization of SMPs. As for shape fixity, other terms such as strain fixity (Tobushi *et al.*, 2001) and shape retention (Lee *et al.*, 2001) also represent the same physical meaning. Shape fixity (R_f) equals the amplitude ratio of the fixed deformation to the total deformation, which is given by:

$$\text{Shape fixity} = \frac{\text{fixed deformation}}{\text{total deformation}} \times 100\% \qquad 10.1$$

Shape fixity is related to both structure of polymers and the thermomechanical conditions of shape memorization. Compared with the structures of SMPs, the thermomechanical conditions play an equal or even more important role in determining the shape fixity and other shape memory properties.

Shape recovery

An SMP holding a deformation at low temperature can restore its original shape by being heated up above T_{trans}. Shape recovery (R_r) reflects how well an original shape is memorized when shape memorization occurs (Tobushi *et al.*, 2001; Kim and Lee, 1996; Lendlein and Kelch, 2002). As with shape fixity, there are diverse and confusing usages not only in terms of notation but also in mathematical expressions for this parameter. Tobushi *et al.* (2001) utilized thermomechanical cyclic tensile tests to evaluate SMPs and entitled the parameter 'shape recoverability' or 'strain recovery'. In their study, the physical meaning represented by this parameter was interpreted as:

$$\text{Strain recovery} = \frac{\text{deformation to sample recovered in this cycle}}{\text{deformation to sample taken place in one cycle}} \times 100\% \qquad 10.2$$

Kim and co-workers also employed thermomechanical cyclic tensile tests to evaluate SMPs and defined the parameter as 'shape recovery' (Kim and Lee, 1996). The definition was given by:

Shape recovery

$$= \frac{\text{deformation to sample recovered in a whole cycle}}{\text{deformation to sample taken place in the first cycle}} \times 100\% \quad 10.3$$

Li *et al.* (1998) utilized the bending test to investigate SMPs. Their definition of recovery rate could be understood as:

Recovery rate

$$= \frac{\text{deformation to sample recovered in reheating process}}{\text{fixed deformation}} \times 100\% \quad 10.4$$

It is evident from equations 10.2–10.4 that the three forms of definition indeed represent different physical meanings even though they can all reflect the shape recovery from a particular angle. As a consequence of the different definitions, different values for the shape recovery will be obtained from the different formulae. For example, in the case of equation 10.2, shape recovery will increase with increasing cycle number and tend to 1, while in the case of equation 10.3 it will simply increase with increasing cycle number. Therefore caution should be taken in calculating shape recovery. Like shape fixity, shape recovery depends both on the structure of the polymers and on the thermomechanical conditions of shape memorization.

Recovery stress

Recovery stress stems from the elastic recovery stress generated in the deformation process. When SMPs are heated and deformed, elastic stress is generated, and the elastic stress is stored when SMPs are cooled down below T_{trans}. If deformed and fixed SMPs are reheated above T_{trans} the stress stored in them is released as shape recovery stress. In this regard, shape memorization can be looked at as a thermomechanical cycle consisting of stress generation, stress storage and stress release.

SMPs are considered to be promising in the development of smart actuators. The characterization of shape recovery stress is therefore essential. However, few attempts have been made to investigate shape recovery stress. Tobushi *et al.* (2001) investigated shape recovery stress through a specially designed thermomechanical cyclic tensile test on a shape memory material testing machine consisting of a tensile machine accompanied by a constant temperature chamber. Tey and co-workers studied the recovery process of the shape memory polyurethane foam MF-5520 produced by MHI (Tey *et al.*, 2001). The shape memory polyurethane foam was compressed at 83°C and was cooled to room temperature (about 30°C) to keep deformation. Then the deformed foam was heated gradually and the change of recovery stress as the temperature increased was investigated. Liang and co-workers investigated the recovery stress of the MM-4500 SMP from

Mitsubishi Heavy Industries (MHI) (Liang *et al.*, 1997). A heat gun was used to heat the specimens (short-duration heating, so there was no apparent creep), and the temperature was estimated in the range of 70–80°C.

The dilemma in characterization of recovery stress of SMPs is chiefly caused by the viscoelasticity of the polymers, especially for thermoplastic SMPs. Owing to the limitations of the equipment and the efficiency of heat transfer, it is practically impossible to heat or cool an SMP to a certain temperature in a sufficiently short time in experiments. Therefore the stress relaxation is inevitable if only the SMP is in a constrained state. As a consequence, the stress generated in deformation must be lost to some extent in the shape fixing and shape recovery processes. Additionally, the rate of stress relaxation alters with temperature change in the whole shape memory process, resulting in its influence on recovery stress being unknown. In other words, the recovery stress may change all the time while undergoing stress relaxation and the progress of change is uncertain. Therefore it is difficult to capture or calculate the recovery stress in absolute terms. Actually, in practical applications, the influence of the viscoelastic behavior will also be inevitable. So if the experimental conditions are constant and the experimental results are thereby reproducible, the measurements of recovery stress in relative terms are valuable. The author carried out some investigations into the recovery stress of SMPs through a specially designed thermo-mechanical cyclic tensile test.

Shape recovery speed

Shape recovery speed describes the speed at which a given SMP recovers from a temporary shape to its original shape when heated. Actually, the parameter has no uniform and clear name. In the study of Li *et al.* (1998) it was called the 'speed of recovery process', while Luo *et al.* (1997) entitled it 'deformation recovery speed'. Here the author proposes the name 'shape recovery speed' as its title, which is in parallel with the other shape memory properties, to make it easy to memorize. The parameter can be measured both qualitatively and quantitatively.

Liu *et al.* (2002) investigated the shape recovery process of some SMPs and qualitatively studied the shape recovery speed using a video camera at a rate of 20 frames per second. It was evident that the SMP restored its original shape in 0.7 s. Li *et al.* (1998) and Luo *et al.* (1997) investigated the shape recovery of SMPs with a constant heating rate. Through the curve of shape recovery as a function of temperature, the shape recovery speed was calculated. The shape recovery speed can be defined as:

$$V_r = \frac{dR_r}{dT} \times \frac{dT}{dt}$$

10.5

where V_r is the shape recovery speed, dR_r/dT is the ratio of shape recovery to temperature, and dT/dt is the heating rate.

10.6.3 Factors affecting shape memory effect

Lendlein and Kelch (2002) pointed out that the shape memory effect was not related to a specific material property of single polymers; it rather resulted from a combination of the polymer structure and the polymer morphology together with the applied processing and programming technology. The shape memory effect (SME) of SMPs is shown through the whole thermomechanical cyclic processes involving deformation, shape fixing and shape recovery. Each process may change the shape memorization and affect the shape memory properties eventually. Changing the conditions in the thermomechanical cyclic processes would lead to variation of the shape memory properties. Different methods and conditions employed in previous reports have resulted in difficulties in comparing the experimental results. It is necessary to characterize the SME with varying conditions in order to provide important information for development and application of SMPs.

The author and her group conducted a number of experiments on the effects of deformation conditions on the SME and found that conditions such as deformation temperature, maximum strain and deformation rate had a profound effect on the SME. In addition the effects of fixing conditions on the SME were also analyzed. Parameters such as fixing rate and fixing temperature were the main conditions that affected the SME. The author also investigated the effects of recovery conditions on the SME. In this case the recovery temperature and recovery time were very important conditions which affected the SME. Shape memory properties are dependent on the special microstructures of SMPs. Therefore the elements affecting the microstructure will definitely influence the shape memory properties. It was supposed that the processing temperature may affect the microstructure, such as morphology, micro-domain structure and phase separation, that determines the shape memory properties of SMPs. Besides processing temperature, the environmental conditions can also play an important role on the shape memory properties. Yang et al. (2004) studied the influence of water on the microstructure and shape memory properties of SMPU MM3520 from Mitsubishi Heavy Industries (MHI). It was found that the SMPU loses its shape fixing capability after being exposed in the air at room temperature (about 20°C) for several days. In addition, for the evaluation of shape memory materials, the conditions of sample preparation have a great influence on the test results.

10.7   Thermal regulation property of phase change materials

10.7.1 Definition and significance of thermal regulation property

Phase change materials (PCMs) are able to absorb, store and release large amounts of latent heat over a defined temperature range when the material changes phase or state. A fabric containing a PCM can act as a transient thermal barrier which regulates the heat flux. The heat absorption by PCMs results in a delay in microclimate temperature and hence a substantial decrease of the amount of sweat produced by the skin of the wearer. Both lead to an enhancement of the wearing comfort and prevent heat stress (Bendkowska et al., 2005).

Fabrics incorporated with MicroPCM are called 'PCM treated fabrics' or just 'PCM fabrics'. The thermal properties of PCM fabrics are dynamic and responsive. That is, their thermal properties are related to the change of temperature and time. For example, when the environment temperature reaches the PCM melting point, the physical state of the PCM in the fabric will change from solid to liquid along with the absorption of heat, while the temperature of the PCM in the fabric keeps constant at melting point, therefore it can regulate its temperature automatically by itself. As the reverse thermal regulation performance occurs during the cooling process, the environment temperature comes to its freezing point. So, the PCM fabric can provide a cooling effect caused by heat absorption of the PCM and a heating effect caused by heat emission of the PCM to the human body.

PCM fabrics are designed to be used under special environmental conditions in which they need to offer a desired temperature lasting for a definite period of time. Hence, thermal properties during phase change are very important because once the PCM fabric's temperature goes out of the phase change range, it is no longer effective as an active thermal wear. In order to evaluate the thermal properties of PCM fabrics, some research has been conducted. Pause (1995) developed the concept of dynamic thermal insulation to measure the transient effect on the insulation value of PCM fabrics, and pointed out that the total insulation of PCM fabrics is comprised of basic insulation and dynamic thermal insulation that was determined from the duration of the temperature variation during the phase change. The dynamic thermal insulation was calculated by comparing the times for achieving the end temperature of the phase change range of the samples with and without MicroPCM and with reference to the basic thermal insulation of the samples. The thermal insulation is given in units of thermal resistance. In 2002, a new test instrument and measurement index were described by Hittle and Andre (2002). The index of temperature regulation factor (TRF) is used to indicate the temperature-regulating ability of PCM fabrics, which is a dimensionless number less than or equal to 1, and the TRF for PCM fabrics will always be less than the TRF for non-PCM fabrics. A test for PCM garments was carried out by Shim and McCullough (2001) in a warm and a cold chamber. The value of heat loss from a thermal manikin was measured and used to quantify the effect of PCMs in clothing on heat flow from the body during temperature transients.

Fabrics containing PCM microcapsules present a unique challenge to the standard test procedures used for determining the thermal properties of fabrics. In the case of traditional fabrics, the thermal properties are investigated by standard steady-state procedures involving the use of guarded hot plate apparatus. A steady state procedure is inadequate in assessing the dynamic performance of fabrics containing PCMs, because PCM is a highly productive thermal storage medium (Bendkowska, 2006).

10.7.2 Primary parameters for testing

By analyzing the physical mechanisms of heat and moisture transfer through textiles incorporating phase change materials, Ying *et al.* (2004) proposed three indices and related test methods to characterize the thermal functional performance of PCM fabrics. They are the static thermal insulation (I_s), the thermal regulating capability (I_d and Δt_d) and the thermal psychosensory intensity (TPI).

The indices of thermal regulating capability (I_d and Δt_d) are used to describe the thermal regulating performance of textiles incorporating phase change materials, and were found to be strongly dependent on the amount of phase change material. From the aspect of thermal comfort of textiles and clothing, the TPI index is used to represent the thermal psychosensory intensity perceived by the body, and the TPI value increases with increase of phase change material level. The static thermal insulation index (I_s) is used to describe the static thermal insulation effects of the textiles. All indices can be measured and calculated by using the testing methods of the Fabric Intelligent Hand Tester.

When there exist gradients of temperature and water vapor pressure across a textile structure, heat and moisture transfer through the structure occur. These two kinds of transportation involve multiple processes under different conditions. By combining the mechanisms of heat and moisture transfer in porous textiles and the phase change process occurring in microcapsules incorporated in the textile, a new mathematical model for the processes has been provided by Li and Zhu (2006). The factors involved in this new model include the heat of moisture sorption or desorption of vapor by fibers, the heat of moisture sorption or desorption of liquid water by fibers, the heat of evaporation of water, the heat change by conduction, the heat related to radiation, and the latent heat which is gained and lost from the MicroPCM. For the process of phase change, the factors of the quantity of PCM and the radius of the MicroPCM spheres have been considered in the model. From the model, it is seen that both factors have a significant influence on the heat energy changes through porous textiles. Thus, the greater the quantity of PCM added, the higher the volume fraction and the more heat energy is gained or lost from the MicroPCM. Therefore, more heat flux is delayed through the PCM fabrics and a stronger thermal regulation performance occurs; also, the smaller the radius of the MicroPCM spheres, the more significant the thermal regulating capability of the porous textile (Li and Zhu, 2006).

With the testing conditions used, no liquid phase is involved and no radiation factors are considered, so the energy balance equations given by Li and Zhu can be simplified. They become

For PCM fabrics:

$$c_V \frac{\partial T}{\partial t} = \lambda_v \frac{\partial (C_f \varepsilon_f)}{\partial t} + \frac{\partial}{\partial x}\left[K_{mix} \frac{\partial T}{\partial x}\right] - q(x,t) \qquad 10.6$$

For non-PCM fabrics:

$$c_V \frac{\partial T}{\partial t} = \lambda_v \frac{\partial (C_f \varepsilon_f)}{\partial t} + \frac{\partial}{\partial x}\left[K_{mix} \frac{\partial T}{\partial x}\right] \qquad 10.7$$

where x denotes the coordinate across the porous textile slab; $x = 0$ and $x = L$ indicate the positions at the lower and upper surfaces of the porous slab, respectively; T is the temperature of the flow fields at x in the porous textile; c_v is the volumetric heat capacity of the fabric (kJ/m^3 K); λ_v is the heat of sorption or desorption of vapor by fibers (kJ/kg); C_f is the water vapor concentration in the fibers of the fabric (kg/m^3); K_{mix} is the thermal conductivity of the fabric (W/m K); and ε_f is the volume fraction of fibers.

The relationship between the porosity of the fabric (ε), the volume fraction of fibers (ε_f), and the volume fraction of MicroPCM (ε_m) is expressed by equation 10.8 under the test conditions:

$$\varepsilon + \varepsilon_f + \varepsilon_m = 1 \qquad 10.8$$

In equation 10.6 the first term on the right-hand side describes the heat of moisture sorption or desorption of vapor by fibers, the second term describes the heat change by conduction, and the last term, $q(x,t)$, describes the latent heat which is gained or lost from the MicroPCM. The above analysis of the physical mechanisms of heat and moisture transfer through PCM and non-PCM fabrics leads to the introduction of three indices and test methods which characterize the thermal functional performance of fabrics.

The static thermal insulation (I_s)

Equations 10.6 and 10.7 show that the temperature and moisture gradually reach equilibrium with the environment, and then $\partial T/\partial t = 0$ and $\partial(C_f\varepsilon_f)/\partial t = 0$. For the PCM fabrics, in the equilibrium state, the phase change will have happened, so the term $q(x,t) = 0$. Therefore, in this state, equations 10.6 and 10.7 are identical and the heat flux value becomes constant, i.e.

$$K_{mix}\frac{\partial T}{\partial x} = \text{const.} \qquad\qquad 10.9$$

Defining this constant to be the index I_s, which represents the static thermal insulation effects of the fabric, given in units of heat flux (W/m²), the defining equation is written as:

$$I_s = K_{mix}\frac{\partial T}{\partial x}\bigg|_{x\in[L-\delta L, L]} \qquad\qquad 10.10$$

The static thermal insulation can also be expressed by equation 10.6:

$$I_s = -\frac{h_t}{\varepsilon}(T - T_{ab}) \qquad\qquad 10.11$$

where T_{ab} is the constant environment temperature (K), and h_t is the convection heat transfer coefficient.

The thermal regulating capability (I_d and Δt_d)

The thermal regulating capability of PCM fabrics is dependent on temperature and time. It takes place only during the temperature range of the phase change and terminates when the phase change in all the MicroPCM contained in the fabrics is complete. Comparing equations 10.6 and 10.7, it can be seen that the thermal regulating capability is related to the term $q(x,t)$. Suppose the duration time of the phase change process in the PCM fabrics is from t_{d1} to t_{d2}, hence, the indices Δt_d and I_d may be defined as:

$$\Delta t_d = t_{d2} - t_{d1} \qquad\qquad 10.12$$

and

$$I_d = \frac{\int_{t_{d1}}^{t_{d2}} q(L,t)\,dt}{t_{d2} - t_{d1}} \qquad\qquad 10.13$$

I_d is the mean of the heat flux delayed by phase change during the phase change period, and the total heat energy change related to phase change is expressed by $\Delta t_d * I_d$.

Thermal psychosensory intensity

From the aspect of thermal comfort of textile and clothing, Ring and de Dear (1991) proposed that the intensity of thermal sensations (termed the psychosensory intensity, PSI) is proportional to the cumulative total impulses from stimulus onset at the thermoreceptor until such time as the receptor firing rate has decayed to within one impulse per second of the poststimulus steady state. On this basis, Wang *et al.* (2002) derived equation 10.14 to calculate the impulse frequency of the thermoreceptors responding to heat flux and temperature profile in the skin:

$$Q(y, t) = C + K_s T_{sk}(y, t) + K_d \frac{\partial T_{sk}(y, t)}{\partial t} \qquad 10.14$$

where K_s and K_d are the static and dynamic differential sensitivity, C is a constant, y is the location of the thermoreceptors (depth from the skin surface), and T_{sk} is the temperature of the skin. In equation 10.14, the values of C, K_s and K_d are taken as 28.1 (s^{-1}), 0.72$(s^{-1} \cdot {}^{\circ}C^{-1})$ and $-50({}^{\circ}C^{-1})$, respectively.

For the PCM fabric, the index of thermal psychosensory intensity (TPI) is defined to express the thermal perception by the body. In the test conditions, the temperature measured at the surface of the fabric, $T(L,t)$, is taken as equal to T_{sk}; therefore, equation 10.14 becomes

$$Q(L, t) = C + K_s T(L, t) + K_d \frac{\partial T(L, t)}{\partial t} \qquad 10.15$$

From the curves generated from equation 10.15, TPI is the integral of $Q(L,t)$ over the time from t_{d1} to t_{d2}, i.e.

$$\text{TPI} = \int_{t_{d1}}^{t_{d2}} Q(L, t) \, dt \qquad 10.16$$

The above three parameters can be effectively utilized to measure the thermoregulatory response of a fabric treated with a PCM material. However, different treatment methods and fabric constructions may require additional tests to be conducted to fully understand the thermal functional performance of the fabrics.

In an attempt to determine the thermal regulating ability of fabrics containing phase change materials, Bendkowska *et al.* (2005) developed an instrument to measure the temperature regulating factor (TRF) (see Fig. 10.4). Determination of the TRF of apparel fabrics was done by means of the instrument, which used a dynamic heat source. This instrument simulated an arrangement of skin–apparel–environment. The fabric sample was sandwiched between a hot plate and two cold plates, one on either side of the hot plate. These cold plates at a constant temperature simulated the environment outside the apparel. Sinusoidally varying heat input to the hot

10.4 General view of apparatus for testing thermoregulation properties of fabrics containing PCMs (Bendkowska *et al.*, 2005).

plate simulated human activity. To measure the steady-state thermal resistance of the fabric (*R*) the controlled heat flux was constant and the test proceeded until a steady state was reached. To assess the temperature regulating ability, the heat flux was varied sinusoidally with time and the TRF was determined.

10.8   Self-cleaning testing

The development of permanent self-cleaning cotton textiles with a life cycle of 25–50 washings or more is an objective sought by the textile industry in the framework of new products classified as intelligent textiles. These self-cleaning fabrics have a nanofilm coating of titanium dioxide nanoparticles, which can break down dirt molecules, pollutants and microorganisms when exposed to visible and UV light. Clothes made using this method could be cleaned by simply exposing them to sunlight.

10.8.1 Self-cleaning effect

The self-cleaning fabrics work using the photocatalytic properties of titanium dioxide, a compound used in many new nanotechnology solar cell applications. The fabric is coated with a thin layer of titanium dioxide particles that measure only 20 nanometers in diameter. When this semi-conductive layer is exposed to light, photons with energy equal to or greater than the band gap of the titanium dioxide excite electrons up to the conduction band. The excited electrons within the crystal structure react with oxygen atoms in the air, creating free-radical oxygen. These oxygen atoms are powerful oxidizing agents, which can break down most carbon-based compounds through oxidation–reduction reactions. In these reactions, the organic compounds (i.e. dirt, body odor, smoking odor, etc., bacteria, color stains, harmful organic materials such as formaldehyde, other carbon-based molecules, pollutants and micro-organisms) are broken down into substances such as carbon dioxide and water. Since the titanium dioxide only acts as a catalyst to the reactions, it is never used up. This allows the coating to continue breaking down stains over and over.

As the technology of self-cleaning is relatively new and not yet well established among researchers, the test methods for testing these fabrics are not yet standardized. The researchers who have been developing these fabrics adapted their own techniques to analyze the effectiveness of the coatings and self-cleaning capability of the fabrics. Several techniques have been reported. In one of the studies on the development of self-cleaning cotton, Qi et al. (2006) used a surface characterization technique to assess the performance of the cloth. The structure and morphology of these coatings were investigated using field emission scanning electron microscopy (FESEM, JSM-6335F at 3.0 kV, JEOL, Tokyo, Japan). The crystallinity of solid powder extracted from the sols was studied by X-ray diffraction spectroscopy (XRD, Bruker D8 Discover X-ray diffractometer) operating at 40 kV and 30 mA. The crystallinity of the titania coatings formed on the cotton fabric was studied by low angle X-ray diffraction (XRD, Philips Xpert XRD system) with incident beam at 3 u using Cu Ka radiation and detector scan mode operating at 40 kV and 30 mA. The UV absorption of coated cotton substrates was measured according to the Australian/New Zealand Standard AS/NZS 4399:1996 using a Varian Cary 300 UV spectrophotometer. In the absence of standard test methods, the following test procedures have been suggested by the researchers.

10.8.2 Test for static bactericidal activities

One of the important tests that is carried out for assessing the self-cleaning ability is the testing for static bacterial activities. The test may be done on fabric swatches for each test specimen (woven cotton and TiO_2 coated cotton fabrics) with a diameter of 5 cm, stacking up and placing in separate 150 ml wide-mouth jars. 1 ml of inoculum is added to each jar. The tops are immediately screwed tightly to prevent evaporation. For the '0 h' contact time sample, 100 ml of sterile phosphate buffer may be added immediately after inoculation and the jars are shaken for one minute. 1 ml of the resulting solution is transferred from each jar to a test tube containing 4 ml of double distilled water ddH_2O and plated out in duplicate. For the '5 h' contact time sample, the jars are incubated for 5 h in a 37°C incubator before the plate count technique is performed. All the Petri dishes are incubated for 24 h in a 37°C incubator. After incubation, the number of viable cells (colonies) is counted manually and the results after multiplication with the dilution factor can be expressed as mean colony forming units (CFU) per ml after averaging the duplicate counts (Qi et al., 2006).

10.8.3 Test for decomposition activities of colorant stains

The decomposition activities of colorant stains are assessed by analyzing the decrease in concentration of the colorants during exposure to UV irradiation. To carry out this test, a 3 g coated woven white cotton fabric is cut into 1 cm × 1 cm pieces. These pieces are placed in a 250 ml beaker containing 75 ml Neolan Blue 2G aqueous solution (0.2 g l^{-1}), 100 ml Cibacron Blue F-R (0.2 g l^{-1}) and uncoated white cotton fabric as well as a comparison. Then the beakers are exposed to UV irradiation provided by Philips UV lamps (20 V) while vigorously shaken (IKA KS260 Basic Orbital Shaker). The light intensity on the top of the beakers is 1.2–1.3 mW cm^{-2}. Prior to UV irradiation, the colorant solution with white cotton pieces is kept in dark conditions for 4 h while shaking to establish the absorption–desorption equilibrium. UV-Vis absorption spectra of irradiated samples are recorded on a UV-Vis spectrometer (Perkin Elmer UV-Vis spectrometer Lambda 18). The colorant solution is centrifuged to precipitate cotton fibers at the bottom of the tube and the upper clear colorant solution is used. The changes in concentration of colorants are estimated by the concentration at the absorption peak for Neolan Blue 2G at 630 nm and for Cibacron Blue F-R at 610 nm (Qi et al., 2006).

10.8.4 Test for degradation activities of colored stains

This test may be carried out to assess the effectiveness of removal of colored stains on the fabric. An example of the evaluation of degradation activities of a red wine stain and a concentrated coffee stain is presented. The test involves using cut pieces of dimensions 4.5 cm × 6.5 cm pieces of woven white cotton and S60 coated woven white cotton fabrics respectively. A red wine stain and a coffee stain are dropped respectively on coated white cotton fabrics as well as pristine white cotton for comparison. The irradiation of all samples may be carried out in the cavity of a Suntest solar simulator on Xenotest Alpha LM light exposure and weathering test instrument (air-cooled xenon arc lamp, irradiance 45–95 mW cm22, Xenotest Alpha LM, Heraeus Industrietechnik, Hanau, Germany). The red wine and coffee stained cotton fabrics may be irradiated for 8 h and 20 h respectively (Qi et al., 2006). The discoloration is observed before and after the treatment and may be quantified using spectrophotometric studies.

10.8.5 Test for tearing strength

The tearing strengths of pristine woven cotton and titania coated cotton before and after 20 h of light irradiation may be measured by an Elmendorf Tearing Tester (Thwing-Albert Instrument Co.) in accordance with ASTM D1424-96. The tearing strength results are analyzed before and after irradiation treatment for coated samples. Further, in an interesting study on self-cleaning of wool–polyamide and polyester textiles by TiO_2-rutile modification under daylight irradiation at ambient temperature, Bozzi et al. (2005) have discussed the self-cleaning evaluation of wool–polyamide and polyester textiles by elemental analysis and dust test. Elemental analysis of the TiO_2-loaded textile fabrics was carried out by atomic absorption spectrometry using a Perkin-Elmer 300 S unit. A flow of dust was applied on the textiles and the dust adherence was estimated from color changes of the textile surface in a relative scale of 1–5.

An irradiation procedure and evaluation of textile cleaning performance by gas chromatography has been reported by Meilert *et al.* (2005). The photochemical reactor consisted of 80 mL cylindrical Pyrex flasks containing a strip of textile of 48 cm² positioned immediately behind the wall of the reactor. The irradiation of the samples was carried out in the cavity of a Suntest solar simulator (Hanau, Germany) air-cooled at 45°C for 24 h. The Suntest lamp had a wavelength distribution with 7% of the photons between 290 and 400 nm and was provided with a cut-off filter at 310 nm. The profile of the photons emitted between 3100 and 800 nm followed the solar spectrum with a light intensity of 50 mW/cm² corresponding to 50% of AM 1, the light intensity of the midday equatorial solar radiation. The CO_2 produced during the irradiation was measured in a gas chromatography apparatus (Carlo Erba, Milano) provided with a Poropak S column. In addition, the authors used attenuated total reflection infrared (ATR-IR) spectroscopy, elemental analysis techniques and transmission electron microscopy techniques to analyze the loading of TiO_2 on the textile fabrics and to understand the cross-sectional behavior of treated fabrics.

10.8.6 Applications of self-cleaning fabrics

In addition to suits, the new self-cleaning coating could be applied to hospital garments, sportswear, military uniforms and raincoats. Other possible applications include awning material for outdoor campers, fabrics for lawn furniture and convertible tops for cars. The coating could appear in consumer products within five years, the researchers estimate.

10.9   Electronic responsive testing

Electronic textiles or e-textiles lead to an interdisciplinary field of research which brings together specialists in information technology, microsystems, materials and textiles. The focus of this new area is on developing the enabling technologies and fabrication techniques for the economical manufacture of large-area, flexible, conformable information systems that are expected to have unique applications for both consumer electronics and the military industry.

Electronic textiles combine the strengths and capabilities of electronics and textiles. E-textiles, also called smart fabrics, have not only 'wearable' capabilities like any other garment, but also local monitoring and computation, as well as wireless communication capabilities. Sensors and simple computational elements are embedded in e-textiles, as well as built into yarns, with the goal of gathering sensitive information, monitoring vital statistics and sending them remotely (possibly over a wireless channel) for further processing. Possible applications include medical (infant or patient) monitoring, personal information processing systems, and remote monitoring of deployed personnel in military or space applications.

Conductive fabrics are gaining popularity amongst electronic textiles. In general, two main methods are followed for rendering a textile material electrically conductive: (a) by applying a conductive coating on the surface of a non-conductive textile after it is formed, or (b) by incorporating conductive fibers (e.g. via interweaving or embroidery) into the textile structure. Any textile structure, including knitted, woven and non-woven textiles, can be thus made electrically conductive. The choice of the textile structure and of the conductive mechanism determines the efficiency of the textile as an equivalent electroconductive material and assures its durability for its intended lifetime. The state of the art in conductive fibers is highly conductive metal wires or plated fibers. Inherently, conductive polymers are becoming closer in performance to metallic conductors and may be suitable for the next generation of applications in electrotextiles.

As these fabrics are new inventions in the field of intelligent textiles, complex approaches have been adapted to integrate these fabrics with conductive wires (conductive fabrics), wearable electronic devices and microcomputers. Accordingly, the testing of these fabrics varies among different researchers. In the absence of standard test methods and procedures, a brief review of the testing approaches adopted by different researchers is presented in the following sections. Some textile products with electrical properties have already found application in the field of EMI shielding, static dissipation and resistive heaters. For these products, the whole area of the fabric has to be conductive, whereas data transmission requires separate conductor lines. For data transmission it is important to have high conductivity.

In order to evaluate the electrical and electronic response of conductive fabrics, Kirstein *et al.* (2002) measured the high-frequency properties of conductive fabrics to predict the electrical properties of different textiles and to optimize the fabrics and the signal line configurations. They analyzed a plain woven fabric with copper filaments. The electrical characterization was performed by measuring material properties such as dielectric permittivity, transmission line configuration and impedance measurement. To investigate the frequency characteristics of textile transmission lines they measured the transmission properties with a vector network analyzer (VNA) up to 6 GHz.

A means of characterizing electrotextiles using a unique waveguide measurement using imbalanced coupling has been reported by Ouyang and Chappell (2005a, 2005b). By using waveguide measurement, the authors measured the change in quality factor, Q, and the resonance frequency to determine the loss tangent and the dielectric constant of the material. In order to characterize the conductive property of various conductive fibers, a fiber fixture for the waveguide cavity measurement was created. With the fiber fixture, conductive threads were arranged parallel to each other with controllable equal space over the aperture of the waveguide in the measurement setup. In this way, the effective surface resistance of different classes of conductive threads was evaluated at high frequency without the pattern affecting the result. Two methods of measuring effective surface resistance and conductivity of different classifications of electrotextile were presented. A waveguide technique is applicable for materials that are metal-plated after forming the textile. A high-Q waveguide cavity-based measurement technique is desired to enhance the sensitivity of the effect of conductivity in order to characterize the electrotextile properties. Measuring Q of a cavity to obtain metal conductivity is well established for traditional measurements with symmetric input and output ports. In addition to the waveguide cavity measurement, the authors utilized another measurement technique for characterizing the electrotextile based on the microstrip resonator.

In general, to evaluate the electronic response of fabrics, the researchers used the following electrical parameters to describe the behaviour of signal transmission in such fabrics.

10.9.1 Transmission line configuration

This transmission line configuration is similar to conventional coplanar waveguides (CPW) on printed wire boards.

10.9.2 Impedance measurement

This parameter is used to investigate the characteristic impedance of the textile transmission lines. It is expected that the textile geometric variations influence the impedance. The signal reflections along the transmission line can be measured with time domain reflectometry, as the metal fibers incorporated in conductive fabrics show different impedance characteristics and signal transmission effects.

10.9.3 Frequency characterization

In order to determine the bandwidth of textile transmission lines, the frequency characteristics of textile transmission lines are investigated and transmission properties with a network analyzer working at up to 6 GHz are measured. The extracted frequency characteristics reveal information such as dielectric and ohmic losses and the line insertion losses.

10.9.4 Digital signal transmission

Testing of digital signal transmission with a line length of 20 cm and a clock signal with a frequency of 100 MHz may be carried out to understand the signal integrity of different line configurations. The more signal lines, the better the signal integrity, but the more energy is needed for transmitting the signal.

10.10 Applications

In the last few years, smart/intelligent materials and structures have found a wide application as novel smart products in aerospace, transportation, telecommunications, homes, buildings and infrastructures. Today smart textiles are being used in the space programme, in skiwear, shoes, sports helmets and insulation devices as heat generating/storing fibers/fabrics. In addition, smart fabrics integrated with optical fiber sensors have been used to monitor the soundness of major bridges and buildings. It is possible to incorporate sensors inside the garments and these fabrics can detect information regarding injury to and health of the wearer. Textiles in fiber, film and foam forms made of shape memory polymers result in a range of high-performance fabrics and garments in a variety of applications. Fiber sensors, which are capable of measuring temperature, strain/stress, sensing gas, biological species and smell, are typical smart fibers that can be directly applied to textiles. Conductive polymer-based actuators have achieved very high levels of energy density. Textile scaffolds can be made from biodegradable fibers and can be used as biological tissues and organs such as ears and noses. It is possible to impart very high-energy absorption capacity to textiles integrated with nanomaterials and they can be used for stain proofing, abrasion resistance, light emission, etc.

10.10.1 Smart textiles/apparel

The term 'smart textiles' is derived from intelligent or smart materials. The concept of 'smart material' was first defined in Japan in 1989. The first textile material that, in retroaction, was labelled as a 'smart textile' was silk thread having a shape memory effect. The continual shrinkage of the textile industry in the Western world has amply raised the interest in intelligent textiles. Smart textile products meet all the criteria for high added-value technology allowing transformation to a competitive high-tech industry: from resource-based towards knowledge-based; from quantity to quality; from mass-produced single-use products to manufactured-on-demand, multi-use and upgradable product-services; from 'material and tangible' to 'intangible' value-added products, processes and services (Van Langenhove and Hertleer, 2004). An unlikely alliance between textile manufacturers, material scientists and computer engineers has resulted in some truly clever clothing. The interaction between woven fabric and electronics is finding favor in the world of interior design as well. Elsewhere, garment manufacturers are focusing on functional benefits rather than aesthetics. The simplest of these so-called 'smart clothing' items are made by adding the required circuitry, power sources, electronic devices and sensors to standard fabric garments. Batteries can be sewn into pockets, wires fed through seams, and wireless antennae attached to collars and cuffs. The design of such clothing items is still important, although appearance is not the sole criterion. Embedded sensors control conductive material on the back of the jacket to keep the wearer warm should identify the temperature drop, while electroluminescent wires are fixed to pockets and hems to light up in the dark as a safety feature. Such clothing doesn't exist simply to look good, or to attract attention, nor does it simply meet needs without regard to aesthetics.

A totally new vision could be possible with the advent of wearable electronic textiles, where functionality is incorporated into the fabric. More sophisticated prototypes for smart clothing use conductive threads to weave switches, circuits and sensors into the fabric itself. These threads can be made from very finely drawn conductive metals, metallic-coated or metal-wrapped yarns, or conductive polymers. Ideas touted to date include jacket sleeve keypads for controlling cellphones, pagers or MP3 players, and sportswear with integral fabric sensors and display panels, ideal for monitoring heart rate and blood pressure during a gym workout or morning run. Clothing fitted with textile Global Positioning System technology could also be suitable for locating skiers or mountaineers in bad weather or even for keeping a watch on young children (Gould, 2003).

Smart textiles are widely used in medical and hygiene-related products. Antimicrobial and odor-preventing fabrics find wide applications in active-wear and underwear fabrics. These fabrics are hygienic and increase the performance of athletes. Another example of smart textiles that is an integral part of our daily life is 'airbag' fabrics in automobiles. The inflation/deflation mechanism of an airbag is a good example of sensible and intelligent response to mechanical shocks and vibrations.

10.10.2 Military garments

Textile-based materials equipped with nanotechnology and electronics have a major role in the development of high-tech military uniforms and materials. Active intelligent textile systems, integrated with electronics, have the capacity of improving combat performance by sensing, adapting themselves and responding to a situational combat need, allowing the combat soldiers to continue their mission. Meantime, smart technologies aim to help soldiers do everything they need to do with less equipment and a lighter load.

Wearable computers – devices that are attached or integrated into an individual's clothing – were once considered useful mostly as bulky maintenance devices, but are now considered to be the electronic heart of the soldier in the future. The first wearable computers were developed for navigation and maintenance tasks. There are also military applications as body-worn computational resources for soldiers. Intelligent textiles can be used in uniforms to protect soldiers in extreme weather and to constantly monitor the wearer's physical condition. Some even claim automatic healing of wounds. Other outfits have the ability to adapt their color to the surroundings for camouflage purposes. The technology is also helping to improve law-enforcement and anti-terrorism efforts. Computers are being developed that are wired into clothing and have the capability to track enemy targets, network the soldier with air, land and sea forces, monitor his or her physical health, and even translate native languages. Devices are in production that will aid law enforcement and military personnel in the global war on terrorism.

A soldier on the battlefield, an athlete on the training track, a patient in the home – every aspect of their physiology can be remotely monitored by the clothing they are wearing. In short, the ability to weave new-technology threads of electronic wizardry into fabrics means that the IQ of textiles is rising sharply. First-generation commercial techniques for producing electronic textiles included impregnating textiles with carbon, incorporating metal filaments. But this new generation of 'gifted garments' gets its brains from conductive polymers, a special class of organic polymers capable of conducting electricity. For soldiers in the field wearing such a 'smart garment', the weight they carry is significantly reduced as much of their communications technology is integrated into their uniform. That same intelligent textile technology can also be used to provide warmth in cold weather, and to monitor constantly their physiology and the air they breathe – and even to render them invisible to radar.

Much of the smart-fabric, 'soldier of the future' research is centered at the US Army Soldier Systems Center in Natick, Massachusetts. The scientists and technologists are tackling a variety of textiles that can transport power and information. One example is a soldier sticking his or her intelligent glove finger into water to see if it is safe to drink. The soldier could communicate with others by a fabric keyboard that might be either unrolled from the pocket of a uniform, or simply sewn or woven in as part of the uniform's sleeve.

If electronics and optical technologies could be integrated successfully into textiles, there could be a striking improvement in battlefield communications. One such project, the Battle Dress Uniform, gives soldiers camouflage and environmental protection, but it also may become a wearable electronic network to send and receive data.

10.10.3 Medical uses

Smart textiles are widely used in medical and hygiene-related products. Antimicrobial and odor-preventing fabrics find wide applications in activewear and underwear fabrics. These fabrics are hygienic and increase the performance of athletes.

At present, there are some applications in the medical field whose thermal performance can be significantly improved using textiles treated with PCM microcapsules. Such applications include the following.

Surgical clothing

Surgical clothing such as gowns, caps and gloves is often worn for several hours at a time in the operation room. Because these materials are designed primarily to prevent permeation of particles and liquids, which carry bacteria, the thermophysical comfort of such garments is usually poor. But the thermophysical comfort of these garments can be improved substantially by coating the inside of the garments with PCM microcapsules. This effect results from the PCM reacting to any temperature change in the microclimate by either absorbing or emitting heat so that thermoregulation occurs. This thermoregulating effect is responsible for the garments' enhanced thermophysical comfort.

Bedding materials

PCM may also be used in bedding materials such as mattress covers, sheets and blankets, as well as for improving comfort in hospitals, for example. Using materials treated with PCM microcapsules to stabilize the thermoregulation of the patient's microclimate may also support the healing process. Blankets made of acrylic incorporating PCMs and mattresses covered with a PCM microcapsule coating are already available in the marketplace.

Materials used in intensive care

A further development of this technology may center on products used in intensive care units with which a sustained and intensive cooling benefit can be realized. This would serve to remove excess heat generated by the patient's body after an operation. Another possible PCM application may be a product which supports efforts to keep the patient warm enough during an operation by providing insulation tailored to the body's temperature.

10.10.4 Protective garments/uniforms

Application of PCM and shape memory polymers in textiles has opened up a number of applications in protective garments. In protective garments, PCMs absorb surplus body heat to keep the body cool. They provide the insulation effect caused by heat emission of the PCM into the fibrous structure and the thermoregulating effect provided by these materials maintains the microclimate temperature nearly constant. Shape memory polymer coated or laminated materials can improve the thermophysical comfort of surgical protective garments, bedding and incontinence products because of their temperature-adaptive moisture management features (Hu and Mondal, 2006). PCMs incorporated in non-woven protective garments can be used to control or to treat hazardous waste. Safety helmets with PCMs can have very good thermal resistance and the heat generated by the wearer can be dissipated only by convection, which results in substantial reduction of the microclimate temperature in the head area. In the case of chemical or biological protective clothing, a conflict between the protective function of the clothing and the physiological regulation of body temperature can be solved by PCMs which provide significant microclimate cooling for 1–3 hours under usually high heat conditions. This cooling apparel uses macro-encapsulated PCMs uniformly distributed within lightweight vests, helmet liners, cowls and neck collars (Bendkowska, 2006).

10.11 Future trends

Intelligent or so-called 'smart' clothing represents the future of both the textile/clothing industry and the electronic industry. As the convergence between these two industries brings great opportunities and challenges, it draws attention and investment from many organizations in different fields, e.g. academic institutions, government organizations, private companies and laboratories. Currently, none of the applications of smart clothing is considered a full integration of high technology and fashion design, since most research attempts are focusing on solving technical problems such as integrating microchip and computer systems into clothing or overcoming wash-and-care issues. Consequently, the current applications are unable to attract the mass market. The unbalanced contribution from the electronics and fashion industries and the users is the main problem for smart clothing development at this stage. Moreover, the research reveals that the strategic thinking which helps in defining true benefits or core values of smart clothing that may lead to better outcomes is still lacking. As the approach has recently changed from the technical one to a user-centered one, strategic thinking and a new product development process addressing key elements from both industries are required. As a result, the key question is identifying and optimally balancing these key elements in the new approach and the new product development process. Nevertheless, fashion design and product design are established fields of their own and it is difficult to adopt or switch to the other's work methods. Therefore, the new approach and the new product development process should encourage the development team to think in a different way and go beyond their current creative boundaries.

10.11.1 Integrated testing and standardization

The technology of intelligent fabrics and clothing is growing fast. The application of intelligent fabrics in technical textiles, protective clothing (fire brigade, police, military clothing) and medical textiles is considered to be very promising. Current developments in R & D of intelligent fabrics are resulting in applications of these fabrics in sports and automobiles. As the need for these textiles is increasing, there is an absolute need to develop an integrated testing system to evaluate the performance of these textiles. Intelligent fabrics and clothing do not belong to just one particular field but result from combined research in several areas, such as electronics, electrical, textiles, thermodynamics, computing, etc. Hence the characteristics of these fabrics differ and the testing of these fabrics should address these issues. Test methods must be adapted to evaluate several properties of fabrics such as heat regulation, thermal comfort, shape memory effect and sensory properties in addition to the regular textile properties. The R & D efforts in this direction must aim at developing testing methods to integrate several characteristics of these smart fabrics so that a complete analysis may be possible. In addition, efforts to develop testing standards for several properties of these fabrics and clothing would be a great move for the benefit of industry and research organizations.

10.11.2 Interdisciplinary approach

Intelligent textiles provide rich evidence of the enormous wealth of opportunities still to be grasped by the textile industry. These opportunities appear equally abundant in the clothing and fashion sector of the industry and in the technical textiles sector. The development of a full intelligent textile product needs a strongly interdisciplinary approach. Such an approach would combine the efforts of several people involved in disciplines other than textiles and clothing. In particular, future developments will arise from active collaboration between people representing a whole variety of backgrounds and disciplines, including electrical, electronics, mechanical, textile, biotechnology, non-technology engineering streams, materials science, process development, design, commerce and marketing. Such an effort will provide the intelligent clothing industry with the opportunity to work together with innovative partners from various other disciplines and thus contribute to a further switch or increased demand for these innovative fabrics for technical, medical, safety, military and fashion sectors.

10.12 Conclusions

Intelligent fabrics will undoubtedly feature strongly in textile developments over the next decade and look set to become more and more part of everyday life. They represent a variety of different types of fabrics and garments incorporating specially constructed polymers, electronic devices or even some types of colorants. Smart clothing is perceived as the next generation of both fashion and electronic products. Its influence is rising dramatically, as indicated by the rapid increase in research and development projects in the last five years. It offers a wide range of possibilities and opportunities for new business and new product lines. As a result, research and product developments have been carried out by multinational companies and leading academic institutes. Testing the various characteristics of these fabrics becomes an important aspect of everyone concerned in the field. Research and development is growing fast in this direction to evaluate these fabrics for successful application in the industry. Although several research groups are working in this area, there is an urgent need for them to coordinate their efforts to set new testing methods and standards to analyze the various characteristics of these fabrics in order to fully utilize them for the said purpose. An understanding between fashion designers and manufacturers of intelligent fabrics would help in designing more and more new products for consumers. This demands a totally new approach for testing and evaluation of these fabrics, to include testing several properties, not just one. Such an integrated approach would greatly help in understanding these new materials in order to increase their scope of application. In this direction, all the research groups should work in conjunction to set up new standards and test methods to assess the performance of these new products which are going to significantly influence our everyday lives.

10.13 Sources of further information and advice

Intelligent textiles and clothing have attracted several research groups working in different countries and organizations. They are all aiming at developing new products to be eventually used in the industry successfully. However, many are in the infancy stage and some patented technologies are available to be used by the industry. Important research groups are North Carolina University, the Institute of Textiles and Clothing at Hong Kong Polytechnic University, Shinshoo University in Japan, Tampere University in Finland, the Indian Institute of Technology at Delhi, the Wearable Computing Lab at Zürich, Switzerland, and many other individuals in various scientific organizations and industries.

10.14 References

AATCC (1998), Wrinkle Recovery of Woven Fabrics: Recovery Angle, AATCC test method 66-1998.

AATCC (2001), Appearance of Fabrics after Repeated Home Laundering, AATCC test method 124-2001.

AATCC (2001), Retention of Crease in Fabrics after Repeated Home Laundering, AATCC test method 88C-2001.

ASTM (1972), Standard Methods of Test for Breaking Load and Elongation of Textile Fabrics, ASTM test method D1682-1972.

Bendkowska W (2006), Intelligent textiles with PCMs, in *Intelligent Textiles and Clothing*, ed. H R Mattila, Woodhead Publishing, Cambridge, UK, 34–62.

Bendkowska W, Tysiak J, Grabowski L and Blejzyk A (2005), Determining temperature regulating factor for apparel fabrics containing phase change material, *International Journal of Clothing Science and Technology*, **17**, 3/4, 209–214.

Boczkowska A and Leonowicz M (2006), Intelligent materials for intelligent textiles *Fibres and Textiles in Eastern Europe*, **14**, 5(59), 13–17.

Bozzi A, Yuranova T and Kiwi J (2005), Self-cleaning of wood-polyamide and polyester textiles by TiO_2-rotile modification under daylight irradiation at ambient temperature, *Journal of Photochemistry and Photobiology* A, **172**, 1, 27–34.

Gniotek K and Krucińska I (2004), The basic problems of textronics, *Fibres and Textiles in Eastern Europe*, **12**, 1(45), 13–16.

Gould P (2003), Textiles gain intelligence, *Materials Today*, October, 38–43.

Hashem M, Hauser P and Smith B (2003), Wrinkle recovery for cellulosic fabric by means of ionic crosslinking, *Textile Research Journal*, **73**, 9, 762–766.

Hittle D C and Andre T L (2002), A new test instrument and procedure for evaluation of fabrics containing phase-change material, *ASHRAE Transactions*, **108**, 175–182.

Hu J and Mondal S (2006), Temperature sensitive shape memory polymers for smart textile applications, in *Intelligent Textiles and Clothing*, ed. H R Mattila, Woodhead Publishing, Cambridge, UK, 104–123.

Karthik T (2006), Intelligent textiles: an overview, *Indian Textile Journal*, October, 31–42.

Kim B K and Lee S Y (1996), Polyurethanes having shape memory effects, *Polymer*, **37**, 5781–5793.

Kirstein T, Cottet D, Grzyb J and Tröster G (2002), Textiles for signal transmission in wearables, *MAMSET 2002, Proceedings of Workshop on Modeling, Analysis and Middleware Support for Electronic Textiles*, San Jose, CA, October, 9–14.

Lee B S, Chun B C, Chung Y C, Sul K I and Cho J W (2001), Structure and thermomechanical properties of polyurethane block copolymers with shape memory effect, *Macromolecules*, **34**, 6431–6437.

Lendlein A and Kelch S (2002), Shape-memory polymers, *Angewandte Chemie International Edition*, **41**, 2034–2057.

Li F, Chen Y, Zhu W, Zhang X and Xu M (1998), Shape memory effect of polyethylene/nylon 6 graft copolymers, *Polymer*, **39**, 6929–6934.

Li Y and Zhu Q Y (2006), A mathematical model of the heat and moisture transfer in porous textiles with phase change materials, *Textile Research Journal* (in press).

Liang C, Rogers C A and Malafeew E (1997), Investigation of shape memory polymers and their hybrid composites, *Journal of Intelligent Material Systems and Structures*, **8**, 380–386.

Liu C, Chun S B, Mather P T, Zheng L, Haley E H and Coughlin E B (2002), Chemically cross-linked polycyclooctene: Synthesis characterization, and shape memory behaviour, *Macromolecules*, **35**, 9868–9874.

Luo X, Zhang X, Wang M, Ma D, Xu M and Li F (1997), Thermally stimulated shape-memory behaviours of ethylene oxide–ethylene terephthalate segmented copolymer, *Journal of Applied Polymer Science*, **64**, 2433–2440.

Meilert K T, Laub D and Kiwi J (2005), Photocatalytic self-cleaning of modified cotton textiles by TiO_2 clusters attached by chemical spacers, *Journal of Molecular Catalysis A: Chemical*, **237**, 101–108.

Murthy V S, Asai R G and Jaybhaye P B (2003), Intelligent textiles: an overview, *Indian Textile Journal*, April, 29–35.

Ouyang Y and Chappell W (2005a), Characterization of the high frequency properties of electrotextiles, *XXVIIIth General Assembly of International Union of Radio Science (URSI)*, October.

Ouyang Y and Chappell W (2005b), Measurement of electrotextiles for high frequency applications, *IEEE MTT-S International Microwave Symposium Digest*, June, 1679–1682.

Pause B (1995), Development of heat and cold insulating membrane structures with phrase change material, *Journal of Coated Fabrics*, **25**, 59–68.

Qi K, Daoud W A, Xin J H, Mak C L, Tang W and Cheung W P (2006), Self-cleaning cotton, *Journal of Materials Chemistry*, **16**, 4567–4574.

Ring K and de Dear R (1991), Temperature transient: a model for heat diffusion through the skin: thermoreceptor responses and thermal sensation, *Indoor Air*, **4**, 448–456.

Shim H and McCullough E A (2001), Using phase change materials in clothing, *Textile Research Journal*, **71**, 6, 495–502.

Tao X (2001), *Smart Fibres, Fabrics and Clothing*, CRC Press, Woodhead Publishing, Cambridge, UK.

Tey S J, Huang W M and Sokolowski W M (2001), Influence of long-term storage in cold hibernation on strain recovery and recovery stress of polyurethane shape memory polymer foam, *Smart Material Structures*, **10**, 321–325.

Tobushi H, Okumura K, Hayashi S and Ito N (2001), Thermomechanical constitutive model of shape memory polymer, *Mechanics of Materials*, **33**, 545–554.

Uddin F (2006), Electronic textiles, *www.fibre2fashion.com*

Van Langenhove L and Hertleer C (2004), Smart clothing: a new life, *International Journal of Clothing Science and Technology*, **16**, 1/2, 63–72.

Wang Z, Li Y and Kwok Y L (2002), Mathematical simulation of the perception of fabric thermal and moisture sensation, *Textile Research Journal*, **72**, 4, 327–334.

Yang B, Huang W M, Li C, Lee C M and Li L (2004), On the effects of moisture in a polyurethane shape memory polymer, *Smart Materials and Structures*, **13**, 191–195.

Yang X B and Huang X B (2003), Evaluating fabric wrinkle degree with a photometric stereo method, *Textile Research Journal*, **73**, 5, 451–454.

Ying B, Kwok Y-L, Li Y, Yeung C-Y and Song Q (2004), Thermal regulating functional performance of PCM garments, *International Journal of Clothing Science and Technology*, **16**, 1/2, 84–96.

11

Key issues in testing damaged textile samples

R V M GOWDA,
Bannari Amman Institute of Technology, India
and K M BABU,
The Hong Kong Polytechnic University, China

Abstract: This chapter deals with key issues in testing damaged textile samples. It provides comprehensive information on different types of damage, causes for their occurrence, stages of damage occurrence, practical significance of damage analysis, methods of damage analysis, factors affecting accurate testing, applications of textile damage analysis, and future trends.

Key words: damaged textile, damage occurrence, damage analysis, sample preparation, forensic application.

11.1 Introduction

Textile damage is defined as an injury caused to a textile material. The damage may be caused in the form of abraded portions, change in colour, cuts, harsh cleaning spots, holes, nicks, pitting, scratches, soiling, stains, streaks, tears, etc., which lower the value of an item. Damages are often referred to as faults or defects, which represent some kind of deviation from prescribed requirements that lower the usefulness of goods. Damage is the disadvantage arising from faults, which may be visible on the surface, easily recognizable defects or those hidden inside the structure, and thus damage analysis is a wider-ranging term than analysis of faults.

Textile damage analysis is a special area in testing of textiles and has significant practical relevance as textiles can be damaged at various stages during their manufacture, processing, distribution, transportation and usage. Determining the exact cause of damage can be of great help for all those involved in the manufacture, distribution and usage. It can greatly help in forensic science examination to detect evidence and punish the culprits. Damage analysis demands a wide knowledge of textile fibres, the process of their conversion to yarns and fabrics, chemical treatment, garment production, and typical application of textiles. In addition, it also requires the knowledge of methods of analysis using microscopy, chromatography, infrared spectroscopy, thermal analysis, image analysis, etc. Additional requirements for successful damage analysis are information regarding the processes and machines used, stages in storage, transportation, conditions of usage, etc.

Analysis of damage to textiles is not usually an exact science although it does use scientific methods. In many cases several different tests are necessary. Their results can sometimes be contradictory. These then have to be evaluated and weighed up against each other very critically, whereby comparison of samples, experience with similar cases and information about the circumstances of the damage can be useful. In many ways this process is similar to a court trial when only circumstantial evidence is available, but fortunately in the laboratory the damage can often be imitated and the evidence thus verified.

The present chapter intends to provide broad information on the importance of textile damage analysis, causes of damage and types of damage to textiles, and the various methods and techniques used to analyse the damages.

11.2   Causes of damage

Inappropriate treatment of textiles during production, distribution and usage can cause chemical, mechanical, thermal and biological damages. Each type of damage has different effects and greatly reduces the serviceability of the textile material. The causes of damage may be attributed to different people, in particular to textile manufacturers, dyers and finishers, garment producers, distributors or consumers. Although there are interrelations between the above causes, it is usual to attribute mechanical damage to the textile manufacturer, chemical damage to the dyer and finisher, and biological damage to the distributor and end user. When a garment or other textile article is damaged in use or in the care process, a determination of the cause can often be made because of the obvious nature of the damage. Once the cause is identified, responsibility can usually be assigned to the consumer, manufacturer or drycleaner (Mahall, 1993).

The manufacturer is responsible for offering a product that will perform satisfactorily for its normal life expectancy when it is refurbished by the care process specified by the care label instructions. Damage such as severe general colour loss and colour bleeding in the care process, shrinkage that makes an item unwearable, colour fade from the decomposition of fluorescent brighteners, and failure of trim and decorations to withstand the care process are examples of manufacturer's responsibility (www.drycleandave. com).

The consumer is responsible for damage that occurs during use and home care. This includes failure to follow care instructions, further complicating a stain by using a home remedy such as water or soda, chemical damage from spillage of alcoholic beverages, medications, perfumes, after-shaves, hair dyes, perspiration and shrinkage of garments due to improper washing techniques.

It may be difficult to determine responsibility for some types of damage. In cases where the cause of damage is uncertain, a garment can be examined by laboratory methods to analyse the nature of the damage and investigate the probable responsibility.

The various causes of damage are of great importance from the point of forensic science investigations. There are various forms of physical damage which may be found on textiles, for example 'normal wear-and-tear' resulting from normal use of textiles. This usually takes the form of a thinning of the fabric prior to a hole forming, but seams may also come undone, threads can catch and be pulled out from the fabric, or the fabric may even be torn. It must also be remembered that the fabric will probably have been cut in order to make the textile item. In forensic investigations, these forms of 'normal' physical damage must be distinguished from other forms which may be related to the crime.

In a violent scuffle, a fabric may be torn, and the seams often fail; the structure of the fabric may also be distorted. Fabrics may be neatly cut, either with scissors or by slicing with a knife. They may also be punctured by relatively sharp (for example, a screwdriver) or blunt (for example, a hammer) objects, and the nature of the damage will depend on the supporting material (if any) beneath the fabric. The stabbing action of a knife may have features of both puncturing and cutting. Pure tensile failure may occur, especially in ropes and webbing (such as seat belts and slings), although this can often be precipitated by some other form of damage which has weakened the textile.

Abrasive damage, normally considered to be due to 'normal wear and tear', can also be of forensic importance. For instance, a seat belt may fail to protect a passenger in an automobile accident if it has been previously caught in the door and allowed to drag along the road. Damage may also be inflicted by insects, such as moths and carpet beetles, which bite the fibre and digest the fibre pieces internally. Microorganisms, such as some forms of bacteria and fungi, can inject enzymes onto the fibre to break it down.

There are many chemicals which can weaken, modify or completely dissolve some textile fibres. The exact nature of the damage depends on the chemical structure of the fibre and the local conditions, such as temperature and presence of other agents such as oxygen (Johnson, 1991). Textiles may also be damaged by excessive heat, for example in fires or ovens. Heat damage could be localized if inflicted by a cigarette or blow-torch. Thus, when examining a damaged textile item, the textile technologist is usually confronted by a wide range of possible general causes. This range must first be narrowed before any particular scenario can be evaluated.

11.3   Types of damage

Damages in textiles, in general, have been classified into six different classes, namely chemical damage, mechanical damage, thermal damage, biological damage, damage by light and heat, and damage due to presence of defects and contaminants. Each of these damages is described below.

11.3.1 Chemical damage

Chemical damage to textiles can occur when the fibres, yarns, fabrics and garments are exposed to the action of acids, alkalis, oxidizing and reducing agents, solvents, etc. Exhaust gases, especially nitrous gases, can cause yellowing and loss of strength and elasticity for some of the textile fibres. Fibres such as elastane fibres can demonstrate loss of strength and elasticity when they absorb oils and fats such as mineral oil, paraffin wax, unsaturated fatty acids, cosmetic oils and sun protection agents. Hence chemical damage can be discussed with reference to different fibres.

Chemical damage to cotton can occur during the wet processing stages such as desizing, scouring, bleaching and finishing. Cotton is most frequently damaged during bleaching. Reducing agents such as bisulfites can attack the disulfide bonds in wool and cause destabilization of the structure. The cystine links in wool can also be cleaved hydrolytically with hot water, steam and especially alkali. A common damage to silk can occur in the degumming process by the action of alkali. Chemical damage to synthetics is much less frequent than damage to natural fibres because they are more resistant to chemical influences during finishing. The most frequently observed chemical damage to synthetics is acid damage to polyamide and alkaline damage to acetate filaments (Schindler and Finnimore, 2005).

11.3.2 Mechanical damage

Mechanical damage to textiles is generally caused by abrasion during mechanical processing while manufacturing a textile material, resulting in greyed spots and light streaks. The other damage effects are tension breaks and cuts, punctures, setting effects and insect attack. This damage is often only recognized in a later finishing stage when large quantities of material have already been damaged. Natural fibres and cellulose regenerated fibres are more susceptible to mechanical damage than synthetics.

The damage caused by abrasion is usually limited to events which occur to fibres in yarn or fabric. This damage can occur at any stage from spinning to the final product and is often not noted until the final product is examined. Yarn abrasion is sometimes accentuated by an increase in tension as a package is wound, which can lead to patterns of damage visible in fabric.

Abrasion to warp yarns in weaving can occur in drop wires, needle eyes and in the reed. This type of abrasion will cause a line in the fabric which follows a warp yarn. Improper machine settings, damaged guides, dirty machine surfaces and damaged needles can give rise to abrasion of yarns in knitting machines. In garments and other textile end products, damage analysis is often directed at assigning responsibility to manufacturer, consumer or professional launderer or drycleaner (Merkel, 1984).

Tension breaks in fibres usually result in a torn, frayed appearance of the ends of the fibres in contrast to the sharp, clean end of a cut fibre. Cracked selvedges can be caused by excessive tension on filling yarns. This type of damage occurs at a bobbin change and is usually accompanied by an indentation in the selvedge at the site. Holes in knit fabrics are very often associated with excess tension in the yarns during knitting. The excess tension can arise from bad cone winding, improper machine settings, improper operation of positive feed units or tension devices, or dirty yarn guides. Weak places, slubs and knots in yarn can also create holes in fabric during knitting.

Punctures in fabric are often associated with a repeat pattern and may be caused by a rough spot on a metallic roll or a piece of metal embedded in a soft roll. Microscopic examination will reveal cut or crushed fibres at the site. Many finishing plants maintain a list of circumstances of the rolls over which fabric passes so that the source of this type of damage can be easily located.

As previously stated, fabrics may also be punctured by relatively sharp (for example, a screwdriver) or blunt (for example, a hammer) objects, and the nature of the damage will depend on the supporting material (if any) beneath the fabric; the stabbing action of a knife may have features of both puncturing and cutting; pure tensile failure may occur, especially in ropes and webbing. Fabric is not often damaged during processing by tension in the warp direction but set marks can occur in warps made of thermoplastic fibres. A set mark is a line across a fabric parallel to a filling yarn that is caused by secondary creep in the warp yarns which occurs when a loom stops with warp yarns under tension.

11.3.3 Thermal damage

Thermal damage is most frequent in synthetic fibres. It is severe in synthetic fibres having a relatively low melting point. Thermoplastic fibres, which include most synthetic fibres (e.g. polyamide, polyester and polyolefin), change primarily in terms of their physical state as temperature increases (contracting and melting), while chemical degradation (decomposition and burning) occurs only after their melting point has been exceeded (Wąs-Gubała and Krauß, 2006). It results in hardening of handle, yellowing, loss of strength, uneven fabric appearance (light reflection) and uneven dyeing behaviour (spots, streaks and other types of unevenness such as warp splashes).

Thermal damage can occur at many stages in processing. Examples are texturizing, setting, pressing and sewing. During texturing, the original circular fibre cross-sections are usually flattened to polygons. When setting is at too high a temperature or for too long, the yarns are flattened at the interlacing points and tight spots are formed at a higher texturizing temperature in case of PET and nylon. During singeing of staple fibre blends with cellulosics or wool, protruding synthetic fibres can melt to form small balls, which cause a hard handle and which dye more deeply in exhaust processes (small dark spots, deeper dyeing being caused by the high amorphous content and the decrease in relative surface area) and after continuous dyeing are lighter than the undamaged fibres. Pressing at very high temperatures can cause flattening and bonding of thermally sensitive synthetic fibres. Thermal damage also occurs through friction, impact, striking, cutting or puncturing out during textile production and garment manufacture.

Textiles may lose their original colour and strength when exposed to heat. Prolonged exposure to heat can damage fibres like cotton, wool and silk owing to decomposition reactions. Scorched wool fibres turn yellow or brown and blisters are formed on fibres exposed to high heat. Damage to cotton fibres caused by heat can sometimes be detected by the spiral staining of the primary wall exhibited in the Cango Red test. Most artificial fibres melt when exposed to heat, forming ball-like tips and other distortions typical of partially melted material. Long exposure to slight over temperature can lead to oxidative damage in the fibres.

11.3.4 Biological damage

Biological damage to textiles may be attributed to the attack of fibres, yarns and fabrics by microorganisms. Textiles are particularly susceptible to damage through microorganism growth, a problem very common in textile manufacturing units. Microorganisms will grow on any surface when the temperature, humidity and pH of the environment are suitable and a source of food is present. The organisms grow in colonies that have characteristic appearances and deposit stains on fibres that can be observed in visible and ultraviolet light. Actually, microbes secrete enzymes which cause damage to textiles. Generally, fibres made of naturally occurring polymers are subject to biological attack, and cellulose is damaged more commonly than the protein fibres.

Mould and mildew are the common terms used to identify a whole range of microorganisms that survive on organic materials and cause damage. A large number of fungi, bacteria, yeast and algae have been identified as surviving on fabrics. Once a microorganism population is established, it alters the pH of the cloth. This alteration may result in a change of colours in the dyes. There can also be a decrease in the strength of the cloth. Associated with a characteristic musty odour, microorganism growth appears as an irregular stain which generally ranges in colour from grey to black, though yellow, orange and red stains are possible. The attack of microorganisms can result in permanent discolouration of the fabric (Conserve O Gram series bulletin, 1993). The factors affecting biological damage are fibre content, environmental conditions, cleanliness of the textile surface, and acidity/alkalinity.

Control of environmental conditions is by far the most effective method of preventing the biological damage to textiles. It is effective not only for the control of microorganisms but for control of other agents of damage such as insects. Microorganism growth will permanently stain a fabric. The resulting damage will remain as a darkened area and only radical treatment procedures will visually diminish such stains.

11.3.5 Damage by light

Damage caused by exposure to light is associated with the breaking of primary valence bonds with formation of functional groups that can be detected by stains or titration. It is difficult to distinguish light damage from other types of chemical damage, but textiles are often tendered by light in a pattern that helps to distinguish from the back, the edges may have been protected, etc. Delustred fibres are more readily damaged by light than are bright fibres. Fabrics printed with white pigments are sometimes damaged to a greater extent in the printed area than in the remainder of the material.

Light damage on many kinds of fibres is oxidative in nature because of the simultaneous exposure of the textile material to atmospheric oxygen. Particularly, when the exposure to light is outdoors (weathering), simultaneous microbial degradation may occur in cellulosic fibres or acid degradation in cellulosic fibres and nylon.

11.3.6 Damage due to presence of defects and contaminants

Textiles are sometimes damaged by the presence of certain types of defects and contaminants. In cotton fibres, the presence of immature fibres is responsible for generation of more neps and higher hairiness in yarns, barre in knitted fabrics, and colour variation in dyed fabrics. The presence of sericin in silk, and grease and suinte in wool, might result in variation in fineness and other properties of these fibres before and after removal of these impurities. The damage may occur in fabrics due to the presence of non-fibrous content, foreign matter and spun-in-fly in yarns, and oil, grease and paraffin wax deposits. Since these deposits are usually not distributed evenly on the fibres, fabrics contain more or less large stains, spots or streaks. They are one of the most frequent causes of damage. Oils, greases and waxes (e.g. from spin finishes, lubricants, coning oils, sizing waxes and loom oils) which are not removed before dyeing can cause reserving and, as with sizing residues, lead to dyeing unevenness. Precipitated dye or undissolved dye particles cause dye stains. Inappropriate finishes lead to the formation of chalky marks when the fabric is scratched. Typical causes are size residues, printing paste thickeners which have not been washed off, or unevenly distributed finishing agents. Film-like deposits usually cause a somewhat blurry appearance of the surface imprint, for example blurred scale structures in wool. Oligomer and lime deposits result in greying and light stains on dyed and printed fabrics.

Defects in the fabrics such as streaks and bars are another cause of damage occurrence in textiles. They occur in numerous forms, for example parallel or oblique bars in the warp or weft direction, repeat pattern or irregularity, bands, short or long sections of threads, irregular numbers of wales or courses. These defects are caused by irregularities in the yarn in terms of yarn count, twist, diameter, etc. In addition, differences in the yarn tension, hairiness and blend composition can also cause streaks or weft bars in the fabric. Faults during texturization, mercerizing and non-woven manufacture can also introduce objectionable defects in the fabric. The defects arising out of dyeing, printing and finishing such as excessive abrasion in jet dyeing machines, plaiting-down faults, alkali swollen fibres, scums in printing, greasy deposits, etc., can all lead to faulty dyeing and damage the textiles severely (Schindler and Finnimore, 2005).

11.4   Stages of damage occurrence

The damage to textiles can occur at various stages, such as production, processing, storage, transport and distribution, and usage.

11.4.1 Production

Damage may be caused to fibres, yarns, fabrics and garments during their production due to selection of improper processing conditions. Use of improper speeds or tension levels, and poorly maintained machine parts that are worn out, rough, blunt or too sharp will cause damage to textiles in one form or another during their manufacture. The selection of exceedingly high draw ratio and improper temperature during fibre spinning might cause tensile damage to fibres, rendering them brittle and thermally damaged. The use of suboptimum speeds and settings during opening and carding of fibres will cause mechanical damage to fibres, leading to surface damage and even rupture. Improper carding and drafting conditions are responsible for generation of excess imperfections, which will mar the appearance of the fabric and the garment. Yarn passing at very high speed through rough, rusted and worn-out parts on winding, beaming and weaving machines encourages the occurrence of abraded portions, fuzziness and stains, which are potential causes of damage in the resultant fabrics. Damages such as hole formation and missing stitches are expected in the knitting process if the yarn tension and fibre-to-knitting needle friction are not optimized according to the type of yarn used.

Uncontrolled sewing parameters such as highly varying sewing speed, improper thread tension and worn-out needles, will not only damage fibres in the fabric but also lead to partial fusion of sewing thread, depending upon its fibre composition, and ultimately poor seam formation, which will prove detrimental to the garment during usage. Seam puckering and seam slippage are the most commonly occurring defects in many garments. Appreciable damage to fibres takes place if the suitable shape of needle, barb size and stroke are not used in needle punching, resulting in considerable reduction in strength, extension, thickness and compressional resilience of non-woven fabric.

11.4.2 Storage

Textiles may be unexpectedly damaged during storage if improper conditions of storage are resorted to. Cellulosic based textiles and sized fabrics are prone to attack by moths, mildew and microorganisms if they are stored for extended periods in places with unfavourable conditions of temperature and relative humidity. Insects generally like a warm damp environment. Discolouration of the cotton takes place much more rapidly in a damp environment.

The growth of moulds produces bright stains on fabrics and these stains are extremely difficult to remove, because they are often insoluble. Damage caused by mould attack can be devastating. Moulds digest and break down the materials they feed on. In the process, textiles, paper and wood become weak and eventually crumble away.

There are myriad insects that can, and will, eat anything edible in our collection. Carpet beetles will feed on a wide variety of materials, of both animal and plant origin (wool, silk, cotton). Clothes moths feed only on proteinaceous materials (wool, silk, leather, feathers) and blends of synthetic and proteinaceous fibres, but will damage synthetic and plant fibres (cotton, linen, canvas) when feeding on sizing, starch, or food spills in the fabric. Furniture beetles, powder-post beetles and termites will tunnel in and feed on wood.

Synthetic fiber based textiles are prone to soiling by attracting dust and particulate matter if they are stored in open conditions. This phenomenon is often referred to as 'fog marking' and leads to a build-up of quite a large mass of dust on the fabric. Dust absorbs moisture readily, so that areas with a large build-up of dust can have quite high local humidity even when the environment surrounding the object is completely stable at 50% RH.

Particulate matter can be generated within a building in which textiles are stored. In new buildings, concrete and cement can give off very fine dust particles for up to two years after initial pouring. These particles are extremely alkaline and will damage objects they settle on; for example, they will discolour cloth and attack alkali-sensitive materials such as silk and wool.

11.4.3 Transport and distribution

Improper care exercised during transport and distribution will cause severe damage to textiles. For instance, if a truck carrying several bales of cotton in an uncovered condition from a ginning factory to a spinning mill is met with heavy rain, the cotton might be rendered matted, sticky and unprocessable. Similarly, if care is not exercised during transport of dyed and printed goods, there is great risk of discolouration due to exposure to light, rain and dusty weather. Further, improper loading and positioning of goods on a platform can cause damages such as undue abrasion, compressing and deshaping to textile materials. Use of unsuitable packing materials may well damage textile goods during their loading and unloading.

11.4.4 Textile usage

During usage, textiles in various forms may be subjected to unnoticed wear and exposed to sunlight, dirt and particulate matter. All these situations may cause damage to textiles in one way or another. When light and UV radiation fall on a textile object, they deliver a large amount of energy and, as a result, various chemical reactions can take place, depending on the amount of energy delivered. These reactions are called photochemical reactions. It is very easy to see the effects of these reactions, for instance, if a sample of dyed fabric is left in sunlight for just a few hours. The fabric becomes discoloured (yellowed). However, most changes caused by photochemical reactions are not as quick as this nor as obvious; nevertheless their effects can be devastating and highly damaging.

Light causes extreme and irreversible damage to many textiles, most notably organic textile materials. For example, UV radiation and visible light set off chemical changes in textiles, which weaken and discolour them and cause inks, dyes and pigments to fade, and hence seriously affect the aesthetic quality of many items.

Due to continuous usage, textile materials need to be washed frequently. The use of improper washing conditions, ignoring wash, wear and care instructions, might lead to progressive deterioration of textiles. The use of excessive detergents during washing, inadequate rinsing conditions and prolonged drying will lead to cumulative damage to textile materials.

11.5   Practical significance of damage analysis

The analysis of damage to textiles is a fascinating area of textile testing. It has significant practical relevance and considerable charm but also many difficulties. Determining the exact cause of damage can often be a real challenge. Those who carry out damage analysis need wide-ranging knowledge, imagination and adequate experience, but also intuition and the ability to reason and weigh up evidence like a detective.

It is a common practice in the textile industry that the textile manufacturers quite often dispute with dyers and finishers about the occurrence of faults or damages and questions are simply raised as to who is responsible for the fault and who has to bear the cost of damage. In the case of large lots or continuous production lines, it is often important to locate the cause very quickly to avoid the recurrence of faults as soon as possible. In this respect, damage analysis plays a vital role in quality assurance. In spite of the extensive use of modern process control and optimization techniques, the occurrence of faults cannot be avoided and the costs caused by the damage can be very high. Although some disputes are settled legally by the court, most are resolved through mutual agreement between the producer and the customer. Depending on the importance of the business relationship, fair dealing and price discounts play an important role. In order to investigate the complaints, most of the industries in developed countries have set up their own testing laboratories to carry out the testing of damaged textiles using their own experts in the field, and some companies resolve complaints by simply offering price reductions without carrying out any laboratory investigations. The experts in damage analysis at the fibre, dyestuff and auxiliary agent producers, as well as those at the testing and research institutes, still have their hands full, dealing with many cases of damage where the costs caused by the fault can be quite high (Schindler and Finnimore, 2005).

Determining the exact cause of damage can be of great help for all those involved in the manufacture, distribution and usage of textiles. It can greatly help in forensic science examination to detect evidence and punish the culprits.

Analysis of textile damage is often regarded as a technical service in order to promote and enhance supplier–customer relationships. Its economic importance can be observed from the expenses incurred by the producers of fibres, yarns, fabrics, apparel and dyestuffs and auxiliary selling agents in dealing with complaints and analysing the damage. These expenses sometimes could be very high due to the cost of equipment and the complex procedure involved in analysing a damaged sample. Hence it is highly important to understand the wide-ranging complexity of damage analysis in textiles and the challenges corresponding to the demands made on damage analysis in terms of broadly based, thorough knowledge, great experience and the right combination of logical intuitive approaches depending on the problems (Schindler and Finnimore, 2005).

11.6   Textile damage analysis: sample preparation

Preparation of specimens for textile damage analysis is very interesting and often challenging in nature. Depending upon the nature of damage and the amount of sample available, one has to ingeniously select a specimen of suitable size and the number of specimens to analyse the damage and gather maximum information. As the nature of the damage varies from one type to another, there cannot be a single standard procedure of sample preparation for damage analysis, hence the sample preparation may be very specific to the nature of the damage and the technique employed for damage analysis. Also, there cannot be any established procedure as to exactly how many specimens may be prepared, as this depends upon the size of the damaged textile and its nature. In the event of availability of a large sample, one can resort to the standard procedures highlighted in various test standards as applicable to testing of normal textiles. A standard size of a specimen may be obtained from a larger specimen available, in which for instance the yarns have been subjected to abrasive forces or experienced a surface damage. In addition, an adequate number of specimens, namely up to five, may be chosen in case of availability of sample, for determining the extent of damage caused in different portions of the textile and to gather cumulative knowledge on the causes and severity of the damage. On the other hand, if a textile material has been damaged by a cut or the presence of a hole, it is quite obvious that selection of a standard size of specimen is very difficult as the information regarding the area containing that damage is alone important rather than the standard size of the specimen. In such cases, the person carrying out the damage analysis may opt for selection of an appropriate size and number of samples depending upon the nature of the damage, its shape or size.

Further, besides the selection of a suitable specimen size, some damages require specific preparation of specimens prior to damage analysis. For instance, analysis of cross-sections of fibres or yarns in a damaged textile requires thorough preparation of the specimen using standard impregnation techniques, image development, photography and analysis. Similarly, other techniques, such as thin-layer chromatography (TLC) and surface imprint techniques, require preparation of specimens using the standard procedure for that type of analysis. The sample preparation procedures for various damage analysis techniques are described in the respective sections below.

11.7   Methods of textile damage analysis

Perception and description of the fault are the first steps in damage analysis. Damage can be perceived visually, macro- and/or microscopically, often only with a specific type of illumination, such as reflected, oblique or transmitted light, or with a specific light source, for example ultraviolet (UV) or polarized light. Some faults are detected by other senses, usually in the form of a handle assessment, or they can be registered by measurements. Examples of the latter are colour measurement, tensile and abrasion strength, extensibility, shrinkage and fastness properties. These technical properties can be supplemented by thermo-physiological comfort properties and care requirements, where significant deviations from the agreed or specified values can be claimed as faults. It is important here to describe the fault as exactly as possible. All the typical characteristics and peculiarities, their frequency and possible regularity have to be noted and also whether they can be localized to individual fibre or thread systems. This makes it easier to determine the cause of the damage.

Unfortunately, there are no hard and fast rules on how to proceed with textile damage analysis. The variety of cases and causes is too great for this. Nevertheless, some companies and institutes use their own preprinted forms with long lists of tests for this purpose. This has the advantage that none of the rare tests is overlooked but also the disadvantage of inflexibility and unnecessary work on certain types of damage. Such preprinted forms may be of help to less experienced testers, but experienced testers tend to have their own specific procedures depending on the case and are often guided by their intuition. As a general rule, preliminary tests are made followed by more painstaking specific tests (Schindler and Finnimore, 2005).

The usual steps involved in an investigation of damage are discussed below:

1. Visual examination of the damage with a description of the type, appearance, distribution and possible causes of occurrence.
2. Microscopic observation of the damage with appropriate magnification.

3. Preliminary tests such as solubility and staining tests.
4. Isolation of the substance causing the stains.
5. Comparison and identification, usually by means of thin-layer chromatography and/or infrared spectroscopy. Comparison is made with the
   blank sample (extract from unstained areas) and with authentic substances which could have caused the stain. If the stain cannot be
   extracted, IR spectra from stained and unstained areas can be compared
   and the spectra subtracted in order to identify the stain substance.
6. Reproduction, if possible, of the damage in order to verify the findings,
   for example comparison with authentic stain substances on the same
   textile material, using conditions as close as possible to those used with
   the damaged sample.
7. Further verification, for example, if possible, by means of consultation
   with the persons concerned in the stage of production suspected of
   causing the damage. It is important here to consider alternatives and
   to test the plausibility of the findings critically.
8. Summary of the findings and discussion of the results. If the cause has
   not been clearly identified, the results should be formulated carefully
   and alternatives mentioned.
9. Documentation of findings, including photographs and if possible the
   sample swatches.

11.7.1 Visual examination

As in any other type of testing, preliminary examination for textile damage
analysis is essential and is composed of visual examination and simple tests,
carried out in a short time and with little effort, which give the first clues
in damage analysis.

It is usual to begin with an exact visual examination, if possible in comparison with an undamaged sample. Notice should be made of any peculiarities in appearance. Sometimes abraded and raised areas, holes, thin
places and pressure marks, and changes in colour can be easily recognized
without optical aids. With the use of a magnifying glass they can be seen
more clearly and in more detail. The same is true for many visible deposits
of foreign matter, defects, mildew spots, stains, etc.

As a next step, easily determinable differences between the damaged
sample and the undamaged comparison sample can be sought, for example
handle assessments, wetting behaviour or pH value. Simple tests of mechanical strength, rubbing fastness and wash fastness also belong to this group.

After marking the damaged area, woven fabrics can be separated into
warp and weft threads and knitted fabrics unravelled in order to investigate
the isolated threads more thoroughly. The threads from the damaged area
are analysed for differences in fibre composition, yarn diameter, twist level

and yarn appearance. Determination of the fibre type, including checking of the stated fibre type, is one of the most important preliminary tests. This may involve quantitative or qualitative analysis. As a part of this, identification of blends consists of dissolving out one fibre component and examining the residual fibre to see which of the fibre components are damaged (Schindler and Finnimore, 2005).

11.7.2 Microscopic examination

Microscopy is certainly the most important method used for damage analysis of textiles. Microscopic examinations are used extensively in the textile industry to investigate certain characteristics of raw materials and product features, to analyse competitors' samples, and to check output quality. Microscopy is indispensable in dealing with complaints concerning damages as well as in repudiating unjustified claims. Microscopy is also of great use in textile damage analysis in the sense that it is essential for analysing the fine structure of fibres in damaged areas, variations in surface, shape and size of damage, and presence and distribution of impurities, contaminants, foreign matter, dyes, auxiliaries, etc. Table 11.1 gives the typical applications of microcopy as applied to textile damage analysis.

Stereo microscope

The microscopic investigation of damage usually begins with a stereo microscope at low magnification (about 5×), which can then be increased to about 100×. If necessary the damaged areas can often be marked and thus distinguished from their intact surroundings. The large distance between sample and objective enables individual threads, foreign fibres or deposits to be easily manipulated. Further advantages of the stereo microscope are the spatial, three-dimensional image obtained and the fact that different types of illumination, in particular reflected light falling obliquely from one side or opposite sides, can be readily used, depending on what is to be examined.

In blends containing cotton and polyester fibres, for example, bundles of fibres which resemble cotton neps when examined superficially may on careful examination reveal the presence of both mature and immature cotton fibres held together by partially melted polyester fibres, which manifests itself like thermal damage (Schindler and Finnimore, 2005).

Optical microscope

More detailed information can be obtained from light microscopes with up to 1000× total magnification. Such microscopes are available with transmit-

Table 11.1 Application of microscopy in textile damage analysis

Production stage	Parameter analysed
Fibre spinning	Cross-sectional shape
	Fineness
	Delustrant
	Drawing
	Core–sheath structure
Yarn manufacture	Fibre identification in yarn
	Fibre distribution in blend yarn
	Yarn twist and hairiness
	Slubs, neps, contaminants
	Yarn structure
Weaving preparation	Increase in neps, hairiness
	Distribution of size, pick-up
Weaving and knitting	Weave/knit pattern
	Fabric faults/damages
	Deposits, contaminants
Pretreatments	Degree of scouring, bleaching
	Mercerization, lustre
Dyeing	Even dyeing, macro- and micro-levelness
	Dye penetration, migration
	Deposits, colour difference
	Bi-colour effects with blends
Printing	Clarity and sharpness of print
	Print penetration
	Distribution of binder
Finishing	Distribution of finish
	Cross-linking
	Mechanical effects
	Fastness

Source: adapted from Schindler and Finnimore (2005)

ted light for single-fibre investigations or thin cross-sections, and with reflected light for non-transparent objects. Bright field illumination is usually used with transmitted light. Dark field illumination allows edge structures such as projecting fibres from yarns, scales on animal hairs and delustrants to be more easily observed. Polarized light is used for the determination of the melting point of synthetic fibres by means of birefringence. Differences in tension and drawing as well as fibres deformed under pressure can be seen more easily between crossed polars. An examination for a pattern in the damage may be very informative. Abraded yarn has more surface fuzz than normal and contains fibres which appear cut, bruised or frayed when examined microscopically (Schindler and Finnimore, 2005).

Fluorescence microscopy

In damage analysis the most important microscopic contrast method is selective staining of structures of interest. If selective staining with fluorescent dyes is successful, fluorescence microscopy can be a useful method of investigation for the analysis of damage to textiles. It is more or less independent of the original dyeing of the textile material and it can create strong contrasts and thus mark substances present in low concentrations. In rare cases the natural fluorescence of the material can be sufficient for the investigation, for example with wool where it occurs in direct relation to 'yellowing'.

Hesse and Pfeifer (1974) described the fluorescent detection of oil stains, the analysis of the distribution of optical brighteners in polyester/ cotton blends and the detection of photolytic damage by means of macroscopic and microscopic fluorescence techniques. The damage to fibres arising from different degrees of optical brightening could also be elucidated with the aid of fluorescence microscopy. Using natural fluorescence, many kinds of fungi and biological damage to textiles could be analysed. The distribution of applied spin finishes has been studied with fluorescent staining.

Scanning electron microscopy

In scanning electron microscopy (SEM), because of the very short wavelength of the electrons used, much greater magnification is possible than with light microscopy. More importantly for damage analysis, SEM images have a large focal depth and appear strongly contrasted and spatial. For example, with SEM the examination of surface cracks and damages and the presence of impurities, deposits of silicone, silicates or calcium salts can be easily analysed.

11.7.3 Image analysis

Image analysis is the extraction of useful information from images, mainly from digital images by means of digital image processing techniques. Surface imprints of fibres, yarns and fabrics, textile severance morphology (e.g. knife or scissor cuts), and fabric defects, which are of great importance for damage analysis, are usually not evaluated microscopically. Here digital image analysis, which has now reached a high stage of development, should be used, since it employs high quality images and allows easy evaluation, archiving and distribution of microphotographs. They should show the object of interest as sharply as possible and with high contrast; as a rule they should not be manipulated. Using image analysis, the cut or broken cross-sections of fibres, yarns and fabrics can be easily processed for fibre identification, fault classification, or estimating the penetration of size materials and dyes into fibres, filaments, yarns and fabrics. Upon analysis of the images of yarns cut mechanically, one can notice a clean, sharp end if they have not been exposed to vigorous handling in subsequent operations. These methods are also used to investigate hollow and multi-component fibres, the build-up, adhesion and evenness of coating layers, and the analysis of other textile composites. Image analysis can also be used for analysing fabric pilling, drape, wrinkle/crease, texture analysis, etc. All of these can be useful for damage analysis.

11.7.4 Thin-layer chromatography

Chromatography comprises an important group of separation methods in which mixtures of substances are separated into their components using a mobile and a stationary phase. With textile damage analysis the possibility of identifying the separated substances by comparing them with authentic samples is often as important as the separation itself. This identification is successful when the separation behaviour in one or preferably more separation systems is the same and when additional findings such as the same staining or reaction behaviour show that the substances are identical. A prerequisite for such chromatographic identification is that the identity of the substance is already suspected so that the relevant substances can be chromatographed at the same time for comparison. An even more fundamental prerequisite is that the substances to be analysed are soluble in the mobile phase.

Of the many chromatographic methods used in analysis, the one preferably used in textile damage analysis is thin-layer chromatography (TLC). The reason for this is that TLC delivers results quickly, simply and cheaply, with usually sufficient accuracy for elucidation of damage cases. Dyestuffs, optical brighteners, soluble textile auxiliaries and fibre finishes are especially suitable for TLC. Many pigments are also sufficiently soluble.

11.7.5 Infrared spectroscopy

Infrared spectroscopy (IRS) is often a useful supplement to TLC, especially in the analysis of insoluble or macromolecular substances, cross-linked finishes, fibres, coatings, etc. However, with mixtures of substances, the superimposed IR spectra are often so complex that they can hardly be interpreted. A previous separation including that with TLC is very useful for IRS.

Table 11.2 Suitability of IR bands for identifying silicone stains in different fibrous materials

Wave number (cm⁻¹)	Cotton	Viscose	Wool	Nylon 6,6	Polyester
770–800	++	++	++	++	0 (+)
1020–1120	–	–	+	+	–
1260	+	+	0 (+)	–	–
2965	0 (+)	0	–	–	0 (+)

Note: ++ indicates very good suitability (single, non-overlapping bands); + good suitability; 0 means that because of superimposition silicone can only be detected by the markedly higher intensity of the bands; – means that no increase in intensity of the superimposed bands is recognizable.
Source: adapted from Schindler and Finnimore (2005)

Molecules can be excited by absorption of IR rays to give stretching vibrations (in the direction of the bond) and also the somewhat less energetic bending vibrations with three or more atoms in the molecule. Thus as a rule spectra with many bands and containing a high degree of information are formed. The position and the shape of the IR bands are characteristic of the particular molecular structure which has been excited. IRS in the intermediate IR range from 2.5–25 μm corresponding to 4000–400 cm⁻¹ (wave number) enables the identification of functional groups and other structural parts of molecules. If all the essential bands in the IR spectra of the sample and the authentic substance correspond, including those in the so-called fingerprint range of carbon backbone vibrations at about 1500–1000 cm⁻¹, the two substances are identical.

This shows that IRS is a particularly powerful method for damage analysis. With this method, fibres, coatings and other deposits, textile auxiliaries and substances causing stains, and also blend proportion in fibre mixtures can be identified. Table 11.2 shows the suitability of IR bands for identifying silicone stains, depending upon the type of fibre. Chemical damage to fibres can also be detected by means of specific structural changes. All states of matter can be investigated with IR spectroscopy. Thus in damage analysis, the composition of mostly liquid extraction residues is of particular interest (Schindler and Finnimore, 2005).

11.7.6 Thermal analysis

Thermal analysis (TA) is the comprehensive name for a group of analytical methods in which physical or chemical properties of a sample are measured as a function of temperature and time. The sample, contained in a defined atmosphere (usually air or nitrogen), is subjected to a controlled temperature programme, for example it can be tested isothermally or with a constant rate of heating or cooling. In the area of textiles, TA has increased its importance owing to the fast-growing market segment of technical textiles, for example in quality control and in analysis of products, competitors' samples and damage analysis. In damage analysis using TA, a variety of techniques such as differential scanning calorimetry (DSC), thermo-gravimetric analysis (TGA) and thermo-mechanical analysis (TMA) may be used (Schindler and Finnimore, 2005).

Differential scanning calorimetry

Using differential scanning calorimetry (DSC), the temperature range for melting (T_m) and decomposition (T_d), and during cooling that of crystallization (T_c) can be determined along with the corresponding enthalpies. Furthermore, the characteristic temperature for the amorphous areas, the glass temperature (T_g) and the so-called effective temperature or middle endotherm peak temperature (MEPT) can be determined. The determination of melting point is used as a basis for fibre identification. By comparing the measured heat of fusion with the theoretical value, the purity or content can be determined. Similarly the measured heat of reaction can be compared with the theoretical value in order to calculate the extent of reaction, for example with cross-linking reactions.

For damage analysis of textiles made from polyester, the MEPT is especially interesting because it gives insight into the thermal prehistory of the fibres. The MEPT is the maximum temperature of a small endothermic peak between the small endothermic stage of glass transition and the large endothermic melting peak. The position of this MEPT peak is variable and depends on the temperature of thermal pretreatments (T_p). Its size depends on the intensity, and thus mainly on the duration, of the thermal treatment and also on the tension, for example during setting. The measured MEPT usually lies several degrees above the temperature of a preceding thermal treatment (MEPT > T_p). It gives useful information for damage analysis. For example, it is possible to determine from this temperature whether polyester goods were dyed at the boil (with carrier), under high temperature (HT) conditions or using the thermosol process. Conclusions about setting temperature are also possible, in particular differences in setting conditions can be determined exactly (Schindler and Finnimore, 2005).

DSC is also useful for characterizing bicomponent fibres, film-forming finishes and coatings. For example, with polysiloxane a very low glass temperature (about −120°C) is characteristic, followed by a crystallization peak (at about −100°C) and a large melting peak (−40°C). As a matter of interest the enthalpies obtained by integration of these peak areas enable the crystalline ratio to be calculated: $(H_m - H_c):H_m$.

Thermo-gravimetric analysis

With thermo-gravimetric analysis (TGA), the change in weight of the samples on heating is determined (usually possible up to 1000°C). In a nitrogen atmosphere, the decomposition of the sample can thus be studied; in air the ability to be oxidized is additionally determined. In this way fibres modified to be flame-resistant can be distinguished from standard fibres. In fibre composites, for example fibre-reinforced rubber, it is thus possible to determine the proportions of the components with relatively little effort: moisture and softeners in the first stage of weight loss up to about 220°C, then the fibre and rubber components up to 500°C and finally after changing from a nitrogen atmosphere to air the carbon used as a filling burns and above 700°C the non-burnable inorganic filling remains. The first derivative of the weight loss curve, the derivative TG (DTG), enables a more exact determination. By coupling TGA with a mass spectrometer or a FT-IR spectrometer, the decomposition products can be analysed. Because of the higher costs such methods are only used in exceptional cases for textile damage analysis (Schindler and Finnimore, 2005).

Thermo-mechanical analysis

Thermo-mechanical analysis (TMA) investigates the changes in the dimensions of a sample as a function of the temperature, for example shrinkage or extension of fibres. It is easier to work here with filaments than with staple fibres. Fibre composites and other materials are also analysed by dynamic loading. This dynamic mechanical analysis (DMA) enables, for example, the glass temperature of elastomers to be determined exactly.

11.7.7 Physical analysis

Physical analysis includes, for example, identification of fibres in a damaged textile by determining its density, moisture regain, strength, elongation, refractive index, etc. The physical analysis also includes observation of yarn packages, fabrics and garments under suitable illumination for presence of defects or damage.

As regards fibre identification, from the damaged portion fibres in relatively good condition need to be collected and analysed for density using density gradient column, for moisture regain using the environment chamber, and by gravimetric methods. The measurement of strength and extension of fibres, using a single fibre tensile tester, often serves as a means of fibre identification and thus prediction of conditions of damage. Determination of refractive index or birefringence of fibres using a polarized light method can be of great utility in understanding the extent of structural changes the fibre has undergone during processing or usage *vis-à-vis* the structural characteristics of the original sample. The various methods of physical analysis discussed above are very simple and easy to carry out as compared to other expensive techniques, and thus they may be resorted to quite often in textile damage analysis.

Inspection of cones of yarns or even fabric surface using UV light in a dark room is quite natural in many spinning mills for identification of unintended fibre mixing, presence of contaminants or foreign matter that constitute defects, rendering the packages to be rejected. The physical inspection of knitted fabrics under fluorescent white light is also popular in mills to inspect the presence of barre and explore the causes and remedial measures for its non-recurrence.

11.7.8 Chemical analysis

Chemical analysis of damage to textiles is a broad subject and involves extensive knowledge of chemical testing methods. Technological faults due to using mistaken material or caused by foreign fibres can often be most simply clarified by chemical identification of the fibres. In practical damage analysis, physical methods are often combined with chemical analysis, for example microscopic staining, swelling and dissolution reactions or colour reactions and derivatization in chromatography. IR spectroscopy, a physical method, requires chemical knowledge for the identification of fibres, textile auxiliaries and stains.

Chemical analysis is popularly carried out to determine the extent of acid to cellulosics, nylon, polyester and other fibres. Detection of acid damage with Fehling's solution is very common in case of cellulosic fibres, which are sensitive to acids and can be easily damaged by the acid catalysts used in easy-care, silicone, fluorocarbon and flame-retardant finishes as well as by drops of concentrated acid or faulty dyeing of cellulose/wool blends. Further, chemical analysis includes determination of the extent of molecular damage to fibres, damages caused by chemical weak spots, hydrolytic degradation, analysis of unwanted deposits on textiles, presence of oil, grease, paraffin and wax deposits, analysis of solvent extracts, etc. It is also used for end group analysis and determination of critical dissolution time, etc.

11.8   Further methods of textile damage analysis

There are a large number and variety of methods which can be used for damage analysis of textiles. In addition to the important methods of damage analysis described above, three further methods will be briefly described here (Schindler and Finnimore, 2005).

11.8.1 Techniques for surface imprints

Imprint techniques have been a proven and important method in damage analysis of textiles for a long period of time. It is often advantageous not to investigate the original object under the microscope but rather the negative imprint of its surface:

- In an imprint, it is often possible to see if a fault in a coloured textile was caused during textile manufacture or during dyeing and finishing. Spinning faults, such as use of different fibre counts or differences in yarn twist, and faults in fabric production, can be seen in the imprint as well as in the original (and in the same location). On the other hand, faults arising from dyeing or printing are eliminated in the imprint.
- The imprint is transparent and the colour of the sample does not interfere. Thus with dark-dyed wool fibres, the cuticle scales can only be easily recognised in the imprint. The same applies to abraded places and other types of mechanical damage to the surface of dark dyeings.
- Since the surface imprint is very thin (about 0.02 mm), the depth of focus is usually much better than in direct microscopy of the uneven, three-dimensional textile surface and possible fibre lustre and transparency do not interfere. In direct microscopy with reflected light, the image is usually not sharp because the fibre interior and the underside of the fibre also reflect light.
- Since the transparent imprints are examined in transmitted light, it is not necessary to have a microscope with reflected light. In addition, the original sample remains unchanged.

There are two different but widely used imprint methods in damage analysis, namely imprints on gelatine-coated plates and on thermoplastic films, usually polypropylene or polystyrene. In Table 11.3, the most

Table 11.3 Comparison of the most important surface imprint methods

Imprint with gelatine-coated plates	Imprint with thermoplastic films
No thermal influence on the sample	No swelling of hydrophilic fibres
No false indication of structural differences, arising from diffusion of grease, oil or wax deposits	Detection of grease, oil or wax deposits possible due to diffusion into and dulling of the film
No special equipment necessary	Special instrument is essential

Source: adapted from Schindler and Finnimore (2005)

important advantages of these complementary imprint methods are compared.

Gelatine-coated plates method

Two grams of gelatine are swollen in 40 ml of distilled water for one hour at about 35°C. One part of this swollen gelatine is diluted with three parts distilled water, and 1020 ml of this solution is sufficient for about 100 cm^2 of glass plate surface. The solution is put evenly on clean glass plates, for example microscope slides, with a pipette. After drying for 24 hours in a dust-free place, the plates are ready for storage or use. Before use they are dipped briefly into water and the adherent water is then removed by shaking. The fabric is then laid on the swollen gelatine and covered with a filter paper and a further glass or metal plate. A weight is added, which should cover the upper plate evenly. For glass plates that have the same size as microscope slides, the weight should be about 500 g. After pressing for 30 min, the textile side is dried briefly with a warm hair dryer and then carefully removed from the imprint. The slides can now be observed under a microscope using appropriate magnification for imprint analysis (Schindler and Finnimore, 2005).

Surface imprints with thermoplastic films

The thickness of the films (usually with polypropylene 30–40 μm or with polystyrene 100–200 μm) depends on the thickness of the individual fibres and yarns or the structure of the fabric. Polystyrene films are preferred for large-scale imprints (up to about 20 × 30 cm) in the Streak Analyser. For small-scale imprints, a piece of film cut to a suitable size is pressed firmly together with the textile sample between two polished metal plates, for example of the size of microscope slides, with two screw clamps. The assembly is placed in a drying oven at 105°C in the case of polypropylene for 30 min and with polystyrene for 45 min. After cooling as rapidly as possible with cold air the sample is separated from the film. The film can then be examined in transmitted light in the microscope without the use of embedding agent (Schindler and Finnimore, 2005).

11.8.2 Extraction methods

Extraction methods using Soxhlet apparatus are quite routine and standard in most textile laboratories. During extraction, substances soluble in organic solvents or water are removed from the textile, then, as a rule, concentrated by distillation, and the extraction residue is analysed qualitatively and/or quantitatively. Examples of extracted substances are stains, fibre spin finishes, lubricants, residual grease in wool, residues of surfactants and other chemicals such as acids, bases or thickeners, soluble finishes, dyes and optical brighteners, pesticides and other biocides, carriers, heavy metal salts and formaldehyde. Stepwise extraction using solvents of increasing polarity (for example first hexane, then methylene chloride, then absolute alcohol and finally water) can give a first indication of the nature of the extracted substances. Different extraction methods and apparatus can be used. Miniaturized versions of the Soxhlet extractor are preferred for very small samples such as stains. As alternatives to the Soxhlet extractor, automated apparatus computer-aided finish analysers, like the ALFA 200, have been developed.

11.8.3 Determination of average degree of polymerization of fibres

Many types of damage, including biological, chemical, photolytic, thermal and some types of mechanical damage, are based on degradation of the polymer chains in the fibre. Thus determination of the average degree of polymerization (DP) gives a direct scale for assessing the extent of such damage but not its cause. The time and cost of determining DP are, however, so great that whenever possible simpler but less accurate methods are preferred. Examples of these are loss of tensile strength and abrasion resistance or the pinhead reaction. An advantage of DP determination is that it allows quantitative estimation of the damage. For example, Eisenhut (1941) defined a damage factor s for cellulose fibres based on the decrease in DP, which allows comparative assessment of damage for cotton and regenerated cellulose fibres:

$$s = \log[(2000 : P_1) - (2000 : P_2) + 1] : \log 2$$

where P_1 = DP before damage and P_2 = DP after damage.

Damage factors of $s < 0.5$ are said to be acceptable after bleaching treatments, but with $s > 0.75$ the goods are said to be badly damaged. It must be borne in mind that the damage factors are also dependent on the method used; they are higher with the cuoxam method than with the EWNN method (Schindler and Finnimore, 2005).

11.8.4 Detection of streaks and barriness in woven and knitted fabrics

Barre is defined as an 'unintentional, repetitive visual pattern of continuous bars or stripes usually parallel to the filling of woven fabric or to the courses of circular knit fabric'. Streaks and bars are second only to stains as one of the most common manifestations of damage. They occur in numerous forms, for example:

- Parallel or oblique to the warp or weft direction
- With a repeat pattern or irregularly
- In bands or bars
- Running along short or long sections of thread or across differing numbers of wales or courses.

As a rule, streaks and bars are caused by faults in textile production. Examples of this are:

- Mistaken material, usually use of the wrong yarn
- Differences in yarn count, yarn bulk, yarn twist, thread tension, plying, pile opening, hairiness, inhomogeneous blends
- Faults during texturizing or mercerizing
- With pile fabrics, more deeply incorporated tuft rows or differences in needling
- Wet abrasion and other types of mechanical damage in jet dyeing machines
- Plaiting-down faults in cotton pretreatment: squashed fibres, notches, cracks and splits in the fibres, which occur when the goods, swollen with alkali, are packed down too densely
- Greasy deposits and resinated mineral oil, which have a carrier effect on polyester, leading to deeper dyeing (Schindler and Finnimore, 2005).

The presence of barre can usually be judged by mere inspection of a fabric under proper illumination and careful observation. Barre analysis methods that help to discriminate between physical barre and barre caused by dyeability differences include flat table examination, light source observation, and the Atlas Streak Analyser. The cause of the fault can be readily clarified with the aid of a microscope, fabric dissection and film imprint analysis (www.geocities.com/vijayakumar777/barre.html).

11.9   Factors affecting accurate testing

The accurate testing and analysis of damage to textiles is affected by factors such as limited sample size, sample preparation, presence of contaminants, impurities, etc. As already discussed in Section 11.6, the extent of information gathered from analysis is highly dependent upon the availability of a sample containing the damage and the sample preparation techniques employed. In most cases, adequate and correct information can be obtained if sufficient quantity of sample containing the damage is available. Otherwise, the availability of limited sample size leads to improper preparation in case of special types of analysis, which ultimately influences the validity of the information gathered.

Further, the presence of impurities might often mislead the scientist carrying out the analysis and offer altogether different information as against the original one. The impurities may subvert the reality and come in the way of drawing inferences. For instance, the presence of titanium dioxide particles in a synthetic textile material used as a window screen may often subvert the fact that the continuous exposure of the screen has rendered it dull and caused surface fading. Also, the presence of impurities in dyes or finishes will influence the characteristics of treated textiles and eventually the causes of damage or deviation from requirements, if any, in the characteristics of such products. Hence one has to be very careful to see the presence of contaminants and/or impurities and their influence on the damage analysis before arriving at appropriate conclusions. In case of difficulty of arriving at appropriate conclusions, one can resort to adopting damage simulation techniques, predictive damage analysis, intuition, practical knowledge and logical thinking. Hence damage analysis demands a wide knowledge of textile fibres, the process of their conversion to yarns, fabrics and garments, chemical treatment, and typical application of textiles, besides the knowledge of various methods of analysis and their utility.

11.10 Applications of textile damage analysis

11.10.1 Forensic applications

Forensic science and its application in criminal investigative technology are used to help clarify criminal cases. Textiles can play an important role here, usually in the form of clothing but also including household and automobile textiles, furnishings and in rare cases also technical textiles, for example strings and ropes used to bind, mangle or hang victims. Textiles used by criminals can also include stocking masks, gloves, bags, sacks and adhesive tapes. Sometimes it is possible to solve cases of murder unequivocally with the aid of a few typical fibre traces transferred from the murderer to the victim and/or vice versa.

Forensic application of textile damage analysis methods will help to extract information from photographs taken at crime scenes using photographic and computer techniques. This can include the matching of items taken from suspects. Analysis can reveal shoeprints in earth, mud, sand or carpet, and hand markings on textile surfaces. This enhances the utility of photographic information and the digital images to match clothing, people, firearms, vehicles, etc. It is also possible to detect prior damage to repaired surfaces. Extensive literature on forensic analysis is available from Grieve (1990, 1994, 2000) and from www.canesis.com.

11.10.2 Other applications

Damage analysis of textiles is highly useful in determining whether the fibres used in a consignment are made up of particular type of pure polymer or not. It is also of great use in identifying whether the correct proportions of fibres are used in a blend or not and thus in helping to avoid disputes that may arise between the manufacturer and the customer. Damage analysis is often sought to resolve issues concerning whether the right type of chemicals are used on textiles or not and whether they are skin-friendly, eco-friendly, hazardous to the wearer, etc.

Textile damage analysis is helpful in identification of fibre type and thus in detection of happenings of accidents, fire hazards or any other type of risks the wearer of that particular textile was subjected to. Damage analysis of fibres containing blood stains, or any secretions or remnants collected from selected parts of a human body, will help in detection of murders, rape cases, suicides, etc.

A special case of fibre identification involves vehicle accidents with fatal injuries to the occupants. The high pressure of impact causes such a high frictional heat that fibres are embedded in plastic surfaces which are momentarily softened and the fibres are retained there after the plastic cools down. With these traces, known as fusion marks, it is possible to reconstruct where the passengers were sitting and to determine who was driving the vehicle (Schindler and Finnimore, 2005).

11.11 Future trends

Future trends in damage analysis may include the demand for use of sophisticated instruments that can provide quick and accurate information about the damages and their occurrences. One can resort to damage simulation techniques to compare and contrast the real damage with the stages of occurrence and thus find out the causes. Modelling and computational techniques can be increasingly used in stress analysis, predicting structural degradation, impact damage, reduction in strength and other mechanical properties in various textiles and textile-reinforced composites. Online monitoring of various processes, data collection, analysis and storage will be of great use in looking back at the processes to identify the causes of damage. Progressive damage analysis techniques consisting of simulating damage growth, data collection and analysis may find extensive application in damage analysis in the years to come. Use of finite element analysis, digital image analysis, neural network techniques and fuzzy logics, creation of special test methods, inspection programmes, standard formats of testing, and certification and regulatory actions will make damage analysis much more sound and authentic.

11.12 Sources of further information and advice

In addition to the references cited at the end of this chapter, readers may refer to the following books, websites or sources of literature for further information on testing of damaged textiles and related aspects:

Atlas of Fibre Fracture and Damage to Textiles, second edition, J W S Hearle, B Lomas and W D Cooke (eds), Woodhead Publishing Limited, ISBN-13: 978 1 85573 319 0, July 1998.

Caring for Collections Across Australia, Heritage Collections Council, www.amol.org.au

Chemical Testing of Textiles, Qinguo Fan (ed.), Woodhead Publishing Limited, ISBN-13: 978 1 85573 917 8, September 2005.

Data from the materials property tests are being used to develop and calibrate a computational model for simulating impact and penetration of fabric targets, http://www.sri.com/psd/poulter/air_safety/material_properties.html

Deterioration of textiles, http://www.ashmolean.org/departments/conservation/deterioration#textiles

Environmental Impact of Textiles: Production, Processes and Protection by K Slater, University of Guelph, Canada, ISBN-13: 978 1 85573 541 5, June 2003.

Microbiological testing of polymers and resins used in conservation of linen textiles, http://www.ndt.net/article/wcndt00/papers/idn002/idn002.htm

Multi-scale modelling of composite material systems, in *The Art of Predictive Damage Modelling*, C Soutis, Sheffield University and P W R Beaumont, Cambridge University (eds), ISBN-13: 978 1 85573 936 9, August 2005.

Novel tests and inspection methods for textile reinforced composite tubes, W. Hufenbach, L. Kroll, M. Gude, A. Czulak, R. Böhm, M. Danczak, *Journal of Achievements in Materials and Manufacturing Engineering*, Volume 14, Issue 1–2, January–February 2006.

Sixth European Commission Conference on Sustaining Europe's Cultural Heritage: from Research to Policy, Queen Elizabeth II Conference Centre, London, 1–3 September 2004.

What could be observed about image properties by looking at the damage from the fire of 1532?, http://www.shroudstory.com/faq/turin-shroud-faq-14.htm

11.13 References

Conserve O Gram series bulletin (1993), *Causes, Detection, and Prevention of Mold and Mildew on Textiles*, National Park Service, Washington, DC.

Eisenhut O (1941), Zur Frage der Bestimmung des Schadigungswertes von Fasern, Garnen oder Geweben aus Zellulose und von Zellstoffen, *Melliand Textilberichte*, **22**, 424–426.

Grieve M C (1990), Fibres and their examination in forensic science, in *Forensic Science Progress*, ed. A Maehly and R L Williams, Volume 4, Berlin, Springer-Verlag, 41–125.

Grieve M C (1994), Fibres and forensic science – new ideas, developments, and techniques, *Forensic Science Review*, **6**(1), 60–79.

Grieve M C (2000), Back to the future – 40 years of fibre examinations in forensic science, *Science and Justice*, **40**(2), 93–99.

Hesse R and Pfeifer H (1974), Fluoreszenzmikroskopie zur Erkennung von Fehlerursachen an Textilien, *Textilveredlung*, **9**, 82–87.

Johnson N (1991), Physical damage to textiles, in *Police Technology: Asia Pacific Police Technology Conference Proceedings*, ed. J Vernon and D Berwick, 12–14 November.

Mahall K (1993), *Quality Assessment of Textiles, Damage Detection by Microscopy*, Berlin, Springer.

Merkel R S (1984), Methods for analysing damage in textile materials, in *Analytical Methods for a Textile Laboratory*, ed. W J Weaver, AATCC, 33–91.

Schindler W and Finnimore E (2005), Chemical analysis of damage to textiles, in *Chemical Testing of Textiles*, ed. Q Fan, Woodhead Publishing, Cambridge, UK, 145–241.

Wąs-Gubała J and Krauß W (2006), Damage caused to fibres by the action of two types of heat, *Forensic Science International*, **159**(2–3), 119–126.

Websites

http://archive.amol.org.au
http://www.canesis.com
http://www.dermestidae.com/Anthrenusverbasci.jpg
http://www.drycleandave.com, Dryclean Dave and Uptowne Drycleaning: Damage and responsibility.
http://www.flheritage.com/museum/collections/artifacts/acs5/lightfading_lg.jpg
http://www.geocities.com/vijayakumar777/barre.html
http://www.vam.ac.uk/school_stdnts/schools_teach/teachers_resources/conservation/deterioration_decay/index.html

Flammability testing of fabrics

S NAZARÉ and A R HORROCKS,
University of Bolton, UK

Abstract: Every application of textile materials has its own textile fire testing method. In some instances flammability standards are mandatory and specific in legislation and regulations, whereas for some applications flammability testing is at the discretion of the manufacturer and/or the user. Moreover, since most standards are assessed and modified every five years or so, we have attempted to highlight key issues of fabric flammability when tested as single layer fabrics or as a part of a composite, flammability measurements in general and the general principles underlying textile flammability test methods and relevant performance requirements when used in various applications. These are illustrated and exemplified by reference to specific test methods where relevant.

Key words: simple ignition test, flame spread, heat release, composite tests, mannequin and flammability standards.

12.1 Introduction

Apart from apparel clothing, the need for textiles requiring defined flammability behaviour for improved safety spans a range of applications that includes furnishing and bedding (domestic and contract), transport, civil engineering, medical and defence. In some instances flammability standards are mandatory and specific in legislation and regulations, whereas for some applications flammability testing is at the discretion of the manufacturer and/or the user. The first UK standards for the flammability testing of textiles appeared in the late 1950s, i.e. the period when around 1000 people died each year in fires with textiles as the first igniting material.[1] These and subsequent standards mainly quantify burning behaviour and flame-resistance of fabrics in terms of ease of ignition and rate of flame spread, and this is true of textile flammability standards introduced worldwide.

The measurement of fabric flammability is primarily based on classification of the burning hazard of self-extinguishing and flammable fabrics. Self-extinguishing fabrics are those which stop burning within an acceptable and defined time before the sample is consumed after removal of an ignition source applied for a defined time. It is generally understood that flammable fabrics are those fabrics which ignite when subjected to a small flame for durations of up to 12 seconds and continue to burn after the source has been removed. Most of the work on flammable fabrics is therefore concerned and directed towards the observation and measurement of ease of ignition, the rate and extent of flame spread, the duration of flaming, measurement of heat release and heat of combustion and quantitative description of burning debris, such as melt dripping. Rarely if ever does a single test method enable all these parameters to be measured. For self-extinguishing fabrics, such as flame retarded fabrics, test methods include measurement of time of afterflame and afterglow and extent of fire damage in terms of char length, hole size or weakened sample length.

It is beyond the scope of this chapter to describe all the flammability standards and requirements of textiles in different applications. In addition, since most standards are assessed and modified every five years or so, the reader should always contact the major standards authorities (ASTM, BS, EN, ISO, DIN, etc.) in order to be aware of the latest revisions. However, we have attempted to highlight key issues of fabric flammability when tested as single layer fabrics or as a part of a composite, flammability measurements in general and the general principles underlying textile flammability test methods and relevant performance requirements when used in various applications. These will be illustrated and exemplified by reference to specific test methods where relevant.

12.2   Key issues of fabric flammability

Textile materials have very high fibre surface-to-mass ratios and hence tend to ignite easily and burn faster than other materials. Furthermore, because they are thin materials tested in two- and sometimes three-dimensional geometries, textile materials burn differently from other solid and liquid fuels. In addition, other intrinsic differences in material properties include the following:

- The matrix-like, open structure of the textiles which makes it easy for air to circulate between the burning yarns
- Their thermally thin character (see below) which ensures that non-thermoplastic fibrous materials may undergo thermal decomposition or pyrolysis in depth, whereas bulk polymers and solid fuels undergo pyrolysis as surface decomposition–vaporisation processes only
- The presence of fibres protruding well above the yarn-fabric surfaces which often promotes the occurrence of surface 'flashing'
- The thermoplasticity of some fabrics which may cause shrinkage in one or more directions, curling and deviations from their original plane and, if melting occurs, give rise to often flaming molten drips, which may transfer heat and flame to other materials including the body of a wearer.

While all conventional textile fibres burn, the difference between them is only one of degree. The hazard associated with flammable fabrics is dependent on various fire science-related parameters such as:

- ease of ignition
- rate of flame propagation
- heat transfer mechanisms
- rate of heat release
- total amount of heat released.

These parameters individually and collectively are influenced by characteristics of particular textile material, including:

- chemical composition of fibres present
- interactions of different fibre types if fabric is a blend or composite
- fabric structure (open versus close weave, knitted versus woven versus non-woven structures)
- fabric area density and thickness
- fabric orientation (e.g. horizontal, 45° or vertical)
- point of ignition (e.g. top or bottom of sample, edge or face).

The chemical composition of the fibre is the important fibre characteristic affecting flammability of fabrics. For example, the more thermally stable is the fibrous polymer, the higher is the decomposition temperature as well as the endothermic nature of the decomposition reaction.[2] At a less sensitive level, fabric properties such as area density, construction pattern and surface characteristics also affect flammability. The majority of textile fabrics behave as thermally thin materials with the exception of heavy industrial fabrics, geotextiles and fabrics for thermal insulation which are usually >10 mm thick and have higher densities, often >1 kg/m^2. However, textile fabrics for major conventional end-use applications are thermally thin, with ease of ignition and flame spread rates being of primary importance. The lighter the fabric, the quicker it ignites and the faster it burns compared with a heavier fabric made of the same material.

Besides fabric structural effects, Miller and Goswami[3] analysed the effect of various yarn parameters on the burning behaviour of the fabrics. They found that zero-twist yarns produce higher burning-rate values, whereas yarns with twists from 0 to 1.37 turns per cm decrease the mass burning rate by up to 40%. Wraight et al.[4] studied the effect of fabric area density or weight on heat transfer rate. They found that for cotton the heat emission is inversely proportional to the fabric weight, whereas for 100% polyester the heat emission rate varies in the same direction as the fabric weight but to a much lesser degree.

When determining textile flammability, many terms are used to describe different stages of the process. Selected terms used to define flammability parameters of textile materials described in many British and international standards are given below:

- *Ignition:* flaming of the test specimen for a period of 1 s or more after removal of the igniting flame.
- *Flaming:* combustion in gaseous phase with emission of light.
- *Glowing:* combustion of a material in the solid phase without flame but emission of light from the combustion zone.
- *Smouldering:* combustion of a material with or without emission of light generally evidenced by smoke.
- *Melting:* liquefaction of material when exposed to heat to the extent of forming a hole in its structure, by either shrinking and/or dripping away under the specified test conditions.
- *Flame spread time:* the time taken by a flame on a burning material to travel a specified distance measured from when the igniting flame is applied or after it has been removed.
- *Flaming debris:* materials separating from the specimen during the test procedure and falling below the initial lower edge of the specimen and continuing to flame as they fall.
- *Afterglow time:* the time for which a material continues to glow, under specified test conditions, after cessation of flaming or after removal of the ignition source, ignoring glowing debris.
- *Surface flash:* rapid spread of flame over the surface of a material without ignition of its basic structure.

12.3   Measurement of fabric flammability

Scientifically based measurable flammability characteristics and parameters can be categorised under the following headings:

- Pre-ignition:
  - thermal decomposition (temperature and weight losses)
  - enthalpy changes
  - products of thermal decomposition
  - kinetics of the ignition process.
- Post-ignition:
  - ignition temperature
  - flame temperatures
  - heat release rates during burning
  - flame propagation rates
  - upward mass burning rates
  - extinguishability
  - products of combustion.

However, assessment of the potential flammability hazard of any fabric can only be followed if the source of danger to life and injury is identified at each stage of the burning process. Thus no single laboratory test can determine the complete burning character of a particular textile.[5] Most investigative procedures for assessing textile burning behaviour fall into two groups: the standard test methods and the scientific or research test methods. The research test methods give a fuller picture of burning behaviour and attempt to quantify a number of aspects of the burning process, whereas the standard test methods usually involve 'pass–fail' and/or performance rating criteria.[6] Some of the principal work on development of scientific test methods for measuring flammability of fabrics is discussed in this section.

During the 1970s, the experimental techniques such as Oxygen Index (OI) and flame-temperature methods gained popularity for measurement of flammability of polymeric materials. According to the ASTM,[7] limiting oxygen index (LOI) is defined as the minimum concentration of oxygen, expressed as volume percent, in a mixture of oxygen and nitrogen that will just support flaming combustion of a material. This technique provides a numerical measure of sample flammability, although it does not explain the burning behaviour of the material. Generally, textiles having LOI values of 21 vol% or less burn rapidly, those having values in the range 21–25 vol% burn slowly, and those with LOI $\geq$ 26 vol% exhibit some level of flame retardancy in air, which has an oxygen concentration of about 21 vol%. While oxygen index methods have not achieved formal standard status for textile materials, they are included as part of relevant national and international test methods for polymers in general. Horrocks *et al.* reviewed the whole area of oxygen index as it relates to textiles.[5] With respect to textiles, OI tests are mainly used in determining the effects of different flame retardant treatments and finishes, varying the add-on of finishes, or varying synergistic combinations of flame retardant compounds. However, because LOI values may be influenced by many fabric variables for a textile comprising a single fibre type,[5] there are reasons why the test method is rarely used to define fabric performance by regulatory and commercial bodies. OI methods, however, do find applications as research and development tools.

Various flame spread theories were also developed in the 1970s, which formed the basis of standard test methods designed for measuring fabric flame spread.[8,9] The theoretical models to predict flame spread behaviour were developed on the basis of measured burned length using video photography, for example. Some of the ways of measuring fabric flame spread involving different techniques from those above and related theories are briefly discussed below.

One of the ways of measuring vertical flame spread is to weigh the burning sample continuously. This would enable the rate of loss of weight to be calculated, and from it the vertical flame spread. The details of the apparatus designed for measuring the vertical flame spread on fabrics by the above principle were discussed over 50 years ago by Lawson et al.[10] The vertical flame spread was calculated by using the formula

$$v = \frac{rl}{w_1 - w_2}$$

where v = vertical flame spread in m/g
r = constant
w_1 = initial weight
w_2 = weight of carbon residue
l = initial length of specimen.

The authors also examined the minimum length of the specimen required for measuring the maximum rate of flame spread and concluded that this is not reached until a sample of about 127 cm (50 inches) is burned. An alternative method of assessing flammability was to allow the material to burn at various angles to the vertical and to see at which angle the sample is no longer able to support flaming. A simple apparatus was designed by Lawson et al.[10] wherein the spread over the sample in a semi-circular track was ignited at one end and the distance to which the flame spread reached was noted. It was found that many materials burned completely round the semi-circular track, and they were differentiated from each other by noting the time taken to burn 21 inches (53.3 cm) of specimen. The authors also derived two empirical mathematical equations for determining flame spread rate and concluded that the vertical flame speed is roughly proportional to the square root of the distance of spread ($V = 1.81d^{0.4}$) and inversely proportional to the time of spread ($V = 1655/T^{1.03}$) for the materials that burn completely (for flammable materials). Both of these sets of experiments were incorporated in early, now obsolete British standards of the 1950–60 period.[5]

Furthermore, heat release measurement is also a significant criterion in assessing textile flammability at the scientific level, although less so in the standard test procedure.[3] A heat release parameter of fabrics is also useful to predict burn injury severity in particular. The rate of heat release is determined by measuring combustion product, gas flow and oxygen depletion, and the rise in temperature of the sample or the rise in temperature of air entrained during burning of the specimen. Initial attempts to measure heat release rate and thus predict burn injury severity differed mainly in the manner in which the specimens were held, the number of sensors and the manner in which they were mounted relative to the specimen and the specimen shape. Of all the early heat release measurement techniques studied, the US Textile Research Institute's (TRI) convection calorimeter gained popularity in the 1970s. Miller et al.[11] developed a technique for continuous monitoring of the heat released when a freely suspended fabric burned under natural convection, using the so-called TRI convection calorimeter. The instrument monitored the rate of heat emission by measuring the airflow rate of convective air. The primary response obtained from these measurements included:

- the maximum rate of heat emission
- the time to reach the maximum rate
- the post-maximum time required for the heat emission rate to subside to one-half its maximum value.

Since the late 1990s, however, heat release rate data have been used as a tool for evaluating the hazard level of a fire system in general and individual combustible materials specifically. Developed by Babrauskas,[12] the cone calorimeter measures the hazard level by measuring the rate and amount of heat released, and smoke and toxic gases generated, and this now standard method (e.g. ISO 5660) is used to define material performance including textiles in a number of applications, notably construction and transport. However, because of its complexity and cost, it is rarely applied to the testing of single fabrics and fabric assemblies.

12.4   Standard testing methods

Ideally, all standard test methods should be designed such that the measurement of flammability parameters, such as time to ignition, rate of flame spread, afterglow times, etc., can be acquired in a reproducible and repeatable manner. The flammability principles on which the standard tests are established should be straightforward and easy to transform into a practically simple and easy-to-use test.[13] Some of the principle test methodologies are discussed below along with illustrative standard method examples.

12.4.1 Simple ignition tests

In standard test methods, fabric ignition is measured in terms of minimum ignition time, i.e. how long a flame needs to be applied to a given material so as to achieve ignition, normally to the nearest second. A simple ignition test includes a vertically oriented fabric subjected to a standard gas flame applied to the face or lower edge (depending on the severity of the test required) of the fabric specimen. Ignition is monitored by visual observations and the time taken to ignite the specimen is recorded. This test is used in many standards including BS 5438, EN ISO 6941, FAR Part 25, etc. A schematic diagram of a typical vertical strip test is shown in Fig. 12.1. For horizontal and inclined fabric orientations (e.g. 45°, 60° etc.), edge ignition is often preferred.

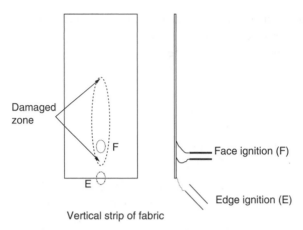

Damaged zone

F

E

Vertical strip of fabric

Face ignition (F)

Edge ignition (E)

12.1 Schematic representation of a simple vertical strip test.

The minimum flame application times determined using Test 1 of BS 5438:1989[14] for ignition of selected different fabrics are given in Table 12.1. Initial analysis[6] of ignition time data in Table 12.1 suggests that the minimum flame application times are very similar for most of the fabrics in spite of different fibre contents, and that the time to ignition is lower for edge ignition in both warp and weft directions than the respective face ignition times. It is also evident from Table 12.1 that the time to ignition is directly related to area density. All fabrics ignite after 1–4 seconds, indicating their respective ease of ignition. This test method appears to distinguish ignitability of fabrics more on the basis of physical factors (relating to area density and specimen orientation) than on fibre chemistry for the examples listed. This can result in significant change in the ranking order of fabrics if the specifications such as physical form and orientation of fabrics are altered. Therefore, for a fuller understanding of the flame initiation process and the response of a material to it, the authors explored experimental methodology to determine the ignition temperature sensitivity of various fabrics. Details of this study are beyond the scope of this chapter, however, and are discussed elsewhere.[15]

For materials with a dripping tendency, the igniting flame burner can be inclined at 45° so as to avoid flame extinguishment by molten drips from the specimen. Such a test method may include a basket with filter paper placed under the vertically mounted specimen (see Fig. 12.2) to judge the hazard of a material burning with flaming drops.

12.4.2 Flame spread

Rate of flame spread is usually calculated by measuring the distance and recording the time taken of the advancing flame front to sever threads

Table 12.1 Minimum ignition times using Test 1 of BS 5438

Fabric sample	Minimum flame application time, s			
	Warp direction		Weft direction	
	Face ignition	Edge ignition	Face ignition	Edge ignition
Light cotton	2	1	2	1
Heavy cotton	4	1	4	1
Poly-cotton (55:45)	3	2	1	1
Poly-cotton (65:35)	2	1	2	1
Polyester	*	*	*	*
Acrylic	2	1	2	1
Light silk	2	2	2	2
Heavy silk	†	3	†	3
Wool	3	3	4	3

*Fabric melted away from the flame.
†Flames extinguished when the flame was moved away.

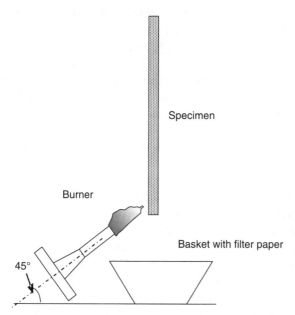

12.2 Vertical strip test for dripping materials with inclined burner.

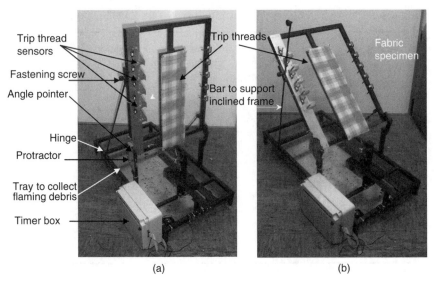

Trip thread sensors

Fastening screw

Angle pointer

Hinge

Protractor

Tray to collect flaming debris

Timer box

Trip threads

Bar to support inclined frame

Fabric specimen

(a)   (b)

12.3 Flame spread test rig: (a) vertical; (b) inclined.

placed at defined distances by the flame front. The flame spread test apparatus is shown in Fig. 12.3. The frame supporting the sample can be hinged (see Fig. 12.3(b)) such that the frame can be moved from the vertical to any other required angle. The frame can be fixed in an inclined position by fastening the screw on the supporting bar.

The upward fire spread is far more rapid than downward and horizontal flame spread and hence is adopted as a better means of measuring the fire hazard of a fabric. Therefore, most standards, including BS 5438:1989, standards for curtains and drapes (see Section 12.5.5) and BS EN ISO 15025:2002, use this type of bench-scale test method for measuring vertical flame spread properties of fabrics in particular.

However, fabrics behave differently in different orientations, depending upon their composition and structure. Table 12.2 shows average rate of flame spread for various fabrics at different inclinations. For each fabric, the rate of flame spread decreases as the angle of inclination reduces to 0° as expected. Further analysis[6] of the data in Table 12.2 has shown that the nature and rate of flame spread vary with the angle of sample inclination, area density and fibre content. This is in agreement with theories of flame spread which state that the phenomenon of flame spread is controlled by the mechanism by which heat is transferred ahead of the burning zone, which in turn is strongly influenced by surface geometry and inclination as well as fibre or material type.

The flame spread test in horizontal orientation is relevant for applications where the textile material is used in flooring, ceilings or any other less hazardous horizontal applications. In this test method, the free end of a hori-

Table 12.2 Average rate of flame spread (m/s) for fabrics at different angles of inclination

Fabric sample	90°	60°	45°	30°	15°	0°
Light cotton	57	40	37	30	18	6
Heavy cotton	27	19	18	14	10	3
Polyester:cotton (55:45)	39	30	27	22	19	8
Polyester:cotton (65:35)	37	27	24	21	13	9
Polyester*	–	–	–	–	–	–
Acrylic	23	15	13	11	8	6
Light silk[t]	–	–	–	–	–	–
Heavy silk	–	–	–	–	–	–
Wool[s]	23	14	12	10	8	–

*The fabric did not ignite.
[t]The flames extinguished on removal of ignition source.
[s]The flames extinguished on removal of ignition source in horizontal orientation of specimens.

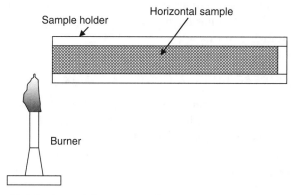

12.4 Horizontal flame spread test.

zontal sample is exposed to the low-energy flame for times up to 15 s in a combustion chamber (see Fig. 12.4). If the sample ignites, then the time to self-extinguish the flame or the time in which the flame passes a measured distance is recorded. This test method is used in various standards for determining the horizontal burning rate of materials used in the occupant compartment of road vehicles, typified by US standard FMVSS 302, BS AU 169a:1992 and ISO 3795:1989.

In the case of flame retarded textiles which are usually tested in vertical orientations, the flame spread is recorded as extent of damage as a hole, char length or weakened (damaged) length measurement of the specimen, and the test is often termed the limited flame spread test. Figure 12.5 shows a Kevlar® (DuPont) aramid and a flame retarded polyester:cotton (70:30) blend fabric specimen after testing in accordance with Test 2 of BS 5438:1989.

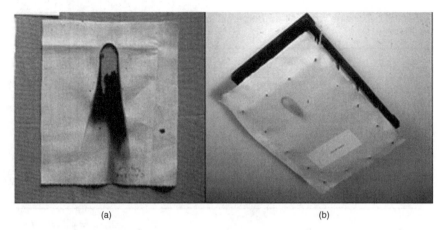

(a)   (b)

12.5 (a) Kevlar® aramid fabric and (b) flame-retarded polyester:cotton (70:30) blend fabric.

Note that the latter is hardly damaged as expected for such a high performance fibre, while the former has formed a hole and significant char accompanied by sooty smoke deposits.

12.4.3 Flame spread under external heat flux

Measurement of flame spread under external heat flux is necessary where thermal radiation is likely to impinge on the textile materials, for example in the flooring material of a building or in transport vehicles whose upper surfaces are heated by flames or hot gases, or both. This situation is usually encountered in a fully developed fire in an adjacent room or compartment. To simulate this scenario, the radiant panel test typically involves a horizontally mounted test specimen positioned at an angle to the radiant heat source shown in Fig. 12.6. The specimen is exposed to radiant heat from an air- or gas-fuelled radiant panel and the textile fabric specimen is at an angle typically of 30° to the panel face. The mounted specimen is thus exposed to a gradient of heat flux ranging from a maximum of 10 kW/m^2 immediately under the radiant panel to a minimum of 1 kW/m^2 at the far end of the test specimen, remote from the panel. The specimen closest to the panel is often ignited by a small flame and the distance burned until the flame extinguishes is converted into an equivalent critical radiant flux in W/m^2 related to the panel intensity at that point. This test method is the basis of that used by the FAA for assessing flammability of textile composites used in thermal/acoustic insulation materials (FAR 25.856(a)) used in aircraft and has also been included by the EU for fire test approval of floorings such as prEN ISO 9239 and BS ISO 4589-1.

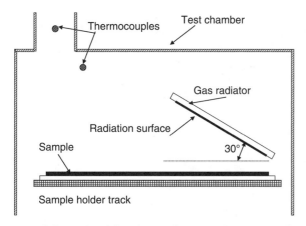

12.6 Schematic of flooring radiant panel test apparatus.

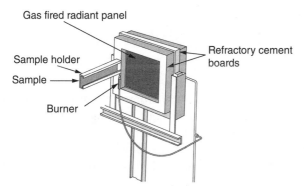

12.7 Radiant panel test for floor coverings as described in BS 476 Part 7.

For textile materials used as interior wall-coverings in UK buildings, including railway carriages, where the fabric could be in a vertical orientation attached to the wall panel, measurement of rate of flame spread under external heat flux is one of the requirements. For such applications, the test method (BS 476 Part 7) essentially requires a vertically oriented specimen (see Fig. 12.7) exposed to a gas-fired radiant panel with incident heat flux of 32.5 kW/m² for 10 min. In addition, a pilot flame is applied at the bottom corner of the specimen for 1 min 30 s and rate of flame spread is measured. The same principle is used in the French test for carpets, NF P 92-506.

In the French suite of test methods (NF P 92-501–506) for testing building materials, the presence of a radiant panel is a significant test feature.

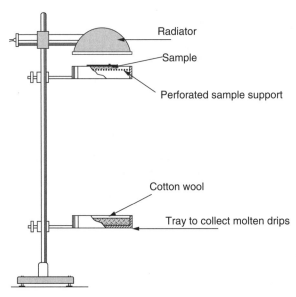

12.8 Dripping test with radiant heat flux.

Normally if the material shrinks away from the ignition source or away from the vicinity of the radiator without burning, in NF P 92-506, for example, then such materials can be further tested for dripping. The dripping test is defined in NF P 92-505 and is a complementary test to determine burning drops which cannot be assessed in the primary test.[16] The test is shown in Fig. 12.8 and the specimen supported on a horizontal grid is exposed to incident heat flux of 30 kW/m². For the specimen to pass the test, the materials should not melt, drip or ignite the cotton wool placed under the specimen holder.

One other particularly important test within this French suite that uses radiant heat flux is the NF P 92-503 Brûleur Electrique or 'M' test for flexible textile materials. The schematic of the test apparatus is shown in Fig. 12.9. The fabric sample is inclined at 30° to the horizontal and is subjected to a radiant heat flux for 5 min and a flaming ignition source is applied to the heated fabric. Time to ignition or time to hole formation, the presence of burning droplets and the length of the damaged specimen are recorded in order to classify materials from M1 to M4, where M1 textiles may be classed as non-flammable, M2 as low flammable, M3 as moderately flammable and M4 as highly flammable. While this test is mainly used in France, Belgium, Spain and Portugal to certify the use of flexible materials in buildings for public use, it affects many UK and other EU manufacturers supplying into EU markets.

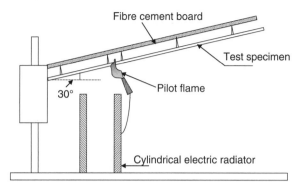

12.9 NF P 92-503 electric burner (brûleur) test.

12.4.4 Heat release tests

When assessing materials for use in buildings and transport, heat release rate is one of the most important parameters characterising the hazard from unwanted fires and is an indicator of:

- rate of fire growth
- size of the fire
- skin injuries from potential fires
- effectiveness of fire suppression agents and their application rates for fire control.

Since textiles may comprise part of a building or transport structure, textile materials, usually part of a building composite, are often subjected to heat release testing. Fire calorimeters to determine the heat release rate during burning of materials operate on a variety of principles, including sensible enthalpy (temperature rise) of the gas stream or enclosure and analysis of the combustion gases for excess carbon dioxide or depleted oxygen.[17] One of the original, successful small-scale calorimeters was the rate of heat release test apparatus developed at Ohio State University (OSU) as shown in Fig. 12.10.[16] The sample is exposed to a heat flux of 35 kW/m² generated by silicon carbide heating rods and a pilot flame is applied on the lower end. The rise in the temperature of the fire effluents is measured using thermocouples and the heat release rate is computed from the temperature rise of the air flowing past a 150 mm × 150 mm burning specimen. This apparatus is defined in the aviation standard FAR 25.853 Part IV Appendix F and ASTM E906-1983 for determining the heat release of internal structural materials in commercial aircraft originally in the US and now worldwide. All decorative textiles fixed to wall panels in aircraft must be tested to this standard.

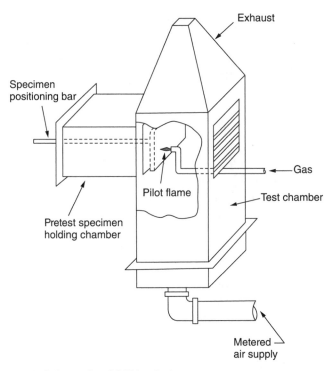

Exhaust

Specimen
positioning bar

Gas

Pilot flame

Test chamber

Pretest specimen
holding chamber

Metered
air supply

12.10 Schematic of OSU calorimeter.

A more sophisticated bench-scale equipment which measures heat release rate by oxygen consumption is the cone calorimeter (see Fig. 12.11). In this test, the fabric or composite specimen mounted over an insulating ceramic blanket is exposed to an external heat flux (0–75 kW/m^2). The volatiles released from the heated specimen are ignited using a spark igniter and the time taken to ignite the gases is recorded as the time to ignition of the specimen tested. Originally designed to study the fire characteristics of building materials which are physically and hence thermally thick, the cone calorimeter can now be used for thermally thin materials such as fabrics,[6] although not yet as a standard test. It has also been used for characterising furnishing fibres which incorporate the samples in a composite fabric/filling form (for example, an upholstery fabric on top of a polyurethane foam).[18] The sample preparation for the cone calorimetric experiment on fabric specimens is shown in Fig. 12.12.

Since textile fabrics are dimensionally as well as thermally thin materials, a number of refractory ceramic blankets are used to form a 13 mm thick layer to separate the lower sample surface and/or backing material from the bottom of the sample holder. However, testing single layer textile materials does pose a major challenge. This is because the fabrics are very light-

12.11 Cone calorimeter apparatus.

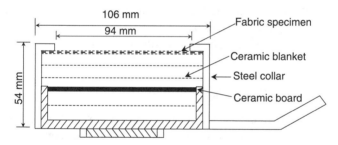

12.12 Sample holder for cone calorimeter.

weight and physically thin, and often tend to shrink and even melt during the course of burning. Even for non-thermoplastic fibres, this shrinkage is estimated[19] at a maximum of 15% of the original surface area and is often accompanied by a distortion of the specimen in the vertical plane. A 3×3 stainless steel (rod diameter 1 mm) crosswire assembly (see Fig. 12.13) can be used for maximising specimen reproducibility, especially in the case of the single-layered specimens.[6]

The cone calorimeter measures the following parameters during a test:

- Sample mass
- Time to ignition
- Flow rate and temperature of the exhaust gases
- Oxygen and carbon oxides (CO and CO_2) in the exhaust gases
- Smoke density.

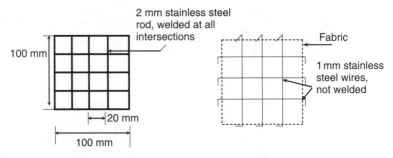

12.13 Sample preparation for cone calorimeter.

By using the above information, calculations are performed to produce the following:

- Heat release rate (HRR)
- Total heat released (THR)
- Mass loss rate (MLR)
- Specific extinction area (SEA), i.e., the amount of smoke produced per unit mass of sample consumed
- Carbon oxides (CO and CO_2) consumed for unit mass of sample.

12.4.5 Mannequin (or manikin) tests

The major means of determining the fabric burn hazard of textiles, and more particularly clothing assemblies, has been by the use of sensored mannequins. Sensor systems used on mannequins have ranked from paper which turns black at certain temperatures[20] to sophisticated sensor systems which indicate the depth of burn.[21,22] The history and development of mannequin tests has very recently been reviewed by Camenzind *et al.*[23] A typical mannequin is equipped with 100–122 individual heat-flux sensors distributed over the surface of the body. The test garment is placed on the mannequin at ambient temperature conditions and exposed to intense fire or flash fire simulation sources with controlled heat flux, duration and flame distribution. The sensors measure incident heat flux upon the underlying mannequin surface during and after exposure. The changing temperature of the mannequin surface temperature simulates damage to human tissue at two skin thicknesses, one representing a second degree burn injury point and the other a third degree burn injury point.[24] The computerised data acquisition system also calculates surface heat fluxes, skin temperature distribution histories and predicted skin burn damage for each sensor location.

These tests offer a more analytical means of assessing apparel burning hazards and predict potential skin burn hazards of garments. They are useful in research but are very expensive and complex for use in standard test procedures,[25] although a new draft standard ISO/DIN 13506.3 is currently being assessed. A major problem has been poor reproducibility, because a major variable is correct clothing assembly fit over the mannequin's torso, even when garment sizes are identical. This is particularly the case in firefighters' clothing where the overlap between top coat and trousers may present a thermal weakness during flame exposure. However, improved reproducibility is being achieved at a level sufficient to enable the test method to be included within the overall performance requirements of firefighter turnout suits within the near future.[23] Notwithstanding these problems, mannequins are considered to be important in simulating the burn injury severity in real-life accidents.

12.4.6 Full product or composite tests

Furniture is a complex product comprising many different materials, including the cover material which is often a textile, a filling or cushioning material which is often polyurethane foam, and the supporting metal or wooden frame. When in contact with an igniting source, for example a smouldering cigarette or a small match flame, the covering fabric can char, melt or catch fire. If the fabric forms a smouldering char, this may generate considerable heat accumulation over a period and subsequently spread into the filling material. The covering fabrics may also melt away from the ignition source, thereby exposing the underlying cushioning foam or filling. Smouldering may have two consequences as it penetrates into the filling: either the intensity of the ignition source can be reduced and oxygen presence diminished thus extinguishing the fire, or the filling material can catch fire and the whole assembly burst into flames. In case of an open flame, if the cover fabric catches fire, it can act as a high-intensity secondary ignition source leading to ignition of the underlying materials. Thus the burning behaviour of the whole composite system depends on the burning behaviour of individual components as well as the interaction between two components, namely the covering material and the filler material.[26]

Bench-scale laboratory tests for individual components are suitable for screening new materials and for quality control purposes in the manufacturing industry and, more importantly, for establishing regulatory performance. However, these component tests cannot assess the fire hazard posed by the upholstered composites, and hence composite flammability tests are essential to measure the fire performance under end-use conditions. Small-scale composite tests present a clearer picture of the now well-established BS 5852 (and its subsequent ISO and EN versions), and small-scale testing for furnishing composites is a prime example.

Table 12.3 Flammability and heat transfer test methods for firefighter's clothing, BS EN 469:1995

Property tested	Standard	Principal performance specifications
Flame spread	BS EN ISO 15025:2002	No flame extending to top or edge, no hole formation, and after-flaming and after-glow times of ≤ 2 s
Heat transfer (flame)	BS EN 367:1992 (ISO 9151)	$HTI_{24} \geq 13$ s
Heat transfer (radiant)	BS EN ISO 6942:2002 at 40 kW/m^2	$RHTI_{24} \geq 22$ s
Residual strength	ISO 5081 after BS EN ISO 6942 Method A at 10 kW/m^2	Tensile strength ≥ 450 N
Heat resistance	BS EN 469:1995 Annex A	No melting, dripping or ignition
Dimensional change	ISO 5077	$\leq 3\%$
Contact heat transfer	BS EN 702:1995 (ISO 12127)	Defined by manufacturer/ consumer
Mannequin (optional)	ISO/DIS 12127	Defined by manufacturer/ consumer

Flammability tests for upholstered furniture using various ignition sources were first developed in 1979 as British Standard BS 5852: Parts 1 and 2 in the United Kingdom. Source 0 is a smouldering cigarette and Source 1 a small butane flame which simulates a lighted match. Sources 2 and 3 are more intense flame sources (see Table 12.9). The test specimen is a composite specimen consisting of fabric and filling material as shown in Fig. 12.14. A similar composite specimen assembly is used in the US standard Cal TB 116 for testing flame retardance of upholstered furniture. In addition to the smouldering cigarette ignition source, BS 5852:1979 and its subsequent variants define the use of a variety of pinewood cribs (Sources 4–7) which match the calorific outputs of increasing numbers of full-size newspaper sheets. Figure 12.15 shows composite test specimens with wooden crib Source 5 as ignition source.

Development of BS 5852 as a small-scale composite test was a breakthrough in realistic model testing that cheaply and accurately indicated the ignition behaviour of full-scale products of a complex nature.[27] Good reproducibility, cost-effectiveness and easy to use features of BS 5852 have led to the establishment of the concept, which was further employed for flam-

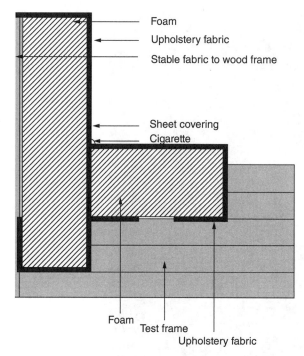

Foam

Upholstery fabric

Stable fabric to wood frame

Sheet covering
Cigarette

Foam
Test frame
Upholstery fabric

12.14 Composite specimen assembly for flammability test of upholstered furniture.

mability testing of bedding and mattresses.[13] The presence of bedcovers, including sheets, blankets, bedspreads, valances and quilts, together with pillows and pillow cases, introduces many interacting variables that affect the fire behaviour of mattresses. Therefore, the test methods developed for determining the ignition resistance of mattresses are composite tests including bed sheets. Assessment of ignitability of mattresses is generally carried out with cigarettes on the bare mattress and sandwiched between two cotton sheets over the mattress. For the mattress to pass the test, flaming combustion should not occur and the char development should typically not be more than 50 mm in any direction from the cigarette. Specific flammability test methods for bedding mattresses are described in Section 12.5.7.

Other textile products that require composite flammability testing are protective clothing assemblies including firefighters' suits, military flight suits, etc. The test methods specific to the flammability of protective garments will be discussed below. The major performance requirement of such clothing demands protection from high-heat flux thermal exposures. Flammability standards and test methods for textile components in protective clothing have been discussed by Bajaj[28] and very recently by

12.15 Composite specimen with wooden crib as ignition source.

Horrocks.[29] Their reviews suggest that standard bench-scale tests for separate determination of flame resistance, thermal insulation/protection and heat resistance may be undertaken on a single fabric or a composite form in a manner that reflects a real application or product requirement. Nowadays, the trend is to provide an overall set of performance specifications for a given type of protective clothing item or assembly. For instance, the performance of firefighters' clothing defined in BS EN 469:1995 (updated 2005) is a composite of tests which measure the different hazards of open flame, hot surface or radiant heat exposure and also offers an optional mannequin test. These are listed in Table 12.3. A similar performance standard exists for protective clothing and other occupations such as workers exposed to heat, BS EN 531:1995.

12.5   Textile flammability standards

Test methods defined in early standards were widely used for general fabrics until 1970 when it was realised that different end-uses required different test methods. Nearly every country had its own set of textile fire testing standard methods. Fabric flammability regulations in different countries are often similar in spirit, but they differ in detail. Since 1990, within the EU in particular, some degree of rationalisation has been underway as 'normalisation' of individual EU member states' testing methods in order to avoid differences in the national standards which could lead to technical barriers to trade. The development of European Standards is thus binding on all community members. The European Committee for Standardisation (CEN) is responsible for publishing European Standards, prefixed by the letters EN which are intended to replace national standards within the European Union, e.g. BS EN, NF EN, DIN EN. Within the US, ASTM is still the overarching standards organisation but increasingly ASTM, ISO and EN standards are becoming similar, if not equivalent, for certain test methods.

12.5.1 Nightwear and apparel

In-depth analysis[6,15,30] of clothing-related fire statistics in the UK show that nightwear is more likely to cause deaths than any other clothing item in spite of the UK Nightwear (Safety) Regulations, 1985.[31] In 1945, legislation regarding flammability of fabrics was brought into force for the first time in California. During the past three decades, governments of the US, Canada, the UK, Australia and several European countries have enacted legislation aimed at reducing the hazards of burning apparel fabrics. Fabric flammability standards and test methods in different countries are summarised in Table 12.4. The Dutch and French standards in Table 12.4 are derived from an ISO standard (ISO 6941) which determines flame spread properties and ISO 6940 which determines ease of ignition. The Swedish and Norwegian regulations are based on the American ASTM 1230 standard test method for flammability of clothing textiles.

This issue of garment flammability has been recognised in the UK since the 1960s when regulations, revised in 1985, were first applied to children's and subsequently to all nightwear.[31] The subsequent UK Nightwear (Safety) Regulations 1985 require the testing of all nightwear, including pyjamas and dressing gowns, and demand that adult and children's nightwear carry a permanent label showing whether or not each item meets the requirements of BS 5722:1984[32] (which uses Test 3 of BS 5438:1976). This latter performance standard defines a maximum permissible burning rate of a vertically oriented fabric. However, it fails to regulate the fabrics according to their heat release properties, which are considered to be more realistic when the burning hazard is to be correlated with the burn injury severity.[33]

In the EU, a new standard that addresses the fire safety hazards associated with children's nightwear has been issued.[34] The flammability testing methods required by EN 14878:2007 are based upon EN 1103 which

Table 12.4 Selected test standards for nightwear (safety) regulations

Country	Test standard	Testing		Type of test	Ignition time, s	Frame V = vertical	Label
		Face (F)/ edge (E)	Flame height, mm				
Germany, France	EN ISO 6940	F-E	40	Ignitability	0–20	ISO, V	–
Germany, France	EN ISO 6941	F-E	40	Flame spread	10	ISO, V	–
	EN1103	F	40	Flame spread Flash Flaming debris	10	ISO, V	–
Ireland	IS 148	F	45	Flame spread	10	BS, V	Yes
UK	BS 5722 method 2	F	45	Limited flame spread	10	BS, V	Yes
UK	BS 5722 method 3	F	45	Flame spread	10	BS, V	Yes
Australia	AS 2755/2	F-E	40	Flame spread	5–15	ISO, V	Yes
Denmark, Finland, Iceland, Norway, Sweden	NT FIRE 029	F	16	Flame spread	1–20	45°	–
Germany	Dutch convenant	F	40	Flame spread Flash Flaming debris	5 spread 1 flash	ISO	Yes
USA, Norway, Sweden	ASTM D1230	F	16	Flame spread	1	45°	–

describes a detailed procedure to determine the burning behaviour of textile fabrics for apparel. The standard covers all types of nightwear, including nightdresses, nightshirts, pyjamas, dressing gowns and bath robes. It also places responsibility on the manufacturer to ensure that any flame retardant chemicals used are effective throughout the life of the garment and do not present a health hazard.

12.5.2 Textiles for protective clothing

As discussed above, tests for specific hazards such as ignition resistance and thermal exposure are well defined individually but may also be grouped together within a single end-use-related performance specification such as BS EN 469 for firefighters' clothing (see Table 12.3). The individual test methods for protective clothing in general are largely based on assessing the resistance of a fabric when tested in a specific geometry (e.g., horizontal, 45° or vertical) and subjected to a small flame igniting source, which usually is a small gas flame applied to the lower edge or face close to the edge of the sample. Parameters measured include time to ignition, rate of flame spread, afterflame and afterglow following a prescribed ignition time (e.g. 10 s), extent of damaged/char or burnt fabric length, or a combination of these. Some of the standards for protective clothing are briefly summarised in Table 12.5. In addition to those in Table 12.3 defined within BS EN 469:1995, an additional example of such an overall performance require-ment is the US NFPA 2112 standard for protection of personnel against flash fires. This is presented in Table 12.6.

The most important aspect of protective clothing testing is the evaluation of burn injury protection and thermal characteristics of clothing systems. Skin burn injury evaluation and subsequent modelling has been studied extensively[35–37] and has been recently reviewed by Song.[38] An overall fire resistance determined under realistic real-fire exposure conditions may be undertaken using an instrumented mannequin (see Section 12.4.4) yielding information regarding a complete clothing system's ability to resist heat flux and protect areas of the torso to first, second or third degree burn injury as described earlier. Resistance to heat transfer by convective flame, radiant energy or plasma energy sources is quantified in terms of thermal protective index (TPI), often related to the time taken for an underlying skin sample with or without an insulating air gap to achieve a minimum temperature or energy condition sufficient to generate a second degree burn (see Tables 12.3 and 12.6).

To evaluate flame and thermal protective performance of a fabric or assembly at a heat flux simulating defined thermal hazards, for example battlefield flame, methods which determine a thermal protective perfor-mance or TPP exist.[29] The instrument measures the time for a thermocouple

Table 12.5 Standards and performance requirements for protective clothing

Standard code	Standard title	Property measured	Performance requirement
ISO 2801:1998	Clothing for protection against heat and flame – General recommendations for selection, care and use of protective clothing	–	–
BS EN ISO 6942: 2002 at 40 kW/m²	Protective clothing – protection against heat and fire. Method of test: Evaluation of materials and material assemblies when exposed to a source of radiant heat	Heat transfer (radiant)	$RHTI_{24}$* ≥ 22 s
ISO 9151:1995 BS EN 367:1992	Protective clothing against heat and flame – Determination of heat transmission on exposure to flame	Heat transfer (flame)	$HTI_{24}^{\dagger}$ ≥ 13 s
ISO 11612:1998 BS EN 469:2005	Clothing for protection against heat and flame – Test method and performance requirements for heat-protective clothing	Heat resistance	No melting, dripping ignition
ISO/DIS 12127:1996 (BS EN 702:1995)	Clothing for protection against heat and flame – Determination of contact heat transmission through protective clothing or constituent materials	Contact heat transfer, manikin (optional)	Defined by manufacturer/customer
ISO 17492:2003 Cor 1:2004	Clothing for protection against heat and flame – Determination of heat transmission on exposure to both flame and radiant heat	–	–

BS EN ISO 15025:2002	Protective clothing – Protection against heat and flame – Test method for limited flame spread	Flame spread	No flame extending to top or edge, no hole formation, and after-flaming and after-glow times ≤2 s
ISO 17493:2000	Clothing and equipment for protection against heat – Test method for convective heat resistance using a hot air circulating oven	–	–
ISO 5081 after BS EN ISO 6942 Method A at 10 kW/m^2	The determination of the breaking strength and elongation at break of woven textile fabrics (except woven elastic fabrics)	Residual strength	Tensile strength ≥ 450 N
NFPA 2112	Standard on flame-resistant garments for protection of industrial personnel against flash fire	–	–
ASTM F 1930, ISO/DIS 13506	Test method for evaluation of flame resistant clothing for protection against flash fire simulations using an instrumented mannequin	Burn injury prediction	–

* RHTI = radiant heat transfer index.
† HTI = heat transfer index.

Table 12.6 Test methods and performance requirements for flame-resistant garments meeting requirements of NFPA 2112

Property tested	Test method	Application of test method	Performance requirements
Thermal protective performance	152 × 152 mm of specimen is exposed to heat and flame source with heat flux of 84 kW/m². The amount of heat transferred through the specimen is measured using a copper calorimeter. The test measures the time taken to transfer amount of heat sufficient to cause second degree burn. This time multiplied by the incident heat flux gives TPP rating.	This test is used to measure the thermal insulation provided by garment materials. The TPP test uses an exposure heat flux that is representative of a flash fire environment.	TPP rating of 3 or more when tested in 'contact', simulating direct contact with skin, and 6 or more when measured 'spaced', simulating an air gap of 6.35 mm between the skin and the garment material.
Flame resistance	76 × 30 mm specimen is held vertically over a small flame for 12 s. After-flame time, char length and length of tear along the burn line are measured. Melting and dripping are also observed.	This test is used to determine ease of ignition and ease of flame spread.	After-flame time ≤2 s, char length ≤102 mm, no melting or dripping.
Thermal shrinkage resistance	381 mm² fabric specimen is suspended in a forced air-circulating oven at 260°C for 5 min to determine amount of shrinkage. The specimen is examined for evidence of melting, dripping, separation or ignition.	The test measures resistance to shrinkage of a fabric when exposed to heat, since this property is considered important in minimising the effects of a flash fire.	Shrinkage ≤10%.

Test	Description	Requirement
Heat resistance	Same as above.	The specimen should not melt, ignite or separate when exposed to heat.
Mannequin testing	Standardised coverall design placed on an instrumented mannequin wearing cotton underwear is subjected to an overall heat and flame exposure averaging 84 kW/m^2 for 3 s. Sensors embedded in the manikin's skin predict occurrence of second or third degree burns. Percentage of body sustaining second or third degree burns is determined using computer program.	This test provides an overall evaluation of how the fabric performs in a standardised coverall design.	Body burn rating ≤50%. Lower body burn ratings indicate greater protection provided by the fabric.
Thread melting resistance	The test involves soaking of the thread used in stitching FR garment in an organic solvent to extract substances that would interfere with the melting of thread. Melting temperature is determined by slowly heating the thread.	Measures the melting temperature of the thread used in flame-resistant garments.	Thread fails the test if melting temperature <260°C.
Label legibility	Sample labels containing product information are subjected to 100 wash/dry cycles and then examined for legibility.	This requirement checks for label durability.	Label must remain legible from 0.3 metres.

placed behind a fabric or assembly to reach a critical temperature equivalent to that causing a radiant energy source.

For fabrics used in flame-resistant garments, a thermal protective performance (TPP) of 3 s or more is required when tested in the 'contact' condition, simulating direct contact with skin, whereas a TPP value of 6 s or more is required when tested in a 'spaced' condition, simulating an air gap of 6.35 mm (0.25 in) between the skin and the garment material.

A similar bench-scale experimental setup is used for test methods described in EN ISO 6942 and EN 367:1992 (see Tables 12.3 and 12.5) for measuring radiant heat transfer index (RHTI) and convective heat transfer index (HTI) respectively. RHTI values of ≥ 22 s and HTI values of ≥ 13 s are required for materials used in protective clothing.

Early full-scale fire-testing of flame-resistant garments initially used a fully dressed mannequin exposed to open-pit fuel fires. However, nowadays, as stated in Section 12.4.4, the military uses a state-of-the-art instrumented mannequin and an environmentally controlled chamber. In the ISO/DIS 13506 standard[23] method for protective clothing, a dressed mannequin is subjected to a full flame exposure with gas burner flames of about 800°C and heat flux of about 80 kW/m² for 8 s. The mannequin test method in ASTM F1930 for flash fire resistance requires exposure of a fully dressed mannequin to a heat flux of 84 kW/m² for 3 s. According to the NFPA 2112 standard (see Table 12.6), the performance requirement for the materials to be used in flash fire-resistant garments is that the body burn rating should be $\leq 50\%$ when tested in accordance with ASTM F1930. This test method is being recommended for adoption as the new military standard to evaluate military flame-protective materials and clothing systems.

12.5.3 Structural fabrics

Tents and marquees are intended to accommodate large numbers of people and, therefore, the fabrics used for such structures must be flame retardant. The primary requirement for the fabrics to be used in tented structures is that they do not readily ignite or produce flaming debris. In the UK the fabrics used for such purposes should either be inherently flame retardant or have a durable flame retardant treatment. The early British standards (BS 6341:1983) for tented structures required the fabrics to be tested in a vertical orientation with bottom edge-ignition as shown in Fig. 12.1. The subsequently modified standard BS 7837:1996 specified mandatory water soaking of the fabric specimen prior to testing. The test method was similar to the one described in BS 6341 except for the addition of filter paper under the test apparatus to catch any flaming debris (see Fig. 12.2).

More stringent standards (BS 7157:1989) use the range of pinewood cribs of various dimensions as ignition sources previously mentioned in Section 12.4.6 and listed in Table 12.9. The fabric specimen is mounted on the test frame in the form of a mini-tent with sloping roof, three side walls, an open front and no flooring element. Depending on the end-use, the ignition source (Sources 4 to 7 as described in BS 5852:1979 Part 2) is placed in one of the corners of the mini-tent at floor level. The test records the progress of combustion and formation of flaming drips. The most recent standard for tented structures, BS EN 14115:2002,[39] requires exposure of the test specimens (600 mm × 180 mm) to radiative heat. Additional hot gases are blown over the surface of the specimen to encourage any spread or propagation of the flame, thus making the test even more stringent. A flame is used to ignite any emitted gases. The effects of ignition and hot flowing gases on the extent of damage are measured.

Nordic countries have specific test procedures for testing fabrics used in the construction of tents. Swedish standard SIS 65 00 82 specifies a test method for testing the ignitability of fabrics for such constructions. The test method is a simple vertical ignition test (see Fig. 12.1) using a diffusion flame of length 38 mm.

In the US, the building code agencies use standards defined by ASTM and NFPA to evaluate the fire-performance characteristics of structural fabrics. The NFPA-701[40] test method is based on the ease of ignition test described earlier (see Section 12.4.1). For the fabric to pass the test, it must self-extinguish within 2 s after removing the flame. The ASTM E84[41] tunnel test evaluates the flammability of a membrane fabric on the underside when in the horizontal position, as would exist for the roof area. The rate of flame spread and smoke formation during a 10 min fire exposure is measured by visual observations. The ASTM E108 test method[42] evaluates the flammability of a membrane fabric and its resistance to penetration by fire on the outer surface. Fabrics for membrane structures also need to be classed as non-combustible materials according to the ASTM E136 test method,[43] which measures flaming, temperature rise and weight loss in a vertical tube furnace at 750°C.

Australian standard AS 1530.2 for flammability of fabrics used as building materials uses the flammability apparatus shown in Fig. 12.16.[44] The test method is quite comprehensive as it measures ignitability, rate of flame spread (expressed as speed factor), temperature of fire gases (expressed as heat factor) and finally the flammability index, which is computed from speed and heat factors. The sample holder is a slightly convex frame inclined at 3–4° to the vertical and the sample is ignited by burning 100 ml pure alcohol in a copper container placed 13 mm below the sample. The temperature of the fire gases is measured by thermocouples positioned in the exhaust hood, and the temperature of combustion gases versus time curve

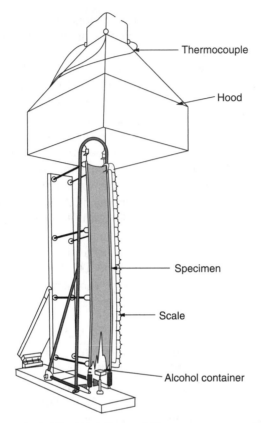

Thermocouple

Hood

Specimen

Scale

Alcohol container

12.16 AS 1530.2 flammability test apparatus for fabrics and films.

is plotted. The average area between the recorded temperature curve and the ambient temperature curve over a period of 180 s is determined to compute the heat factor *H*.

12.5.4 Floor covering textiles

Flammability tests for carpets used in residential as well as commercial buildings are intended to assess the ease with which the textile floor covering will ignite if a burning cigarette, hot coal or similar source of ignition is dropped on the flooring material. To simulate such a phenomenon, the primary UK test for carpets as described in BS 6307:1982[45] uses a methanamine tablet of 6 mm diameter and 150 ± 5 mg weight as ignition source, which models a small flame source such as a lighted match. The tablet is placed in the centre of the specimen and ignited. The damaged zone is measured and the flaming and/or afterglow time is also recorded. This test is carried out in controlled laboratory conditions.

The ease of ignitability test for carpets described in BS 4790:1987[46] uses a 30 g hexagonal hot metal nut to model a burning ember ejected by a domestic coal/solid fuel fire as a source of ignition. A stainless steel nut is heated to $900 \pm 20°C$ in a furnace and placed on a specimen in an enclosed chamber. The times of flaming and of afterglow as well as the greatest radius of the effects of ignition from the point of application of the nut are measured. The radius of the affected area should not be greater than 75 mm. The methanamine tablet test is preferred as an acceptance criterion in industry and is also used in the ISO standard (ISO 6925:1982) for textile floor coverings due to the greater consistency of the ignition source. These tests, however, do not give an overall indication of the potential fire hazard under actual conditions of use.

The US ASTM E648[47] and NFPA 253[48] standards for floor covering textile materials are mainly applicable in the construction industry. These standards use the radiant panel test described in Section 12.4.2 to evaluate the tendency of a flooring system to spread flame when exposed to the radiant heat flux. The test method determines a material's critical radiant flux (measured in W/cm^2), i.e. the minimum energy necessary to sustain flame propagation. Various test methods for textile flooring materials defined in standards of different countries are summarised in Table 12.7.

Carpets are used also in many transportation applications including automobiles, buses, rapid transit vehicles, aircraft and marine vessels. In these applications, carpet is not subject to the traditional carpet test methods described above, but subject to the rules and requirements of respective transportation departments or agencies.[49]

For floor coverings in motor vehicles the BS ISO 3795:1989 as well as FMVSS 302 published by the US Office of Vehicle Safety Compliance of the National Highway Traffic Safety Administration describes a test method which measures horizontal rate of flame spread as described in Section 12.4.2. To comply with these standards, the material must not have a burn rate of more than 102 mm/min (4 in/min).

The flammability regulations for carpets in passenger aircraft and ships are probably the most stringent for any interior textile application. In addition to a radiant panel test, the Federal Aviation Regulation (FAR) 14CFR25 (Airworthiness Standards: Transport Category Airplanes) describes a vertical flame test whereby a carpet sample is suspended vertically in a chamber. A propane/butane flame is applied to the lower edge of the specimen for 12 s. After removal of the flame, the after-flame time, burn length, and any flaming drippings are measured and recorded. The minimum requirements under FAR 25.853(b) for crew or passenger compartments require average flaming and glowing times which are listed in Table 12.8. The IMO (International Maritime Organization) Fire Test Procedures

Table 12.7 Test specifications for textile floor coverings

Country	Standard	Sample size, mm	Replicates	Orientation	Ignition source	Flame application time	Test duration	Performance criteria
Germany	DIN 54332	340 × 104 × original thickness	5	Vertical	Small flame burner (20 mm), inclined at 45°	15 and alternatively for 5 s.	20 s	
Germany	DIN-4102-14, EN ISO 9239-1	230 × 1050 × original thickness	3	Horizontal	Gas heated radiation panel, inclined at 30° to horizontal		600 s if no ignition occurs, 1800 s max	
USA	ASTM D 2859-96, 16 CFR, Part 1630.4, ISO 6925:1982, BS 6307:1982	230 × 230	8	Horizontal	Methanamine tablet (0.15 g)	–	Until flame extinguishes or reaches edge of the steel frame	Damaged area should be <25 mm from the steel frame

Country	Standard	Specimen dimensions	No.	Orientation	Heat source	Time	Duration
USA	ASTM 648-99, NFPA 253	254 × 1070 × original thickness	3	Horizontal	Gas heated radiation panel, inclined at 30° to horizontal	300 s	900 s including 300 s preheating
Austria	ÖNORM B 3810	200 × 800	3	Horizontal	Gas heated radiation panel, inclined at 60° to horizontal	Swivelling pilot flame for 120 s	≤1200 s
France	NF P 92-506	400 × 95 × 55 (max)	3	Vertical, Perpendicular to the radiating panel	Gas heated radiation panel, 850°C + gas pilot flame	600 s for radiation	Until extinction of specimen
UK	BS 4790:1987				Hexagonal hot metal nut, 900 ± 20°C		

Table 12.8 Flammability tests for components inside the fuselage

Test	Acceptance criteria						
	Burn length	Rate of flame spread, mm/min	After-glow time, s	Weight loss, %	Heat release rate, kW/m²	Self-extinguishing time, s	Drip extinguishing time, s
Vertical Bunsen burner	≤203 cm	–	–	–	–	≤15	≤5
	≤152 cm	–	–	–	–	≤15	≤3
45° Bunsen burner			≤3			≤15	
60° Bunsen burner	≤76 cm					≤30	≤3
Horizontal Bunsen burner	–	64	–	–	–	–	–
OSU heat release	–	–	–	–	≤65; total heat release in first 2 min ≤65 kW-min/m²		
Seat cushion oil burner	≤432 cm	–	–	≤10	–	–	–
Insulation blanket radiant panel	≤52 cm	–	–	–	–	≤3	–

(FTP) for testing flammability of carpets (46CFR72.05-55) also use the radiant panel test method described in Section 12.4.2.

12.5.5 Fabrics for curtains and drapes

Curtains and drapes are the vertically oriented fabrics used in interior furnishings and hence flammability testing of such fabrics is often carried out by mounting the specimen vertically on the testing rig. The principal UK flammability test uses Test Method 3 described in BS 5438:1976[14] for curtains and drapes, which measures the rate of flame spread of a vertically oriented fabric specimen. For curtains and drapes used for domestic usage, the flame application time is typically 10 s, whereas for curtains used in contract furnishing, the test is more severe with a longer flame application time of 15 s. The flammability test is even more stringent for the curtains and drapes used in more hazardous applications such as hospitals, prisons, etc. For these applications the fabric has to be tested with four flame application times of 5, 15, 20 and 30 s. The flammability requirements for curtains and drapes employed for various applications are given in BS 5867:Part 2:1980[50] which also mentions that 'the fabric complying with the requirements of the present standard may not always withstand exposure to large sources of heat, but it should have some resistance to flame spread following accidental contact with small sources of ignition'. Furthermore, the UK fire statistics[51] have shown that curtains are often ignited by flame spread from other sources such as a fire started in an armchair by a discarded cigarette. In order to address this issue, a test method devised is defined in BS EN 13772:2003,[52] which specifies a method for the measurement of flame spread of vertically oriented textile fabrics using a large ignition source. A heat flux of a defined energy is applied to a specified area of the lower part of the vertical specimen for 30 s. A small propane flame is then applied for 10 s to a small piece of cotton fabric fixed around the bottom edge of the specimen. Flame spread is measured through severance of marker threads.

In the US, the NFPA 701 fire test for flame resistant textiles and films is used for curtains and drapes including all the decorations and trimmings. The test apparatus for measuring flammability of single as well as multilayer fabric assemblies of less than 700 g/m^2 uses a freely hanging test specimen. A methane gas burner is used as the ignition source, which is applied to the bottom edge of a vertically held specimen (150 mm × 400 mm) for the duration of 45 s. Average weight loss and burning time of flaming fragments for 10 replicates are recorded. The National Building Code of Canada (NBC), ULC-S 109, uses a test method which in principle is similar to NFPA 701, except that the intensity of the igniting flame is different. The ULC-S 109 test method also suggests the possibility of applying the ignition flame at an angle of 25° to the vertical. Application of a Bunsen burner flame at an angle would thus prevent flame extinguishment due to molten drips from a dripping specimen.

12.5.6 Pile and fur fabrics

Fancy fabrics, including pile and fur fabrics, are often used in manufacturing filled soft toys, 'fancy dress' or disguise costumes including beards, moustaches, wigs, etc. Flammability testing of toys using hair or protruding fibres is in accordance with BS EN 71-2:2003,[53] which specifies a test method to determine flammability of toys under particular test conditions. The test rig is as shown in Fig. 12.17. The test method uses a butane/propane gas flame (flame height 20 ± 2 mm) as an ignition source. The flame is usually applied for 2 ± 0.5 s to the lower edge of a representative sample. For products with hair and fibres protruding more than 50 mm from the surface, the test flame is applied either vertically or at an angle of 45°. If ignition occurs, the duration of flaming is measured and the length of specimen damaged by flames is noted. For certain types of toys, the rate of flame spread is also measured through severing of marker threads.

In the UK, BS 4569:1983 describes a test method to determine whether or not the pile of pile fabrics and simulated fur fabrics promotes rapid flame spread. The test method in BS 4569:1983 has also been used as the basis for the EN 14878 standard for flammability of children's nightwear as mentioned in Section 12.5.1. The test method is also applicable to assemblies having a pile on the surface. The flammability testing rig for pile fabrics is shown in Fig. 12.18. The test specimen is held vertically in a draught-free enclosure and a flame is moved across the surface of the pile at a known speed to determine whether flame from the ignition source flashes over the

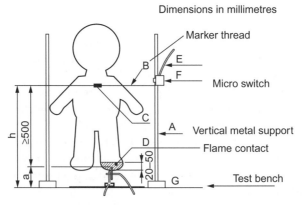

12.17 BS EN 71-2:2003 test setup for measuring flammability of fur toys.

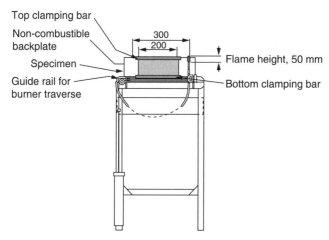

Top clamping bar

Non-combustible
backplate

Specimen

Guide rail for
burner traverse

300
200

Flame height, 50 mm

Bottom clamping bar

12.18 BS EN 4569:1983 flammability test rig for pile fabrics.

surface of the pile. If there is no surface flash on either side of the test specimen, the fabric is considered to have passed the test.

12.5.7 Upholstery and bedding fabrics

Upholstered seating furniture

Flammability testing of upholstered furniture including mattresses is a complex process and fully reviewing the history, development and detail of these test methods is beyond the scope of this chapter. Section 12.4.5 has discussed the general principles behind the current composite testing methodologies used, and work on the development of standards, test procedures and reproducibility of test methods has been published in abundance.[2,54–56] Post-ignition flaming behaviour of full-scale upholstered furniture in general can be classified into four main phases: flame spread, burn through, pool fire and burnout. Among these processes, flame spread occurs mainly through the fabric component of the furniture and ends when the fabric from the underside of the seat cushion has fully burnt out.

The majority of upholstered furniture fires result from ignition by smoking materials with a strong emphasis on cigarettes[54,55] and hence most flammability tests for upholstered furniture use a smouldering cigarette as the primary ignition source, defined as ignition Source 0 in BS 5852 (see Table 12.3 and Section 12.4.5), for example. To simulate the open flame of a burning match, a butane gas flame (gas flow rate of 45 ml/min, applied for 20 s), referred to as ignition Source 1 in BS 5852, is designed to give a similar calorific value. However, in the UK, the upholstered furniture used in contract environments, for example public buildings, restaurants, hospitals, prison cells, etc., is not covered by the Furniture and Furnishings (Fire) (safety) Regulations but is required to meet requirements of the performance standard BS 7176:1995.[57] Requirements of this performance standard are classified for four categories of hazards depending on the type of building in which the furniture is used. For buildings with very high hazard levels, combinations of ignition sources are required to be used for evaluating fire performance of upholstered furniture. For example, upholstered furniture in sleeping accommodation in certain hospital wards and offshore installations has to be tested using three different ignition sources: the smouldering cigarette (Source 0), the match flame equivalent (Source 1) and the pinewood crib (Source 7). The small-scale UK and EN composite test method for upholstered seating is briefly described in Section 12.4.5 and elsewhere.[13]

In California, the Cal TB 133 test method uses oxygen depletion calorimetry for testing flammability of seating used in public occupancies such as public auditoriums, hotels, hospitals, etc. This test method uses a 250 mm^2 tube burner with heat generating capacity of 18 kW. The burner is placed on the seating area for 80 s. This full-scale furniture item is tested in a room and the pass/fail criteria depend on the measurements used, i.e., temperature increase at the ceiling thermocouple or peak heat release rate.

In 2006, the US Consumer Product Safety Commission (CPSC) revised draft standards for performance tests for major upholstery materials.[58] The revised standards include requirements for cigarette and small open flame (SOF) ignition performance of fire barriers and filling materials. The open flame ignition test is not mandatory for the fabrics in upholstered furniture unless the fabric is part of the composite in the end-product test and could contribute to mass loss. All fabric materials, however, have to pass the cigarette test.

Mattresses

Bedding materials have traditionally been combustible materials. Requirements of comfort, convenience and cost will continue to demand the substantial use of textile materials. Fire behaviour of bedding materials has been thoroughly reviewed by Hilado.[59] The California Test Bulletin CAL TB 106 test method for testing ignition resistance of mattresses uses a burning cigarette, whereas in the Cal TB 603 test method the mattress is exposed to large open flames from a gas burner to replicate localised heat flux from burning bedclothes. Two burners with heat flux of 19 kW each are placed on top of the mattress for 70 s and the side burner with 10 kW heat flux is applied for 50 s. The performance criterion is based on peak heat release with maximum allowable peak heat release

Table 12.9 Types of ignition sources specified in BS 5852

Source number	Ignition source	Description	Energy input, kWh	Combustion type	Time of application	Time limits for flaming of composites
0	Cigarette	–	–	Smouldering	Throughout the test	No ignition or progressive smouldering
1	Burner	45 ml/min	0.001	Flaming	20 s	After-flame time up to 120 s
2	Burner	160 ml/min	0.004	Flaming	40 s	After-flame time up to 120 s
3	Burner	350 ml/min	0.016	Flaming	70 s	After-flame time up to 120 s
4	Crib	8.5 g	0.04	Flaming	Throughout the test	Total flaming time up to 10 min
5	Crib	17 g	0.08	Flaming	Throughout the test	Total flaming time up to 10 min
6	Crib	60 g	0.28	Flaming	Throughout the test	Total flaming time up to 13 min
7	Crib	126 g	0.59	Flaming	Throughout the test	Total flaming time up to 13 min

of 200 kW and total heat release rate of 25 MJ for 10 min from the start of the test.[60] In the UK, the standard BS 6807:1996 is used to assess ignitability of mattresses and uses ignition sources specified in BS 5852 (listed in Table 12.9), while BS 7177:1995 specifies various combinations of ignition sources for four different hazard classifications: low, medium, high and very high. The 0/NS (cigarette plus non-smouldering insulation) ignition source is also described in Annex B of BS 7177:1996 to provide guidance for users on the ignitability behaviour of mattresses when covered with bedding. These test methods are not mandatory except in contract furnishings, but they are used for quality control in industry or for development of new products.

12.5.8 Textiles in transportation

Fire and heat resistant textiles find applications in the transportation industry (e.g., seat coverings in commercial aircraft) and do so without sacrificing aesthetic and comfort properties. Despite no-smoking regulations, smouldering fires in transport vehicles often occur as a result of carelessly discarded matches or cigarettes in bins, apertures or hidden spaces.

Upholstered fabrics used in seating of public transport are often the first item to ignite. The risk of fire incidents and hence the flammability standards for textile materials used in road, rail, air and water transport vehicles are discussed in this section. Specific flammability test methods and performance requirements are also mentioned in brief. Most of the textile materials used in mass transport vehicles may also be regulated for smoke and toxic gases but this topic is beyond the scope of this chapter.

Motor vehicles

The flammability testing of fabrics used in motor vehicles has not yet been made mandatory due to the fact that fire incidents in motor vehicles are rare and, moreover, such fire spreads relatively slowly. Flammability standards for textiles used in motor vehicles are usually governed by individual countries, although with globalisation of the automotive industry, major manufacturers subject internal textile materials such as seating covers and carpets to the US FMVSS 302 tests described previously in Section 12.4.2. The Federal Motor Vehicle Safety Standard (FMVSS 302) was brought into force in the US in 1972 and specifies a simple horizontal flame spread test as described in Section 12.4.2. It measures rate of flame spread over a horizontally placed specimen subjected to a Bunsen burner flame for 15 s. This test method has also been adopted by the German (DIN 75 200), British, Australian (BS AU 169) and Japanese (JIS D 1201) automotive standards.

In 1982, specific technical fire protection requirements for motor coaches with more than 16 passengers were introduced in France.[61] The Specification Technique ST 18-502 specifies that the materials used for curtains and blinds must be tested in a vertical orientation to the specifications described in ISO 6940. ST 18-502 also defines a dripping test for the head lining material in coaches to be tested in accordance with NF P 92-505. The test apparatus (see Fig. 12.8) specified in NF P 92-505 uses an electric burner test as described in Section 12.4.2.

The European Union directive 95/28/EC, defining requirements for the fire behaviour of upholstery of seats including the driver's seat, interior lining material and any textile material used for thermal and/or acoustic insulation in certain categories of motor vehicles, was released in 1995. Appendix IV of the directive 95/28/EC describes a test method to determine the horizontal burn rate of the materials and is similar to the FMVSS 302 test method. The lining material is tested for burning drips using the test apparatus described in French standard NF P 92-505 (see Fig. 12.8). For curtains and blinds and/or any other hanging materials in the motor vehicle, a test method similar to ISO 6941 to measure the vertical burn rate is specified.

Rail vehicles

Textile flammability standards are extremely severe because fires in railways can spread very quickly and can result in significant losses. Besides fires caused by technical defects, passenger-induced hazards are also very common.[62] In European countries the Union Internationale des Chemins de Fer (UIC) Code, Sheet 564.2, harmonises test procedures and performance criteria for flammability and smoke production. The Annex of UIC 564.2 describes a test method for textiles in particular. The test measures after-flame time, extent of charred surface and observed dripping when a fabric specimen inclined at 45° is exposed to a Bunsen burner flame for 30 s. Flooring textiles are tested using a fishtail burner with a 42–48 mm flame width. In the UK, the standard for flooring textiles in railways uses the radiant panel test described in BS ISO 4589-1.

Furnishing fabrics used in railway seatings are tested as a complete assembly with a 100 g paper cushion used as an ignition source (Annex of UIC 564.2). In the US, complete seat assemblies are tested according to ASTM E 1537-98. In the UK, ceiling lining materials are tested according to the test method described in Part 6 of BS 476 (see Section 12.4.2, Fig. 12.8). This fire propagation test, described in Section 12.4.2, measures the contribution of the lining material to the growth of fire. If textiles are used as lining materials for walls and ceilings of a passenger car, the flame spread along the surface of the specimen is tested according to Part 7 of BS 476 with the radiant panel described in Section 12.4.3 employed to measure the rate of flame spread (see Fig. 12.7).

Aircraft

Approximately 900 kg of combustible textile material in the form of seat upholstery, decorative textiles, wall coverings, carpeting, tapestries, blankets, curtains and seat belts are used in a modern commercial passenger aircraft. Flammability testing of such textile materials has been regulated by the US Federal Aviation Administration (FAA) under Federal Aviation Regulations (FAR) and these latter extend to all commercial airliners operating across the world internationally.

Test methods for measuring ignitability and flammability of fabrics used in aircraft are, in principle, similar to the ones already described in Section 12.4 of this chapter. However, the performance criteria are often more stringent than for other applications. Tests involving a small ignition source simulate the start of fire in the cockpit or cabin due to electrical faults or overheating during the flight. However, catastrophic fires occur as a consequence of crash landings or crashes when taking off. All textiles present in an aircraft have to pass the test requirements defined in FAR 25.253(b) in which a 10 mm diameter Bunsen burner flame impinges upon the bottom edge of a vertically oriented 75 mm (3 in) wide by 305 mm (12 in) long sample for either 12 or 60 s depending on the requirement. Maximum permeable flame-out and melt drip times are defined following removal of the ignition source, e.g. ≤5 s for flameout after 12 or 60 s ignition and ≤3 s or ≤5 s drip times respectively. Average burn lengths should not exceed 152 mm (6 in) for 12 s and 203 mm (8 in) for 60 s ignition times. Textiles required to be subjected to 12 s ignition include carpets, curtains, upholstery outer fabrics, seat cushions, padding, blankets, coated leather fabrics and galley furnishings. Textiles are required to be subjected to the 60 s condition only if they are part of a composite such as interior wall panels, ceilings, partitions, etc.

For textiles used in seat harnesses and webbings, the above test is used with the sample held in a horizontal orientation and with a 15 s flame application time. The above vertical strip test is used as a prior requirement of any textile used in a composite or product requiring a subsequent, more intense flammability requirement. In this respect, measurement of ease of ignition under high heat flux and post-ignition behaviour of textile materials may be of crucial importance. For example, the oil burner test for seat cushions described in FAR 25.853(a) Appendix F Part II uses a kerosene burner calibrated to produce a heat of 12 kW/m^2 at the seat surface and with flame temperature of around 1000°C. The flame is applied to the complete seat assembly as shown in Fig. 12.19. The flame is applied for 2 min and average percentage weight loss is measured. The pass requirements are that the total average weight loss is 10% or less and average burn lengths of the cushions do not exceed 43 cm. For textiles attached to internal wall and ceiling panels or partitions, they are tested as a composite for heat

12.19 FAR 25.853(a) Appendix F Part II kerosene burner test for aircraft seating.

release using the OSU calorimeters as defined in ASTM E 906-1983 which specifies the test method for measurement of ease of ignition and associated heat release of textiles used in commercial aircraft when exposed to high heat flux (see Section 12.4.3). The overall method is contained within FAA 25.853 Part IV Appendix F in which the vertically oriented composite is exposed to a heat flux of 35 kW/m^2. A maximum peak heat release rate of ≤65 kW/m^2 and average heat release rate over the first 2 minutes of the test not exceeding 65 kW/m^2 are required for a textile composite to achieve a pass. For more detailed information, the reader is directed to the recent review by Lyon.[63]

Ships

Flammability standards and tests for furnishing fabrics, bedding and draperies used in ships and submarines have been developed by the International Maritime Organisation (IMO) and the National Fire Protection Association (NFPA) and include Safety of Life at Sea (SOLAS), High Speed Craft (HSC) and Fire Test Procedures (FTP) codes. The Fire Test Procedures (FTP) code describes the flammability tests and performance criteria for combustible materials[64] and Sorathia has recently reviewed this area.[65]

The surface flammability of interior finish materials used in bulkheads, overheads and decks is evaluated through either the IMO Resolution A.653(16) test method or the NFPA 25/ASTM E84 tests, which in principle measure the rate of flame spread under a radiant heat flux (see Section 12.4.2). The performance criteria for textile wall coverings demand a flame spread index (FSI) of ≤75 and a smoke developed index (SDI) of ≤450. In addition to this, materials to be used as deck overlay and finishes are required to be tested in a vertical orientation and should not produce more than 10 flaming droplets. Decking materials also must not have jetting combustion in the presence of adhesives or bonding agents.

Upholstery fabrics used in passenger ships must be tested in accordance with the IMO Resolution A.652(16) test method. The upholstered part of the furniture is placed on the back and bottom on the test seat fame and exposed to cigarette and butane flame for 1 h. The sample fails if any progressive flaming or smouldering is observed during the test period. In addition, the upholstered furniture (one chair) must meet the limited heat release criteria such that when tested in accordance with NFPA 266,[66] the maximum heat released must not be more than 80 kW, and when tested in accordance with ASTM E1537[67] and UL 1056,[68] the total heat released during the first 10 minutes should be less than 25 MJ.

Bedding components, including mattresses, pillows, blankets, quilts and bedspreads, used in ships are tested in accordance with IMO Resolution

A.688(17) according to the SOLAS and HSC Code, whereas NFPA 301 requires mattresses and mattress pads to comply with the NFPA 267,[69] ASTM E1590[70] and 16 CFR 1632[71] test methods. When tested according to the IMO Resolution A.688(17), the bedding should not ignite readily or exhibit progressive smouldering when subjected to smouldering or flaming ignition.

The SOLAS and HSC Code for materials used in hanging drapes specify that they have to meet flammability requirements when tested in accordance with IMO Resolution A.563(14). The test method includes a small swatch of material exposed to a small flame either at the bottom of the swatch or in the centre. The sample must not continue to burn for more than 5 s after application of the flame. In addition to this, the sample must not burn through the edges and should not have a char length of more than 150 mm.

12.6   Future trends

Bench-scale flammability tests are useful in that several material fire properties can be derived and data can be used for relative ranking of textiles. The tests are fairly cost-effective and can be employed for screening of new materials. The data derived from bench-scale tests can also be used for predicting large-scale fire behaviour using mathematical models. It is essential that future bench-scale tests should therefore be scientifically sound and should be a good indicator of large-scale end-use performance.

Computer models can be used to simulate real-scale fire tests using data obtained from bench-scale tests. However, this approach has its own limitation since the results will be applicable to a specific textile or composite geometry with a particular ignition scenario. Modelling of furniture fires and thermal burn injury have been developed by simplifying the problem and overlooking some important details of the burning item, and so cannot replace real testing methods. Reaction-to-fire tests such as the cone calorimeter are more robust and scientific and have greater applicability to realistic end-use applications and fire scenarios. However, they are too complex and expensive for everyday textile testing.

It is most likely, therefore, that future testing methods will be based on simple but increasingly scientifically sound principles, will be normalised across trading areas (e.g. the EU) and consequently will be cost-effective in application. Increased instrumentation will occur, as evidenced by the development of mannequin-type tests, providing costs of purchase, servicing and use remain affordable. At the present time predictive testing has too many unquantifiable variables and so will remain a research tool only, for at least the next 10 years.

12.7   References

1. Eaton P.M., 'Flame-retardancy test methods for textiles', *Rev. Prog. Coloration*, **30**, 51–62 (2000).
2. Kasem M.A. and Rouette H.K., 'Flammability and flame retardancy of fabrics: A review', in *Flammability of Fabrics*, Hilado C. (ed.), Technomic Publishing Co., Westport, CT, 11–24 (1974).
3. Miller B. and Goswami B.C., 'Effects of constructional factors on the burning rates of textile structures: Part I: Woven thermoplastic fabrics', *Text. Res. J.*, **41**, 949 (1971).
4. Wraight H., Webster C.T. and Thomas P.H., 'The heat transfer from burning fabrics', *Fire Research Station, Fire Research Note* No. 340/1957, November 1957.
5. Horrocks A.R., Tunc M. and Price D., 'The burning behaviour of textiles and its assessment by oxygen-index methods', *Textile Progress*, **18**(1/2/3), L. Cegielka (ed.), 1989.
6. Gawande (Nazaré) S., 'Investigation and prediction of factors influencing flammability of nightwear fabrics', PhD Thesis, 2002.
7. ASTM D2863-00: Standard method for measuring the minimum oxygen concentration to support candle-like combustion of plastics (oxygen index).
8. Parker W.J., *J. Fire Flam.*, **3**, 254–268 (1972).
9. Thomas P.H. and Webster C.T., Fire Research Note No. 420, Department of Scientific and Industrial Research and Fire Officer's Committee Joint Fire Research Organisation, March 1960.
10. Lawson D.I., Webster C.T. and Gregsten M.J., *J. Text. Inst.*, **46**(7), T435–T463 (1955).
11. Miller B., Martin J.R., Meiser C.H. and Gargiullo M., *Text. Res. J.*, 530, July 1976.
12. Babrauskas V., 'Development of the cone calorimeter – a bench-scale heat release rate apparatus based on oxygen consumption', *Fire and Materials*, **8**(2) 81–95 (1984).
13. Horrocks A.R., 'Textiles', Chapter 4 in *Fire Retardant Materials*, Horrocks A.R. and Price D. (eds), Woodhead Publishing, Cambridge, UK, 128–181 (2001).
14. BS 5438:1976 British Standard Methods of test for flammability of textile fabrics when subjected to a small igniting flame applied to the face or bottom edge of vertically oriented specimens.
15. Horrocks A.R., Gawande S., Kandola B. and Dunn K., 'The burning hazard of clothing – the effect of textile structures and burn severity', in *Proceedings of the 11th Annual BCC Conference on Flame Retardancy*, Stamford, CT, Business Communication Co. Inc., Norwalk, CT, May 2000.
16. Bonnarie T. and Touchais G., 'National and international fire protection regulations and test procedures: buildings', in *Plastics Flammability Handbook*, Troitzsch J. (ed.), Hanser Gardener Publications, Cincinnati, OH, 297–308 (2004).
17. Filipczak R., Crowley S. and Lyon R.E., 'Heat release rate measurements of thin samples in the OSU apparatus and the cone calorimeter', *Fire Safety Journal* **40**(7), 628–645 (2005).

18. Fire Safety of Upholstered Furniture – the final report on the CBUF research programme, EUR 16477EN, Bjorn Sundström (ed.), *circa* 1994.

19. Toal B.R., Shields T.J., Silcock G.W., Hume J. and Thomas P.H., 'Letters to editors', *Fire Mater.*, **14**(2), 73–76 (1989).

20. Barker S., Tesoro G.C., Toong T.Y. and Moussa N.A., *Textile Fabric Flammability*, Part III: *Ignition*, MIT Press, Cambridge, MA, 1975.

21. Krasny J.F., in *Flame Retardant Polymeric Materials*, Vol. **3**, Lewin M. *et al.* (eds), Plenum Press, New York, 155 (1982).

22. Norton M.J.T., Kandolph S.J., Johnson R.F. and Jordon K.A., 'Design, construction, and use of Minnesota woman, a thermally instrumented mannequin', *Text. Res. J.*, **55**(1) 5–12 (1985).

23. Camenzind M.A., Dale D.J. and Rossi R.M., 'Manikin test for flame engulfment evaluation of protective clothing: Historical review and development of a new ISO standard', *Fire Mater.*, **31**(5), 285–296 (2007).

24. Song G., 'Clothing air gap layers and thermal protective performance in single layer garment', *J. Ind Text.*, **36**(3), 193–205 (2007).

25. Horrocks A.R., *Rev. Prog. Coloration*, **16**, 62 (1986).

26. Fleischmann C.M., 'Flammability tests for upholstered furniture and mattresses', Chapter 7 in *Flammability Testing of Materials used in Construction, Transport and Mining*, Apte V.B. (ed.), Woodhead Publishing, Cambridge, UK, 164–186 (2006).

27. Horrocks, A.R. and Kandola B., 'Flammability testing of textiles', Chapter 6 in *Plastics Flammability Handbook*, Troitzsch J. (ed.), Hanser Publications, Munich, 173–188 (2004).

28. Bajaj P., 'Finishing of technical textiles', in *Handbook of Technical Textiles*, Horrocks A.R. and Anand S.C. (ed.), Woodhead Publishing, Cambridge, 152–172 (2000).

29. Horrocks A.R., 'Thermal (heat and fire) protection', Chapter 15 in *Textiles for Protection*, Scott R.A. (ed.), Woodhead Publishing, Cambridge, 398–440 (2005).

30. Horrocks A.R., Nazaré S. and Kandola B., 'The particular flammability hazards of nightwear', *Fire Safety Journal*, **39**, 259–276 (2004).

31. *The Nightwear (Safety) Regulations*, SI 1985/2043, HMSO, London (1985).

32. BS 5722:1984 British Standard specification for flammability performance of fabrics and fabric assemblies used in sleepwear and dressing gowns.

33. Krasny J.F. and Fisher A.L., 'Laboratory modelling of garment fires', *Text. Res. J.*, **43**(5), 272–283 (1973).

34. Anon., 'New European standard issued on flammability of children's nightwear', in *Regulatory News: Bulletins*, August 2007 (07B-143), http://cps.bureauveritas.com/

35. Mehta A.K. and Wong F., 'Measurement of flammability and burn potential of fabrics', Report from Fuel Research Laboratory, Massachusetts Institute of Technology, Cambridge, MA (1973).

36. Henriques Jr. F.C., 'Studies of thermal injuries V. The predictability and the significance of thermally induced rate processes leading to irreversible epidermal injury', *Archives of Pathology*, **43**, 703–713 (1947).

37. Stoll A.M. and Greene L.C., 'Relationship between pain and tissue damage due to thermal radiation', *J. Appl. Pathology*, **14**, 373–382 (1959).

38. Song G., 'Modelling thermal burn injury protection', Chapter 11 in *Textiles for Protection*, Scott R.A. (ed.), Woodhead Publishing, Cambridge, 261–292 (2005).
39. BS EN 14115:2002, British Standard for burning behaviour of materials for marquees, large tents and related products.
40. NFPA 701: Standard methods of fire tests for flame propagation of textiles and films.
41. ASTM E84-03: Standard test method for surface burning characteristics of building materials.
42. ASTM E108-07a: Standard test methods for fire tests of roof coverings.
43. ASTM E136-04: Standard test method for behavior of materials in a vertical tube furnace at 750°C.
44. Dowling V., 'National and international fire protection regulations and test procedures: Buildings', in *Plastics Flammability Handbook*, Troitzsch J. (ed.), Hanser Publications, Munich, 395–403 (2004).
45. BS 6307:1982, ISO 6925-1982 Method for determination of the effects of a small source of ignition on textile floor coverings (methanamine tablet test).
46. BS 4790:1987 Method for determination of the effects of a small source of ignition on textile floor coverings (hot metal nut method).
47. ASTM E648-99: Standard test method for critical radiant flux of floor-covering systems using a radiant heat energy source.
48. NFPA 253: Standard method of test for critical radiant flux for floor covering systems using a radiant heat energy source.
49. Turner R.C., 'Carpets for US transportation markets: Flammability testing requirements', http://www.tx.ncsu.edu/jtatm/volume3issue4/auto_carpet.htm
50. BS 5867: Part 2:1980 Specification for fabrics for curtains and drapes. Flammability requirements.
51. Fire Statistics United Kingdom 2006, *Home Office Statistical Bulletin* (UK Government Statistical Service).
52. BS EN 13772:2003 Textiles and textile products. Burning behaviour. Curtains and drapes. Measurement of flame spread of vertically oriented specimens with large ignition source.
53. BS EN 71-2:2003 Safety of toys. Flammability.
54. Hirschler M.M., 'Fire tests and interior furnishings', in *Fire and Flammability of Furnishings and Contents of Buildings*, Fowell A.J. (ed.), ASTM Publication, Philadelphia, PA, 7–31 (1994).
55. Paul K.T., Reimann K.A. and Sundström B., 'Furniture and furnishings', Chapter 13 in *Plastics Flammability Handbook*, Troitzsch J. (ed.), Hanser Publications, Munich, 580–607 (2004).
56. Sundström B. (ed.), Fire Safety of Upholstered Furniture – the final report on the Combustion Behaviour of Upholstered Furniture (CBUF) Programme, June 1995, Interscience Communications Ltd.
57. BS 7176:1995 Specification for resistance to ignition of upholstered furniture for non-domestic seating by testing composites.
58. Wakelyn P.J., 'Potential USA textile flammability regulations under consideration', in *Proceedings of the 18th Annual BCC Conference on Flame Retardancy*, Stamford, CT, Business Communication Co. Inc., Norwalk, CT, May 2007.

59. Hilado C.J., 'Fire studies of bedding materials', in *Flammability of Fabrics*, Volume 9, Hilado C.J. (ed.), Technomic Publishing Co., Westport, CT, 140–182 (1974).
60. Technical Bulletin 603, Requirements and test procedure for resistance of a mattress/box spring set to a large open-flame, California Bureau of Home Furnishings and Thermal Insulation, North Highlands, CA, 1986.
61. Antonatus E. and Troitzsch J., 'Transportation: motor vehicles', Chapter 11 in *Flammability Testing of Materials Used in Construction, Transport and Mining*, Apte V.B. (ed.), Woodhead Publishing, Cambridge, UK, 431–442 (2006).
62. Ebenau A., 'Transportation: rail vehicles', Chapter 11 in *Flammability Testing of Materials Used in Construction, Transport and Mining*, Apte V.B. (ed.), Woodhead Publishing, Cambridge, UK, 442–456 (2006).
63. Lyon R., 'Flammability requirements for aircraft cabin materials', Chapter 20 in *Advances in Flame Retardant Materials*, Horrocks A.R. and Price D. (eds), Woodhead Publishing, Cambridge, UK, in press.
64. Lattimer B.Y., 'Ships and submarines', Chapter 15 in *Flammability Testing of Materials Used in Construction, Transport and Mining*, Apte V.B. (ed.), Woodhead Publishing, Cambridge, UK, 361–384 (2006).
65. Sorathia U., 'Flame retardant materials for maritime and naval applications', Chapter 19 in *Advances in Flame Retardant Materials*, Horrocks A.R. and Price D. (eds), Woodhead Publishing, Cambridge, UK, in press.
66. NFPA 266: Standard test method of upholstered furniture exposed to flaming ignition source, National Fire Protection Association, Quincy, MA (1998).
67. ASTM E1537: Standard test method for fire testing of upholstered furniture (1996).
68. UL 1056: Fire test of upholstered furniture, Underwriters Laboratories, Inc., Northbrook, IL (1989).
69. NFPA 267: Standard test method for fire characteristics of mattresses and bedding assemblies exposed to flaming ignition source, National Fire Protection Association, Quincy, MA (1998).
70. ASTM E1590: Standard test method for fire testing of mattresses (1999).
71. 16 CFR 1632: Standard for flammability of mattresses, US Government.

Index